Einstein's Pacifism and World War I

Virginia Iris Holmes

Syracuse University Press

Syracuse, New York 13244-5290

First Edition 2017

17 18 19 20 21 22 6 5 4 3 2 1

∞ The paper used in this publication meets the minimum requirements of the American National Standard for Information Sciences—Permanence of Paper for Printed Library Materials, ANSI Z39.48-1992.

For a listing of books published and distributed by Syracuse University Press, visit www.SyracuseUniversityPress.syr.edu.

ISBN: 978-0-8156-1062-5 (hardcover)
978-0-8156-1085-4 (paperback)
978-0-8156-5360-8 (e-book)

Library of Congress Cataloging-in-Publication Data

Available from the publisher upon request.

Manufactured in the United States of America

Einstein's Pacifism and World War I

Modern Jewish History
Henry Feingold, *Series Editor*

Select titles in Modern Jewish History

American Jewish Political Culture and the Liberal Persuasion
Henry L. Feingold

Assimilated Jews in the Warsaw Ghetto, 1940–1943
Katarzyna Person

Bundist Counterculture in Interwar Poland
Jack Jacobs

The Children of La Hille: Eluding Nazi Capture during World War II
Walter W. Reed

The Downfall of Abba Hillel Silver and the Foundation of Israel
Ofer Shiff

Judah L. Magnes: An American Jewish Nonconformist
Daniel P. Kotzin

Leaving Russia: A Jewish Story
Maxim D. Shrayer

Will to Freedom: A Perilous Journey through Fascism and Communism
Egon Balas

In loving memory of Iris
forever Iris
she led me to boycott
slaughterhouses

Contents

Illustrations

Following page 117

Acknowledgments

I would like to express my gratitude to individuals at Syracuse University Press: Jennika Baines, my earlier acquisitions editor, for her enthusiasm and perseverance regarding my book; Deborah Manion, my later acquisitions editor, for her steady guidance toward the end of the process; editor-in-chief Suzanne Guiod, Kay Steinmetz, Kaitlin Carruthers-Busser, Elizabeth Myers, and others for help in preparing the manuscript; Mona Hamlin and others for marketing; Matthew Kudelka for developmental editing; and Modern Jewish History series editor Henry Feingold and several anonymous outside readers for their insights. Many archives, libraries, and collections were most helpful. I thank Chaya Becker and Roni Grosz of the Albert Einstein Archives at the Hebrew University in Jerusalem for help with documents and photos. In addition, I am grateful to the following archives, libraries, collections, and individuals, both for help with documents, photos, and copyright permissions, and for help with attempting to locate photos: Roland Lüthi, Heike Hartmann, Yvonne Voegeli, and Nicole Graf of the ETH Bibliothek in Zürich; Anina Hug of the Swiss National Library (Schweizerisches Landesbibliothek) in Bern; Tess Hinds and Lucinda Moore of the Mary Evans Picture Library in London; Dr. Thomas Einstein, great-grandson of Albert Einstein and grandson of Hans Albert Einstein; Ursula Cohrs, granddaughter of Heinrich Zangger; Christine Baur of the Zentralbibliothek Zürich; Lara Szypszak of the US Library of Congress; Michael Simonson of the Leo Baeck Institute in New York; Alexandra Kosubek of the Bundesarchiv in Koblenz; Esther-Julia Howell and Alexander Markus Klotz of the Institut für Zeitgeschichte in Munich; John R. Waggener of the American Heritage Center at the University of Wyoming; Ceylan Akturk of Wayne State

University Press; and any other copyright holders whose names I learned too late to include here.

I thank Robert Schulmann for an early reading and feedback on the manuscript, John Stachel for the same generous input on a slightly later version, and Fred Jerome for his comments. For nudges that led to the original inception of the idea for this book, I thank Diana Buchwald and other editors and officers of both the Einstein Papers Project (EPP) and Princeton University Press, who encouraged me and other EPP editors to create "Einstein spin-off books." I also thank Diana Buchwald and the EPP for the allocation of a portion of my workweek to my research and writing during my tenure as senior research editor there, time that contributed significantly to this book among other projects. I also thank the EPP editorial team for a unique dialectical process that helped me to arrive at a stronger understanding of my own authority as a historian. During my time at the Einstein Papers Project, I also appreciated collaboration with fellow editors Ze'ev Rosenkranz, Jozsef Illy, Tilman Sauer, Daniel J. Kennefick, Jeroen van Dongen, A. J. Kox, and Issachar Unna; the invaluable assistance of Osik Moses, Rudy Hirschmann, Rosy Meiron, Barbara Wolff, Nurit Lifshitz, Shady Peyvan, Jenny Nollar, Jordan Boyd-Graber, and Lea Hildebrandt; and helpful input from Peter Dougherty and Alice Calaprice. I want to express my gratitude to Gabriele Wohlgemuth of the Zentralbibliothek Zürich and her staff for their help on the research trip in April 2007 when I discovered Einstein's letter of August 21, 1917, and other documents.

In the acknowledgments of my dissertation, among the outstanding scholars who contributed to my development as a historian, I honored four historians who showed exceptional humanity and integrity while interacting professionally: Kathleen Canning, Robin D. G. Kelley, Carol Faulkner, and Andreas Reinke. I have since had the good fortune to become acquainted with another historian of this caliber, Robert Schulmann, to whom I have enormous gratitude for his generosity with time, moral support, intellectual and editorial insights, and formidable command of the Einstein topic. I am grateful to my dissertation chair, Rabbi Lance J. Sussman, dissertation committee members Gisela Brinker-Gabler, Wulf Kansteiner, and Allan Arkush, as well as other historians and scholars who contributed to my intellectual development, including Todd M. Endelmann, Jean H. Quataert, Geoff Eley, and Avrom Nowersztern. Other historians whose support and useful input have been warmly appreciated include Connie Shemo, Ute Ritz-Deutch, Michelle Kuhl, and Trent Maxey. I am also thankful to Susan Cocalis for giving me a remarkable jump start with the German language, Marianne Schloten for the same, Avrom Nowesztern for the same with Yiddish, and various professors at the University of Michigan for the same with Hebrew.

I will always remain grateful to Woody and Trudi Widrick and other members of a long-ago spiritual growth group led by Woody, for their enthusiastic, decisive encouragement during months when I was awakening to my calling to become a scholar of modern Jewish history. I am thankful to many aficionados of the Quaker tradition, including Frances Crowe, for my early and extended introduction to the richness of pacifism, and for friendship and support. I am grateful for a family that launched me with economic stability and a faith in my intellect, and for some financial help from my uncle and mother, Frank Bequaert and Helen B. Holmes, during and after graduate school. I am deeply inspired by friends who are animal advocates, including Kate Carpenter, Gracia Fay Ellwood, Peter Hartgens, Sheryl Becker, Larry Kuttner, Rachael Gordon, Jamie Kordack, Linda Huebner, Miriam Jones, Andy Doyle, and Anne Flaherty. I have also received abiding love from Jupiter, Phoebe-Bubs, Willow, Sidney, Chambonnaise, Deta, Orlando, Ripinsky, Emmeline, Brownell, and Cady. Along with colleagues, mentors, and friends mentioned above, other luminaries whose encouragement and wide variety of support have been invaluable over the years include Jonathan McClellan and Linda Jenness-McClellan, Hans and Erna Lüdecke, Mary-Elaine Koenig Leake, the late Joan Lockhart, Erica Bergquist, Bob Gettings, Jill Faulkner, Ramón Oliva, Don Campbell, Nancy Slator, Libby Maxey, Heather Davies Williams, Ann Jarvis, Sophie Goodenough, Leila Jojua, Tika Revishvili and Peace Corps Georgia staff, Angela Wen, Nicholas Beck, and many other fellow Peace Corps Volunteers, RPCVs both of other years and countries and from my group, my brother Joseph Holmes for a variety of help across the last decade, and my nieces Emily Avery Holmes, Cara Avery Holmes, and Xiaowen Avery Holmes.

Introduction

Entering the fateful year of 1914 at age thirty-five with long-standing but untested pacifist and internationalist leanings, Albert Einstein found his incipient pacifism and dormant capacity for political engagement awakened by the First World War. His thoughts and feelings in response to World War I evolved across the years of the War, from horrified disbelief; to ironic alienation from both the War's violence and patriotic support for it by the German people, including most of his scientific colleagues (he characterized the "war-enthusiasm" [*Kriegsbegeisterung*] as a peculiar kind of insanity); and finally, toward the War's end, to a kind of bleak endurance with occasional bursts of the old irony. Einstein's responses to the War evolved in that same period from a relatively passive support for the initiatives of other pacifists (e.g., he signed Georg Friedrich Nicolai's "Appeal to the Europeans" and joined the pacifist New Fatherland League); to a more active pursuit of a friendship with the prominent French pacifist Romain Rolland, as well as developing his own philosophy of a postwar order and avenue to lasting peace, the core element of which would be some sort of international or world governmental organization with executive powers of enforcement; and, finally, to being the impetus or co-impetus behind initiatives such as a commission of private German citizens who investigated atrocities committed by the German Army in Belgium and France and published their findings in booklet form, hoping thereby to help the German people better understand the feelings of their former wartime enemies. In the first few years after the War, his responses to postwar developments in Germany and internationally also evolved, from enthusiastic hopes in Germany's new Weimar Republic, trust in the laudatory effects of military defeat and economic hardship on the

German people, and support for the international principles outlined by US president Woodrow Wilson in his "Fourteen Points" speech; to great disappointment in the early Weimar Republic's leaders and their actions, aversion to the culture of violence that rapidly developed in Germany, and offense at the Allies' betrayal of Wilson's principles in the Treaty of Versailles. Yet Einstein's postwar trajectory toward disappointment, like his wartime trajectory toward alienation, was again accompanied by actions that reflected a deeper optimism about the potential for peace between nations—such as his participation in the aforementioned commission investigating wartime atrocities—as well as by pacifist stances in private letters and public statements. This book focuses on the words and statements of Albert Einstein, utilizing them to give us direct access to the awakening of his pacifism and the related development of his thoughts and feelings during the events of World War I and its immediate aftermath.

Many intellectuals, Jews, and pacifists have been inspired by the political thought of Albert Einstein, particularly as he expressed it during the decade after World War II, the last decade of his life (and as a United States citizen). But to understand how his pacifist and internationalist thought matured from a youthful inclination to a source of pragmatic initiatives and savvy insights, one must examine his interactions with the people around him, individually and collectively, in response to World War I, when he was in his late thirties and early forties, living and working at his highest scientific productivity in Berlin, the city at the pinnacle of scientific activity in German-speaking Europe, yet simultaneously the capital city of one of the principal warring nations. This book offers a close perusal of the emergence of Einstein's conscious political philosophy, convictions, and choices of occasions on which to act publicly, highlighting the circumstances and interactions in which his pacifist world view developed.

Although there is evidence of incipient pacifism from Einstein's childhood and youth, World War I was the period when he first articulated an explicit commitment to pacifism, took public stands on behalf of peace and international organization and reconciliation, and used his growing stature and influence to promote political objectives for which he cared. In the autumn of 1914, during the earliest months of the War, he was one of a courageous few to sign the pacifist "Appeal to the Europeans." Around June 1915, almost a year into the War, Einstein joined the German pacifist New Fatherland League (Bund Neues Vaterland). In October 1916, the year of the battles of Verdun and the Somme, he joined in a Dutch initiative, the Anti-War Council of the Central Organization for a Durable Peace, and also engaged in early discussions of a commission to investigate German war crimes. By November 1916, he was pursuing a friendship with the prominent French pacifist Romain Rolland, exiled in Switzerland, whom he had first contacted in March 1915; they would continue the early stages of a decades-long correspondence during 1917. In August 1917, the year following the harsh "turnip winter" (*Kohlrübenwinter*) of hunger in Germany, Einstein wrote for his friend Heinrich Zangger a point-by-point description of his envisioned postwar order of states. In September 1918, the year that started with Wilson's "Fourteen Points" speech in January and ended with the November armistice, he offered thoughtful feedback to proposals by the creative revolutionary pacifist Kurt Hiller, as part of his ongoing interactions with various high-profile pacifists. In 1919, the year of the Paris Peace Conference where the Treaty of Versailles was written, he participated in a commission of private citizens investigating atrocities committed by the German army in France and Belgium during the War; this group published its findings in booklet form. In 1920, extreme economic hardship in Germany and intensifying German resentment against the harsh terms of the Versailles treaty manifested themselves in the brief rightist Kapp Putsch in March, harassment of Einstein and other pacifists through disruptions of their lectures, occasional physical attacks on pacifists such as the statistician Emil J. Gumbel, and politicized attacks on Einstein's scientific contributions. Yet notwithstanding these reactionary and potentially repressive conditions, in May of that year he signed a declaration supporting the republican constitution and in July he wrote a statement of support for Anglo-American Quakers who were feeding German children. Meanwhile, in his rising fame, he showed discernment in declining many solicitations for his endorsement of causes and initiatives that did not meet his standards.

This book tracks the emergence and expressions of Einstein's earliest professed pacifism. It offers a thorough presentation of published writings and private correspondence written by him in response to World War I, and summarizes, with excerpts, the correspondence of friends and relatives who clearly contributed to his thought on the topic, both during and immediately after the War. It offers the everyday thoughts of an extraordinary private citizen in his ordinary interactions during a time of historic crisis. Many selections from Einstein's correspondence have a personal flavor and are obviously drawn from the middle of a conversation; some of that conversational quality has been retained in the excerpts provided, while some has been eliminated in order to maintain a focus on his response to the issue of war and peace. The book explores Einstein's expressions of pacifist convictions, outrage, and irony about the War, along with his sense of alienation from the majority of his academic colleagues, most of whom were swept up in the 1914 war-enthusiasm. Some of Einstein's most interesting activity in response to the War took place in the years immediately after the War had ended, by which time his political engagement had become much more substantial; hence, the book addresses the wartime years, 1914 to 1918, and also the immediate postwar years, 1919 to 1921. (The epilogue addresses his continuing pacifism through 1955.)

As a senior research editor with the Einstein Papers Project for six years, I participated in difficult, careful decisions regarding which documents to deselect from the vast Einstein archive—that is, which documents would not be presented in full in the volumes of *The Collected Papers of Albert Einstein* (CPAE). (All deselected documents were "calendared"—that is, their existence is recorded in the calendar at the end of the volume.) My editorial colleagues and I, aware of documents we had deselected from inclusion in the *CPAE* volumes, and of our decisions to minimize contextualization of Einstein's political engagement in our annotation, were encouraged to fill the resulting gaps by producing our own "Einstein spin-off books." This awareness and encouragement contributed to the initial vision and inspiration behind this book. As a book focused exclusively on the awakening of Einstein's pacifism in the years 1914 to 1921, the present volume offers the most complete presentation available of Einstein documents focused on that theme in that time period (including some documents not presented in the *CPAE*). David Rowe and Robert Schulmann's *Einstein on Politics* and Otto Nathan and Heinz Norden's *Einstein on Peace* are similarly focused thematically, but they address Einstein's full lifespan, and only a small percentage of the documents presented here appear in either of those books. A larger percentage appear in the *CPAE*, but not all; and those that do appear were not all selected for the *CPAE*'s English-language translation volumes. (Also, the percentage of documents appearing elsewhere varies greatly from one chapter of this book to another, a higher percentage during the earlier War years than in the later War years and immediate postwar years.)

Primary sources, originally in German, are presented in English translation, and all translations are mine. The two exceptions to this are the 1914 manifesto "To the Civilized World" and the 1914 "Appeal to the Europeans," for which I present contemporaneous 1919 English-language translations. As mentioned above, not all Einstein documents are published in the volumes of the *CPAE*, and this editorial deselection is extended further in the *CPAE*'s English-language translation volumes, in which some documents presented in the *CPAE* are further deselected from the *CPAE*'s English-language translation volumes. Additionally, in the major published translations of many Einstein documents (except Rowe and Schulmann's *Einstein on Politics*), occasionally entire sentences or clauses, and often crucial words essential to the sense or context of the document, have been omitted. Mainly for these reasons, but also as a matter of intellectual integrity, I have done all of my own translations. My translations occasionally retain some of the rhythms of Einstein's elegant German, as part of a translation style and strategy aimed primarily at precision and completeness. In my own usage and in my translations, I refer to World War I as "the War" (i.e., with a capital W) because that was the English-language practice in the First World War's immediate aftermath, through the interwar period, and sometimes beyond.

The book concisely touches upon the broad contours and major events of the First World War but does not attempt to offer an interpretation or history of World War I or political development in the early Weimar Republic. Hence, the book does not examine the texts of German daily newspapers, except when explicitly mentioned by Einstein or his correspondents. Einstein was well aware that the German press was subject to censorship and tied to government propaganda, and while he did read newspapers, he also acquired his information from other sources, including pacifist sources such as the New Fatherland League, the German Peace Society (Deutsche Friedensgesellschaft), and the main German-language pacifist monthly journal *Die Friedens-Warte*, as well as from the sources from which populations typically obtain information during censorship—soldiers on leave, friends, neighbors, colleagues, etc. The original vision of this book involved "nesting" Einstein's statements in contextualizing passages that would situate his thought in the immediate sequential events of World War I and its aftermath. In addition to events on the frontlines of the War, this material included in the German domestic arena both propaganda and economic hardship. However, early readers found these passages distracting; therefore most of this contextual material has been moved into the footnotes.

Because this is primarily an exploration of Einstein's pacifism, I do not analyze in depth his stances on social justice and anti-oppression issues that are not directly linked with war and peace. However, because questions of war and militarism, including in Einstein's World War I experience, are often interwoven with those of economic deprivation, class divisions, political regimes, and relations between governments and their populations, the book does include Einstein's comments on these matters as they connect to pacifism and internationalism. As I have argued elsewhere, I see an inherent link between pacifism and opposition to social injustice; pacifists reject the outward violence of killing, while proponents of social justice reject the psychological and economic violence of inequality and oppression, often reinforced by the threat of physical violence.[1] Einstein was by nature compassionate on behalf of the underdog, and he was particularly concerned about antisemitism, which directly affected him and his own people. Indeed, unlike many German Jews, he extended his public stances to the plight of East European Jews.[2] Because Jewish men were denigrated as weak and cowardly, in an interweaving of gender and antisemitism, it took exceptional courage for them to embrace pacifism, as I have discussed elsewhere.[3] So I consider it relevant and worthwhile to include Einstein's philosophy and stances in opposition to antisemitism, his advocacy on behalf of East European Jews, his promotion of Jewish pride, and his statements on Zionism, including his attempts to reconcile its nationalism with his pacifism. Pacifists have a long tradition of extending their activism to advocacy of social justice, and Einstein certainly carried this banner as well, as his documents demonstrate.

This expression of Einstein's Jewish identity within his broader commitment to social justice, while interrelated with his pacifism and expressed throughout the same period, is addressed separately in chapter 6; his pacifist activities are explored chronologically in chapters 1 through 5. Chapter 1 has a slightly different "rhythm" than the later chapters in that it offers two long and historically crucial documents from the autumn of 1914 that are rarely published and have never (since that era), to my knowledge, been presented together. These two documents are the German rightist-militarist manifesto "To the Civilized World" ("*An die Kulturwelt*") and the German pacifist "Appeal to the Europeans" ("*Aufruf an die Europäer*"). Both are also presented in the original German in appendices A and B. The former contributed enormously to Allied mistrust of Germans (and to German pacifists' embarrassment) during and long after the War; the latter shows Einstein's early, unhesitating readiness, based on his convictions, to stand publicly among those very few who opposed the War. I judged it essential to devote space for each of these documents in its entirety. Chapters 2 through 6 offer many documents, long and short (and excerpts), but none so central to setting the context for Einstein's awakening pacifism.

In the epilogue, I offer a synopsis of Einstein's pacifism through the remainder of his life. Although very abbreviated, it includes several fascinating and compelling Einstein statements, stances, and interactions over the years. Also in the epilogue, I touch on the secondary literature and range of perspectives on Einstein's pacifism. Although he temporarily withdrew from pacifist activism—in particular from advocating refusal of military service—during the Second World War, when Nazi racial policies and the Holocaust presented a uniquely troubling ethical dilemma for Jewish pacifists, it is valid to consider Einstein a pacifist, as I have argued elsewhere, for a combination of reasons. Most prominently, he continued throughout World War II to declare himself a pacifist and to advocate international organization or world government. Then, after the Second World War, he maintained his pacifist engagement and activism in the

United States, in particular vis-à-vis atomic weapons.[4] There is a significant difference between those who regard war as tragic but inevitable (to be passive is not to be pacifist) and those who, in rejecting war, engage in persistent efforts to eradicate war and establish world peace. The latter are pacifists, and Einstein was one of them.

In this era of the hundredth anniversaries of World War I, this book shows one brilliant thinking individual's experiential angle on the War that was described in its early days as "the war to end all wars."[5] My hope is to enhance our insight into the upheaval in worldwide political culture that took place in response to that historical rupture. The documents in this book highlight Einstein's political discourses in interaction with World War I and its aftershocks, when he found himself situated in the capital city of one of the belligerent nations, then later a defeated nation, also the country of his birth from which he had withdrawn his citizenship in his late teens in favor of Swiss citizenship. These materials demonstrate that Einstein was an independent thinker and instinctive internationalist pacifist, that he was not swayed by mob mentalities or blinding nationalism, and that he was always open to reasoned, meaningful discussion of issues with which he was engaged, ready to add unexpected or unusual factors to the equation. Ultimately, he was motivated by compassion for the oppressed, individually and collectively, and by the desire to alleviate suffering. As the horrors of World War I were fought and endured in his immediate vicinity, Einstein's pacifism was awakened, and his prior inclination developed into a concrete, focused, and enduring outward commitment.

Einstein's Pacifism and World War I

1 Horror at War, Pacifist Affiliation, and Alienation from Colleagues

1914–1915

When Albert Einstein was confronted with the beginning of World War I, he already had pacifist inclinations from earlier in his life. According to his younger sister, Maja Winteler-Einstein, as a child he disliked competitive, physical games, and often served as referee to other children's disputes. He even disliked games often enjoyed by the scientific and technically oriented, such as chess, because of the element of competition, the need for there to be winners and losers. He also felt an aversion to military discipline, coercion, and obedience. On one occasion in his childhood, when observing uniformed soldiers marching in a parade, he told his parents that he did not want to be one of them when he grew up.[1]

Later in his youth, Einstein chose a self-directed path that might have reflected incipient pacifism, besides fostering his future pacifism. When his father and uncle decided to move their business from southern Germany to Italy in response to a failure to be awarded a contract (almost certainly due to antisemitism), Einstein at the age of fifteen decided to move to Switzerland to complete his high school and university education. Thus, he emigrated from Germany just barely in time to avoid the legal requirement for military service.[2] This was not an explicit pacifist stance, and he did not articulate it as such, but it may have indicated a predisposition to pacifism. He later officially renounced his German citizenship and acquired Swiss citizenship. This served to internalize and enhance his internationalism; it also introduced him to republican perspectives and critiques of pan-German expansionism by Jost Winteler, father of the family with whom he first lived in Switzerland.[3]

On June 28, 1914, the Serb Gavrilo Princip assassinated Archduke Franz Ferdinand, heir to the throne of the Austro-Hungarian Empire, and his wife Sophie, in Sarajevo. Between that date and Austria-Hungary's declaration of war one month later, most educated Europeans like Einstein were aware of the posturings and accusations of diplomats. These included Austro-Hungarian officials suggesting the Serbian government had been complicit in the assassination as well as addresses in the British Parliament. Meanwhile, anti-Serb rioting broke out in Sarajevo and Bosnia, and the Austro-Hungarian press used denigrating language against Serbs and frightened their readers with stories about Serbian conspiracies and projected attacks. After sending the Serbs an ultimatum on July 23, Austria-Hungary declared war on Serbia on July 28, 1914, which ignited World War I. At this point, a series of diplomatic and military events took place in rapid succession. On July 31, the French pacifist Jean Jaurès was assassinated. On August 1, Germany declared war on Russia, after Russia mobilized its military in support of Serbia. On August 2, German troops occupied Luxemburg, and a secret treaty was reached between Germany and Turkey to obtain Turkish neutrality. On August 3, Germany declared war on France, Russia's ally. On August 4, German troops invaded neutral Belgium, Britain declared war on Germany for violating Belgian neutrality, and the United Stated declared its neutrality.[4] At this point, in August 1914, there was an outpouring of "war-enthusiasm" (*Kriegsbegeisterung*), also referred to as "the spirit of 1914," in most of the belligerent nations, but especially Germany, England, and France. The people of each nation believed that they had righteousness on their side and that the War would end, with victory for their side, by Christmas of that year.[5]

In March 1914, at age thirty-five, having been recruited to an academic post by physicist Max Planck and others, Einstein moved from Zurich to Berlin with his first wife, Mileva Einstein-Marić, and their two young sons, Hans Albert and Eduard. (In 1917, he would be named head of the Kaiser Wilhelm Institute in Berlin.) In mid-July 1914, Einstein-Marić separated from him, largely in response to the close presence of his cousin and future second wife, Elsa Einstein. She moved with their two sons to another household in Berlin, then returned to Zurich on July 29, the day after Austria-Hungary declared war on Serbia. Einstein and his first wife would remain separated for the duration of the War, with Einstein in Berlin, and Einstein-Marić and their sons in Zurich. Shortly after the War ended, they divorced and Einstein married his second wife.[6]

Between August and December 1914, the War's battles intensified, with one memorable hiatus, the Christmas truce of 1914.[7] Einstein soon discovered that he took a very different view of the War than did most of his academic colleagues in Berlin. On August 19, 1914, in a postcard to his fellow physicist and friend Paul Ehrenfest in the Netherlands, Einstein expressed his earliest thoughts and feelings about the War, making his first allusion to it as a form of insanity:

> Europe in its madness has now embarked on something incredibly preposterous. At such times one sees to what deplorable breed of brutes we belong. I am musing serenely along in my peaceful meditations and feel only a mixture of pity and disgust.[8]

On August 24, he expressed these feelings in more depth to his friend Heinrich Zangger, a physician and medical professor at the University of Zurich and neighbor of his wife and two sons. This letter included an early reference to the sharp difference between his perspective and that of the majority:

> What a horrific picture the world is now offering! Nowhere is there an island of culture where people have retained human feeling. Nothing but hate and a lust for power! The question, "where can justice be found?" is becoming sheer mockery. One lives the life of a stranger on this planet, happy when one isn't killed for untimely sentiments. I feel so strangely drawn to early Christianity and feel acutely as never before how much nicer it is to be anvil than hammer. What galls me most is that today the best talent is also being forced into this senseless butchery and henchman's service. I have blind luck to thank that I was spared this.[9]

In early October 1914, a statement came out of Germany that would have a profound impact on the Allied countries' perception of Germany for years to come, both during and after the War, and that would have great ongoing significance to Einstein and other pacifists in subsequent years. It denied events sometimes called the "rape of Belgium," when the German army killed 23,700 Belgian civilians (about 6,000 of them directly, the rest through deportation and other harsh conditions of the German occupation), caused 10,400 permanent and 22,700 temporary injuries, orphaned 18,296 children, burned and destroyed hundreds of homes and buildings, expelled thousands of Belgian civilians, and stole large amounts of food, equipment, and materials.[10] In response to the Allies' accusations that the German army was committing atrocities during its invasion of neutral Belgium, ninety-three prominent German intellectuals and cultural figures signed a published manifesto "To the Civilized World." This manifesto denied the events and defended German honor as follows (in the almost contemporaneous 1919 translation by Constance A. Grande and Julian Grande):

> As representatives of German science and art, we protest before the whole civilized world against the calumnies and lies with which our enemies are trying to besmirch Germany's undefiled cause in the severe struggle for existence which has been forced upon her. The course of events has mercilessly disproved the reports of fictitious German defeats. All the more vigorous are the efforts now being made to distort truth and disseminate suspicion. It is against these that we are raising our voices, and those voices shall make the truth known.
>
> *It is not true* that Germany was guilty of this war. Neither the nation nor the Government nor the emperor wanted it. The Germans did everything possible to avert it, documentary evidence of which is before the whole world. In the twenty-six years of his reign William II has frequently shown himself the defender of the world's peace, as has frequently been acknowledged even by our enemies. Indeed, this same emperor, whom they are now presuming to call an Attila, was ridiculed for twenty years and more because of his unswerving devotion to peace. Not until our people was attacked from three sides by superior forces, which had long been lying in wait at the frontier, did it rise as one man.
>
> *It is not true* that we criminally violated Belgian neutrality. It can be proved that France and England had resolved to violate it, and it can be

proved that Belgium had agreed to this. It would have been suicidal not to have anticipated them.

It is not true that the life and property of a single Belgian subject were interfered with by our soldiers except under the direst necessity. Again and again, despite all warnings, did the population lie in ambush and fire on them, mutilating wounded men, and murdering doctors even while actually engaged in their noble ministrations. There could be no baser misrepresentation than to say nothing about the crime of these assassins and then to call the Germans criminals because of their having administered a just punishment to them.

It is not true that our troops behaved brutally in regard to Louvain. They were forced to exercise reprisals with a heavy heart on the furious population, which treacherously attacked them in their quarters, by firing upon a portion of the town. The greater portion of Louvain is still standing, and the famous town hall is quite uninjured. It was saved from the flames owing to the self-sacrifice of our soldiers. Every German would regret works of art having been destroyed in this war or their being destroyed in the future. But just as we decline to admit that any one loves art more than we do, even so do we refuse no less decidedly to pay the price of a German defeat for the preservation of a work of art.

It is not true that we disregard the precepts of international law in our methods of warfare, in which there is no unbridled cruelty. But in the East the ground is soaked with the blood of women and children slain by Russian hordes, and in the West the breasts of our soldiers are lacerated with Dumdum bullets. No one has less right to pretend to be defending European civilization than those who are the allies of Russians and Serbians, and are not ashamed to incite Mongolians and blacks to fight against white men.

It is not true that fighting our so-called militarism is not fighting against our civilization, as our enemies hypocritically allege. Without German militarism German civilization would be wiped off the face of the earth. The former arose out of and for the protection of the latter in a country which for centuries had suffered from invasion as no other has done. The German Army and the German people are one, and the consciousness of this makes seventy millions of Germans brothers to-day, without regard to education, rank, or party.

> We cannot deprive our enemies of the poisoned weapons of falsehood. All we can do is to cry aloud to the whole world that they are bearing false witness against us. To you who know us, who, together with us, have hitherto been the guardians of man's highest possessions—to you we cry aloud:
>
> Believe us; believe that to the last we will fight as a civilized nation, to whom the legacy of a Goethe, a Beethoven, and a Kant is no less sacred than hearth and home.[11]

The fact that many of his colleagues in the Prussian Academy of Sciences, including Max Planck, signed this manifesto was a major early source of Einstein's sense of alienation from German intellectuals during the War. Yet his relationships with German scholars were complex, as were their attitudes and stances. Einstein aficionados tend to recall his post–World War II letter to Arnold Sommerfeld declining his invitation to rejoin the Bavarian Academy of Sciences because "the Germans massacred my Jewish brethren in Europe."[12] But long before the unimaginable horrors of the Holocaust, a warm correspondence continued between Einstein and other members of the Prussian Academy, notably Max Planck, Max von Laue, Fritz Haber, and Walther Nernst, all the way through early 1933.[13] When Einstein worked to restore good relations between physicists of different countries in mid-1915, he told Dutch physicist Hendrik Lorentz that he did it with the encouragement of Planck.[14] Also in 1915, Planck helped to defeat a petition in the Prussian Academy to sever ties with French academies.[15] Einstein spoke publicly in praise of Planck at an event honoring Planck's sixtieth birthday in April 1918, and he expressed appreciation to Planck for his friendship as late as April 1933 after Einstein's emigration and the Nazi confiscation of his property in Germany.[16] Einstein believed that Planck and chemist Emil Fischer had signed the manifesto "To the Civilized World" hastily, without reading the text—although they were unwilling publicly to retract their signatures.[17] Overall, Einstein felt gratitude to Planck for promoting his theory of relativity in the physics community and recruiting him to the Kaiser Wilhelm Institute in Berlin, faith that Planck could show a more internationalist side (which was vindicated a few times, such as when Planck signed the Delbrück-Dernburg petition opposing annexations, discussed below), understanding of the social pressure on Planck, and tolerance of Planck's overall timidity or indifference on political-ethical issues that inspired Einstein. On many occasions, mutual friendship was declared on both sides.

However, this manifesto "To the Civilized World," also known as the "Manifesto of the 93," was to become a major source of mistrust of Germans and German intentions among the Allied populations, including intellectuals. Both during and after the War, German pacifist intellectuals and cultural figures were deeply concerned with countering it and thereby vindicating Germany's reputation. Their earliest response to the manifesto was immediate. In mid-October 1914, the pacifist medical doctor and University of Berlin medical professor Georg Friedrich Nicolai wrote a counter-manifesto, "Appeal to the Europeans." But he only succeeded in recruiting three other signatories besides himself, and for this reason he did not publish the appeal until three years later, when he included it in his book *The Biology of War* (*Die Biologie des Krieges*), published in Switzerland in 1917. One of the three other individuals with sufficient conviction and moral courage to step forward and sign this unpopular appeal was Albert Einstein. This was Einstein's first significant public stance and his first move toward linking himself with politically and ethically like-minded people. The appeal read as follows (again in the almost contemporaneous Grande and Grande translation):

> Technical science and intercommunication are clearly tending to force us to recognize the fact that international relations exist, and consequently that a world-embracing civilization exists. Yet never has any previous war caused so complete an interruption of that cooperation which should exist between civilized nations. It may, of course, be that the reason why we are so profoundly impressed by this is only that we were already united by so many ties the severing of which is painful.
>
> That such a state of things should exist must not astonish us. Nevertheless, those who care in the slightest degree for this universal world civilization are under a twofold obligation to strive for the maintenance of these principles. Those who might have been expected to care for such things, in particular men of science and art, have hitherto almost

invariably confined their utterances to a hint that the present suspension of direct relations coincided with the cessation of any desire for their continuance.

Such feelings are not to be excused by any national passions. They are unworthy of what every one has hitherto understood by civilization, and it would be a misfortune indeed were they generally to prevail among persons of culture; and not only a misfortune for civilization, but, we are firmly convinced, a misfortune for the very purpose for which, after all, in the last resort all the present hell was let loose—the national existence of the different countries.

Technical achievement has made the world smaller, and today the countries of that large peninsula Europe seem brought as near to one another as the cities of each individual small Mediterranean peninsula used to be; and Europe—it might almost be said the world—is already one and indivisible, owing to its multitudinous associations.

Hence it must be the duty of educated and philanthropic Europeans to make, at any rate, an effort lest Europe, owing to her not being sufficiently strongly welded together, should suffer the same tragic fate as ancient Greece. Is Europe gradually to be exhausted by fratricidal war and perish?

The war raging at present will scarcely end in a victory for any one, but probably only in defeat. Consequently, it would seem that *educated men in all countries* not only should, but absolutely must, exert all their influence *to prevent the conditions of peace being the source of future wars,* and this no matter what the present uncertain issue of the conflict may be. *Above all must they direct their efforts to seeing that advantage is taken of the fact that this war has thrown all European conditions, as it were, into a melting-pot, to mold Europe into one organic whole, for which both technical and intellectual conditions are ripe.*

This is not the place to discuss how this new European order is to be brought about. We desire only to assert in principle that we are firmly convinced of the time having come for *all Europe to be united together, in order to protect her soil, her inhabitants, and her civilization.*

Believing as we do that the desire for such a state of things is latent in many minds, we are anxious that it should everywhere find expression and thus become a force; and with this end in view it seems to us before all else necessary that there should be a union of all in any way

> attached to European civilization; that is to say, who are what Goethe once almost prophetically called "*good Europeans.*" We must never abandon hope that their collective pronouncement may be heard by some one even amidst the clash of arms, most especially if the "good Europeans of tomorrow" include all those who are esteemed and considered as authorities by their fellow-men.
>
> To begin with, however, it is needful that Europeans should unite, and if, as we hope, there are enough *Europeans in Europe,*—in other words, enough persons to whom Europe is no mere geographical term, but something which they have profoundly at heart,—then we mean to attempt to found such a union of Europeans.
>
> We ourselves wish only to give the first impulse to such a union; wherefore we ask you, should you be in agreement with us, and, like us, bent upon *making the determination of Europe as widely known as possible, to send us your signature.*[18]

In early December 1914, Einstein wrote to Ehrenfest with an early declaration of his identity as an internationalist, again reiterating the themes of war making as insanity and his own alienation from the majority:

> The international catastrophe weighs heavily on me [as an] international person [*internationalem Menschen*]. Living in this "great age" it is hard to understand that we belong to this mad, degenerate species that imputes free will to itself. If only there were an island somewhere for the benevolent and the prudent! Then I too would want to be an ardent patriot.[19]

Naval warfare and some limited use of poison gas and bombing raids began in late 1914 and early 1915.[20] At the start of 1915, Einstein continued to express dismay at the War and its intrusion into intellectuals' lives, as well as his compassion for those who did not share his freedom to stand back from the War. He also speculated about the gendered drive behind war making, and he began to express his incipient faith in a supranational organization. Sometime after December 27, 1914, Einstein wrote to Zangger:

> The world is like a madhouse now. What drives people to kill and maim each other so savagely? I think, in the end, it is the sexual character of

> the male that leads to such wild explosions from time to time, *if it's not prevented by careful organization*. The special calamity of our times, however, is that bestial instincts together with existing tools are leading to veritable destruction. The real betterment of the masses moves so slowly ahead in relation to the rapid development of technology that there is now a disparity of the worst kind. We must therefore, in my opinion, strive toward a large-scale organization that acts toward individual states the way the latter act toward individual thieves. But the *bestia masculina* in particular still resists this. That is how a completely dispassionate person like me can appear *deficient* to others. . . . As long as I am left in peace, I calmly continue my work with the usual pleasure, without letting myself be infected by mass psychosis. What hurts me most bitterly, though, is that people similar to me are prevented by the force of circumstances from acting as I do.[21]

On January 2, 1915, Einstein alluded to the War disparagingly to fellow physicist Edgar Meyer:

> In spite of the tiresome, distressing, atrocious War, I am working quietly in my room.[22]

On January 8, Einstein used the analogy of disease in characterizing the War to Italian mathematical physicist Paulo Straneo. He again noted that he was alienated from many academic colleagues, while urging the sustenance of international ties:

> I love science twice as much in this era, when I feel so painfully the emotional misjudgments of almost all of my fellow human beings and their sad consequences. It is as if an insidious epidemic had confused their minds! All the more must the circumspect, especially we scientists, foster international relationships and distance ourselves from the raw emotions of the mob; unfortunately we have had to experience serious disappointments, even among scientists, in this regard.[23]

On January 11, Einstein praised Zangger for his medical research on the effects of poison gas warfare on soldiers. He continued to express horrified dismay at both the War and people's involvement in it, along with an ironic dig at contradictions in the manifesto "To the Civilized World":

> As a man with faith in society you are seeking to assuage the terrible suffering to which people are now falling victim, and are finding solace in action. The likes of us, however, are so disgusted by the abundance of incomprehensible, puzzling, and unspeakable ugliness that we crawl even more than usual inside our shell of brooding. . . . Here scientific life is almost at a standstill. Everyone is working and suffering for the state, in part voluntarily, in part involuntarily. If only one could do something about it so that the prior relatively harmonious conditions could be reestablished! But in this state of agitation lack of passion is insulting to the others who are more deeply caught in this affair. . . . It's good that among the fine minds strikingly many are considered cripples by the state, especially among theoreticians. [Peter] Debye, [Max] Born, and [Max von] Laue were all deemed absolutely unfit. The latter told me recently that he had insurmountable difficulties learning the rifle drills—a cute illustration of the manifesto published by my colleagues, according to which "we owe our scientific prowess not least to our military education."[24]

Einstein, along with other pacifists, shared with war enthusiasts the belief that the War would end quickly. On January 25, he explained to his almost eleven-year-old son Hans Albert, who lived with his mother in Zurich, why he hoped to go hiking with him the following summer:

> If all goes well, the terrible War will be over by then.[25]

On February 20, Einstein wrote to Nicolai, whose "Appeal to the Europeans" he had signed in October 1914, agreeing, albeit without optimism, to sign another manifesto (and also suggesting the names of four academic colleagues in Berlin and two in Zurich who might also sign). Einstein felt especially pained at the way in which international scholarly collaboration had been stymied by the "Manifesto of the 93":

> Although I am convinced that the voice of a handful of the informed carries little weight against the lust for power of the mighty and the

> fanaticism of the many, I still welcome your manifesto with pleasure and am pleased to be permitted to add my name to it. The manifesto will contribute in any case toward reestablishing the former good relations among researchers of civilized countries, which have already been clouded by some vehement and rash words.[26]

On March 21, 1915, Albert and Elsa Einstein attended their first meeting in Berlin of the pacifist New Fatherland League, or Bund Neues Vaterland (BNV). The BNV was founded in November 1914, sparked by a German governmental shift away from portraying the War as defensive toward early discussions of territorial annexations, which pacifists and BNV members opposed. In addition to opposing annexations, other goals of the New Fatherland League were to create a culturally European concept of nation surpassing national boundaries, to reach political and economic understanding between the European nations, and to base foreign policy on democratic parliamentary participation. The prevention of war through international legal organization, which became part of Einstein's lifelong political philosophy, was also promoted by the BNV. One week before the Einsteins attended this meeting, the BNV sent out a circular urging cultural and political understanding between the European peoples, a new generous spirit in German foreign policy, and participation of the German people in this. The New Fatherland League took the old concept of the gentleman-diplomat, in which aristocrats pursued diplomacy as a vocation, expanding it slightly to include intellectuals in all fields informally self-educating on foreign affairs, then attempting to influence foreign policy through personal contact. German government officials met with BNV members in its first months, probably seeking information toward a negotiated peace with the Allies. However, by mid-1915, as government interest waned, the New Fatherland League became expendable, contact ceased, and in 1916 the BNV was ordered to disband.[27]

The day after he attended this meeting, Einstein clearly aligned himself with pacifism when he initiated contact with the prominent French pacifist Romain Rolland, who was living in exile in Geneva. He expressed his profound admiration for Rolland, characterized the war-enthusiasm as delusion and epidemic, stressed the lack of progress in European history from religious to patriotic mania behind war making, and offered his help to pacifist endeavors:

> From the daily paper and through my connections with the highly commendable "Fatherland" League I have learned of how courageously you have placed your life and person at risk toward eliminating the so ominous misunderstandings between the French and German peoples. I feel impelled to express my complete admiration and respect. May your magnificent example awaken other prominent men from the inexplicable delusion that has gripped like an insidious epidemic even competent men of otherwise steady reason and healthy sentiment! After three hundred years of assiduous cultural activity, must our Europe really be remembered in coming centuries for not having advanced further than from religious mania to patriotic mania? Even scholars of the various nations behave as if their cerebrums had been amputated 8 months ago.
>
> I place my slight powers at your disposal in case you think I can be of service to you as a tool, be it through my residence or through my connections with German and foreign individuals in the exact sciences.[28]

Rolland's views contributed to the development of Einstein's pacifist philosophy and pragmatic world view. At this early point in their correspondence and in the War, Rolland wrote that this "terrible crisis" had been "a harsh lesson for all of us Europeans . . . In the future we have to be better armed against a recurrence of such a scourge (for we cannot delude ourselves that this folly of humanity was the last . . .)." Rolland argued that right at the start of the War some delegates should have met in a neutral country to "shed light on the facts" in a spirit of "tempering the impassioned assertions by both camps and offering the voice of reason," to be "the lucid and steadfast conscience of their nations." For Rolland, "to resist the derangements of the community as a whole" imposed a mission "to organize ourselves later in a more European—that is to say, a truly universal—manner." This would be even more difficult after the War, "because misunderstandings, grudges, and bitterness will last for a long time." He hoped that "the immense sufferings and delirium" would lead to a reaction and a reawakening, and he advised that "one of our most effective tasks must be to disseminate documents that can oppose the spirit of hatred."[29] These ideas are clearly visible in Einstein's later reliance on supranational organization, his focus on anticipating the future, and his involvement after the War in efforts to document its events.

On April 2, 1915, and later, Einstein, his cousin Elsa, and her daughter Ilse sent postcards of support to the embattled pacifist Georg Friedrich Nicolai, coauthor of the "Appeal to the Europeans," then stationed at the garrison town of Graudenz. Nicolai, when called up for military service, had refused to take the oath of allegiance and as a consequence was demoted to a private and medical orderly. He was now attempting to resist the War and propagate pacifism from within the military.[30] Also on April 2, members of the New Fatherland League met, along with Einstein, prior to sending a delegation to a meeting in The Hague on April 7 to 10. At the meeting in The Hague, the Central Organization for a Durable Peace was founded.[31]

Around April 10, Einstein again resorted to the analogy of insanity to characterize supporters of the War. It was clear in his letter to Zangger that Rolland's advocacy of international organization had made a strong impression on him, although he had less faith than Rolland in people's potential to change:

> I am now starting to feel comfortable in the present-day mad turmoil, in conscious detachment from all things that occupy the deranged public at large. Why is it not acceptable to live enjoyably as a member of the madhouse staff? . . . Romain Rolland, who currently lives in Geneva, recently sent me a suggestion, which—to continue the metaphor—amounts to the organization of the sane staff at all asylums for the purpose that they not become deranged as well. Besides, he has high hopes that such an organization would even more or less heal the insane, the optimist! If you have the opportunity, take care of him; he is being persecuted for his international convictions. . . . I hope confidently that there will soon be peace; the insane will soon turn their efforts to a more harmless field again. But they will remain just what they are.[32]

In mid-April, Einstein expressed to Tullio Levi-Civita, an Italian professor of rational mechanics, his frustration at the War's interference with productive intellectual relationships:

> Unfortunately, the attitude of our nonprofessional fellow human beings is incomparably stranger. So it is doubly gratifying getting to know a man like you more closely. I shall happily strive to turn our acquaintance by letter into a personal one: one more reason to cross the Alps at last again one day. It is to be hoped that our fatherlands will not become rebellious against each other as well![33]

Three weeks later, Einstein reiterated to Levi-Civita his desire to make his personal acquaintance.[34]

The spring of 1915 was a time of intensive developments and soaring mortality rates in the War. In April and May, the Turks carried out genocide against the Armenians, poison gas was used for the first time on the western front, the disastrous Gallipoli campaign began, and large numbers of merchant ships were sunk, as well as the passenger ship *Lusitania*.[35] On May 17, 1915, Einstein described to Zangger his mental survival strategies, which included reliance on individual friendships and a partly feigned detachment. He alluded again to the social isolation of pacifists, and he touched on the food supply issue in Switzerland, a particular concern for him because his sons were living there:

> In these times one sees that the only thing in this world really worth striving for is the friendship of exceptional and independent persons who cannot be convinced by any published nonsense to assume the most perverse mental stances. How few such people there are you surely see in Zurich as well as I do here.
>
> . . . People are not as consistent in anything, not even in hatred and unfriendliness, as it may seem. . . . I always get by best with my naiveté, which of course is 20% deliberate. This is simply done if you are indifferent enough to the feelings of your dear fellow men; of course, one can never be as indifferent as they deserve.
>
> Apart from the general misery, of which we learn mainly from the newspapers, I am leading a happy, remarkably peaceful life, more withdrawn than in Zurich. The scarcity of genuinely like-minded people is perhaps not worse than what you experience in Zurich. The likes of us simply have to be content with loneliness in this regard, if no special stroke of luck prevails. . . .

> It is very grave for Switzerland that in addition the Italians are also starting up. Maintaining the food supply will be difficult.[36]

At the end of May, Einstein urged Walter Dällenbach, an electrical engineering student in Switzerland, to be proud he was a citizen of neutral Switzerland.[37] By June 1915, Einstein had become a documented member of the New Fatherland League.[38] On July 7, he wrote to Zangger criticizing the evil effects of unreasoned emotion and its relationship with war. He also praised the few academic colleagues he could identify who were trying to resist the chauvinistic tide:

> The longer this dreadful state of war lasts, the more grimly people become fixed on unreasonable hatred based on nothing. When one is young, one admires fervent emotion and scorns cold calculation. But today I think that the blunders that stem from blind emotion bring much more bitter unhappiness into the world than the most heartless calculating mind could.
>
> But in these times one doubly appreciates the few people who stand entirely above the fray and do not let themselves be carried along in the grim current of the times. One such person is [David] Hilbert, the Göttingen mathematician. . . .
>
> Among my truly profound colleagues, like [Max] Planck and Hilbert, the wish to maintain warm relationships with like-minded scholars abroad is very much alive. Hilbert now doubly regrets, as he told me, through carelessness not having better cultivated his international relationships. Planck is doing everything to rein in the chauvinistic majority at the Academy.[39]

Both Max Planck and David Hilbert, as well as Einstein, were among 140 signatories of the Delbrück-Dernburg petition to German chancellor Bethmann Hollweg. Written on July 9 and dated July 27, 1915, the petition rejected the annexationist war aims declared in the Seeberg address of July 8. That address, also known as the "Petition of the Intellectuals" (*Intellektuelleneingabe*), was drafted by Berlin theologian Reinhold Seeberg under the authority of the Pan-German League. Signed by 1,347 professionals, among them 352 university professors, the Seeberg address urged German annexation of territories on both the eastern and western fronts, both for Germany's military security and as part of a "civilizing" mission in the east.[40] The Delbrück-Dernburg petition, in response, while reasserting that Germany would win the War, opposed territorial annexations by Germany on both principled and pragmatic grounds, out of concern for the future well-being of Germany:

> Germany did not enter the War with the intention of conquest, but rather to maintain its existence, threatened by the enemy alliance, its national unity, and its continuing development. Thus Germany may only pursue in a peace agreement what serves these goals. Your Excellence has received petitions that contravene these goals. We therefore see it as our duty to counter these efforts with total determination and to state openly that we would see in their realization a serious political error and not a strengthening but a disastrous weakening of the German Reich.
>
> Based on purely factual considerations, we support the principle that the annexation of politically independent nations and nations accustomed to independence must be rejected. The German Reich arose from the concept of national unity, of national identity. It has assimilated foreign elements only slowly and incompletely, and we don't want to be pressured, neither through events nor people nor easily fabricated voices, to abandon or alter the leading guidelines of the creation of the Reich or destroy the character of the national state.
>
> It is entirely self-evident that the regions to be evacuated by us according to the stipulation of our peace conditions may not become a bulwark for our opponents, that no rival of Germany may position itself there. It is impossible for hostile feelings of residents to be converted by hostile treatment, which could threaten the peace and security of our borders. Such dangers can be prevented, and we trust that appropriate and effective means will successfully be chosen and implemented. To that end, though, we cannot include such means as would ultimately lead us by a circuitous route to annexation.
>
> We are all firmly convinced, with the entire nation, that this War will end with a complete German victory. After such admirable heroic actions, after such unending sacrifices and efforts, after such glory and such quietly and magnanimously born suffering, the German people will be entitled to a winner's prize, which, as much as possible,

> corresponds with what they sacrificed. The highest winner's prize will always consist in the proudly achieved certainty that Germany also need not fear a world of enemies, and in the unprecedented proof of strength that our people have given to the other people of the earth and to future generations. However, the German people can only conclude a peace that gives a secure basis to the strategic needs, the political and economic interests of the land, and the unhindered activity of its power and its entrepreneurial spirit in its native country and on the open seas. We harbor the trust that Your Excellency and the constitutionally appointed authorities will succeed with composure, should the occasion arise, at the height of our military success to create such a peace.[41]

On July 21, while on the Baltic coast in northern Germany, Einstein wrote to the Dutch physicist Hendrik A. Lorentz, an important mentor in his life, out of concern about the destructive effects of the War on relations among scientists of different countries. He appealed to Lorentz to take action toward international reconciliation among fellow academics:

> I spoke with [Max] Planck recently, and we both reflected sadly on the cruel division that has arisen between us and our highly treasured foreign colleagues as a result of the unfortunate War. Whatever may have been failing in deplorable political agitation on both sides, it is never too late for turning back. It is certain that we scholars are all innocent of the War and that the current sad circumstances should induce us even more to solidarity . . .
>
> . . . If I did not live in Berlin, I myself would write to our close colleagues in France and England . . . to the extent that the old friendship in our circle can be reestablished. I would ask them all to assemble at an appropriate location (Holland or Switzerland) already during these holidays, entirely voluntarily and unofficially, primarily for the cultivation of personal relationships.
>
> But I live in Berlin, have few relationships, and also little skill in communication with people. This is why I am turning trustingly to you in the hope that you can transform all this, of which I can only dream, into reality. Would you not have joy in dedicating some time to this fine task? . . . This is all the more important as there are signs that the official relationships between the societies, as well as the academies,

> could get broken off; because the waves of nationalistic blindness are rising very high. But I notice here that precisely the very best are struggling against it with all their strength, and surely it will be so in other countries as well.[42]

Several days later, Einstein described to Zangger the intellectual and experiential factors in the development of his world view; these ranged from Tolstoy to train journeys with soldiers. He noted with baffled regret that even wartime losses seemed not to convince people—not even the highly educated—of the reprehensible nature of war and the untrustworthiness of national leaders:

> I am sorry for Kaufler concerning his loss. He is not alone in it. These hard fates have an entirely different effect on people than one would have thought. Rather than rising to the recognition that they should not let themselves be led blindly and should discard all unjustifiably instilled trust, in order not to be shamefully abused, they hurl their hatred unreasonably wherever the leaders wish it. It seems that no education and intellectual cultivation can protect against this wretched mania. The upright few surely are already born with the ability to remain upright. Read the moving little book *Christianity and Patriotism* by Tolstoy. Tolstoy knows that what he says will indeed help nothing, although it too ends more optimistically with a hymn to the development and purification of public opinion. He has not traveled with soldiers, as I have, and taken in their stupidity and crudeness.[43]

In early August 1915, Einstein expressed to Lorentz his view that scientists were less chauvinistic than scholars in the humanities. His comments reflected the negative impact of the "Manifesto of the 93," along with his dismay at power politics, narrow nationalism, and the splitting up of the international scholarly community:

> I already had indications of the mood of our colleagues abroad. It is strange in Berlin. Natural scientists and mathematicians as scholars are rigorously internationally minded and guard carefully against any unfriendly measures being taken against their colleagues who live in

> hostile foreign countries. Historians and philologists, however, are for the most part chauvinistic hotheads. The famous and notorious "Manifesto to the Civilized World" is being regretted by all calm-thinking people here. The signatures were given negligently, some of them without prior reading of the text. This is how it was, for example, with Planck and Fischer, who have very resolutely shown a commitment to the preservation of international ties. . . . But I believe that these people cannot be induced to retract their words.
>
> I must confess that the narrow nationalistic sentiment even of people of high standing is a bitter disappointment to me. Furthermore, I must say that my respect for the politically more advanced states has diminished significantly through the awareness that they are all in the hands of oligarchies, which own the press and wield the power and can do what they want. . . . Add to this that the perceptive and powerful have no feeling for the multitude; there one has the sad picture of what is revered as the "fatherland" by those who belong to it. This does not alter with the boundary posts, but rather is everywhere nearly the same. And behind this flimsy ideal, should the relationships between people who have learned to respect one another in personal and scholarly interactions fade? It is too unbelievable and incomprehensible to me. It seems that people always need a fantasy for the sake of which they can hate one another; in the past, it was religious faith, now it is the state.[44]

In August and September 1915, the casualty figures at Gallipoli rose even higher, and the Allies began using poison gas.[45] In August 1915, Einstein reiterated to his other Dutch colleague, Paul Ehrenfest, his passion for the international scholarly community:

> I would so much like to do something to hold together our colleagues from the various fatherlands. Isn't this little cluster of industrious intellectuals the only "fatherland" to which people such as ourselves have any serious attachment? Should *these* people too have fundamental attitudes that are solely a function of their place of residence?[46]

Einstein shared many of the same daily struggles and hardships that most citizens encountered in wartime—for example, difficulties in crossing borders (particularly to Switzerland, where his sons lived) and in obtaining passports for himself and his loved ones. At the end of August, he wrote from his mother's home in Heilbronn to his future second wife Elsa in Berlin to get her help in registering his departure from Berlin, something he was required to do before he could cross the border into Switzerland with his Swiss passport.[47]

He also discussed with Elsa his observations of Germans on the home front, his own self-censorship, and his assessment of how the world works:

> Here people are politically much more confidently and aggressively engaged than in Berlin, no doubt because the food situation is better. I have very much restrained myself in conversations about these matters . . . Also, experience has convinced me that words and ideas do not guide the world, but rather direct, tough experience.[48]

In September 1915, Einstein described to Zangger further wartime travel problems encountered on his return trip:

> In Constance my passport was questioned because *the visa from the German consulate in Zurich was missing*. Therefore, I was transported by honorary military escort . . . to Kreuzlingen. . . . Now temporarily I am stuck in the Hotel Löwen in *Kreuz*lingen . . . I enclose my passport with the request that you get the visa and send it back to me. Should the "powers-that-be" however refuse to give their blessing to the passport, then please summon me back to Zurich by telegraph.[49]

In mid-September 1915, while in Switzerland visiting his sons, Einstein traveled to Vevey to meet Romain Rolland, the French pacifist who was still in exile in Switzerland and about to be awarded the 1915 Nobel Prize for literature.[50] In a letter preceding the visit, he offered Rolland his observations of Germany and recounted the incident that had led him to distinguish between scientists and scholars in the humanities. He also discussed factors influencing levels of confidence that Germany would be victorious and the growing pressure faced by the pacifist organization he had joined:

> I want at least to use my time in Switzerland to send you an uncensored letter. The "New Fatherland" association is going through rather difficult times; it is being harassed by the authorities and condemned (on the whole) by the press. It appears as if the pro-military party, i.e., the pan-Germans, have attained influence on the government through the successes in Russia. But on the other hand, the people of my acquaintance who are most discerning on economic matters do not have especially great confidence; this seems to be connected with the shortage of certain raw materials. Remarkably, in Germany one finds, alongside a strange self-conceit, a love for France and its population, whereas great hatred against England is nearly universal. Among the uncritical masses confidence in victory is almost universal and annexation greed no less prevalent. It is peculiar how the ordinary man can feel recompensed for his heavy sacrifices through stolen territory, from which he personally certainly benefits little.—Hopefully it won't come to that! A decisive victory for Germany would be a misfortune for all of Europe, but especially for this country itself.
>
> One of the most distressing phenomena of this terrible time is that many of the intellectuals have so completely lost their composure. . . . You certainly must have been surprised that so many men who in peacetime were justifiably regarded as level-headed signed the infamous "Manifesto to the Civilized World." Condemnation of this step is now also quite universal in Berlin. Incidentally, the best people in fact had given their signatures *by telephone*, without having read the manifesto! Recently there was great agitation at the Academy, because someone had responded to Voille's Academy speech by submitting a petition to break off relations with the French Academies. Something odd happened then: almost all the historians, philologists, etc., supported the petition, while almost all the natural scientists and mathematicians worked assiduously to maintain the international ties. Thank God, the latter won, if only by a small majority . . .
>
> I wish you good success in your humanitarian endeavors.[51]

Also in September, Einstein's reflections to Zangger on the War made clear his view that Germany was more militaristic than France or Switzerland, as well as his aversion to the many manifestations of militarism, including territorial conquest:

> This printed leaflet is very sweet and there is some truth to it. But why is Delcassé made minister? Why is revenge taught in schoolbooks? Why are parts of Turkey consumed while merrily whistling at the Algeciras Treaty? Why did Jaurès have to be murdered? . . .
>
> And yet how nice France looks when I think of the others! . . . A drive to dominate and a lust for power have poisoned the minds of the upper class everywhere; but in France at least there has been a lack of uniformity and consistency in the evil. Furthermore, I am convinced that the gospel that brutal armed force is the basis of all existence finds its main support in Germany. Even the disgraceful regime in Russia finds powerful support in the German ruling class.
>
> I am very pleased that in Switzerland recognition of the seat of the real danger is dawning, and I confidently believe that this simple recognition will prevail despite the press. The press! The article in *Volksrecht* on the petition of intellectuals on the annexation question has been silently ignored by the bourgeois press. Is it fear? Have scholars of distinction elsewhere lent their support to something like this?[52]

In late September, Einstein rejoiced to Lorentz that fellow scientists Max Planck, David Hilbert, and others had been persuaded to sign the Delbrück-Dernburg petition, opposing annexationist policies:

> Today I had the great joy of seeing many highly regarded local scholars declare their vigorous opposition to annexations in a petition to the Chancellor.[53]

In early October 1915, Einstein sent a printed version of the Delbrück-Dernburg petition to Zangger, commenting in a separate letter:

> It was a very conciliatory manifesto . . . the best people in the country signed.[54]

Also in early October, Einstein complained to the Göttingen physicist Paul Hertz of the destructive effects of a failure to show civil courage:

> . . . this kind of cautiousness, i.e., not standing up for one's rights, is guilty of the entire terrible political mess.[55]

Shortly afterwards, still in early October, he nudged Hertz:

> People like you from the most highly educated class offer the best guarantee for the preservation of the political quagmire (in eternity amen).[56]

Later in October, Einstein wrote to the prominent German pacifist Walther Schücking, a professor of public and international law at the University of Marburg and member of the New Fatherland League, who had contacted him through another league member, Lilli Jannasch. Einstein accepted Schücking's invitation to serve on a German national committee of the Central Organization for a Durable Peace, a group created by the Dutch Anti-War Council (Anti-Oorlag Raad):

> From Miss Jannasch I hear that you would like to have me included in the Anti-Oorlag Raad's Grand Council. I have absolutely no experience nor am I a competent person in political affairs. Nonetheless, I am happily ready to support this excellent cause. So do include me, if you consider it beneficial.[57]

In late October or early November, Einstein agreed to submit a statement of his views on the War to the Berlin Goethe League (Berliner Goethebund). He titled it "My Opinion on the War," and in it he identified himself as an internationalist. He argued that male aggression and subconscious patriotism, even in peacetime, were the roots of war, whose eradication would require both a supranational organization and efforts by individuals to eliminate their combative vices:

> The psychological root of the War lies in my opinion in a biologically grounded aggressive characteristic of the male creature. . . . This aggressive tendency comes into force pervasively whenever individual men are placed side by side, and even more so when relatively close-knit societies have to deal with each other. Almost without fail these dealings with one another will turn into disputes that escalate into quarrels and murder, if no special precautions are taken to prevent such incidents. . . . Who could doubt that vendetta and dueling spring from this feeling? I

even believe that the *honor* that is so carefully cultivated among us gets its main nourishment from this source.

Modern organized states of course had to push these expressions of primitive virile characteristics decisively into the background. But wherever two nation states lie next to each other without jointly belonging to an all-powerful organization, this feeling from time to time generates an enormous tension in the emotions, which leads to catastrophes of war. Meanwhile, I consider the so-called aims and causes of war to be quite trivial; they are always found when passion needs them.

The more refined intellects of all times have agreed that war is one of the most terrible enemies of human development, and that everything must be done to prevent it. Despite the unspeakably sad conditions of the present, I have the conviction that in the not too distant future a state-like organization in Europe will be formed, which will suspend European wars, just as the German Reich now does with war between Bavaria and Württemberg. No friend of spiritual evolution should fail to stand up for this most important political goal of our time.

One can consider the question: Why does the human being in peacetime, when the community of the state suppresses almost every expression of tough rowdiness, not lose the characteristics and motivations that enable him to commit mass murder during war? The reasons seem to me as follows. When I look into the mind of the well-intentioned average citizen, I see a moderately illuminated cozy room. In one corner stands a well-tended shrine, of which the man of the house is very proud, and every visitor is loudly alerted to the presence of it; on it in huge letters the word "patriotism" is written. However, to open this shrine is usually frowned upon. Moreover, the owner of the house barely knows, or does not know at all, that this shrine holds the moral requisites of bestial hatred and mass murder, which he obediently takes out in case of war in order to serve them. This shrine, dear reader, you will not find in my little room, and I would be happy if you would adopt the perspective that in that corner of *your* little room a piano or a small bookcase would be more fitting than a piece of furniture that you only find bearable because you have been accustomed to it since your youth.

I have no intention of making a secret of my international sentiments. How close I feel to a person or a human organization depends solely upon how I judge their intentions and capabilities. The state to

> which I belong as a citizen does not play the slightest role in my emotional life; because I regard this affiliation with a state as a matter of business, something like the relationship with a life insurance company. That I probably must try to be a citizen of a country that presumably does not force me into participation in a war is self-evident from these statements.
>
> But how does a powerless individual contribute to attaining this goal? . . . I actually think that the intellectually more mature people of Europe have sinned through the neglect of general political questions; yet I do not see the cultivation of politics as the greatest effectiveness of the individual in this matter. I believe, rather, that each individual should act personally to steer those feelings . . . as much as possible on a course that could no longer become a curse for humankind as a whole.
>
> . . . Lust for power and greed should, as in the past, be treated as despicable vices, likewise hatred and combativeness. . . . in my view we unfortunately have not made progress on this important point, but rather have fallen backward. Every well-intentioned person should work to improve himself and his personal circle in this respect. Then the heavy plagues that currently afflict us in such a horrible way will also vanish.
>
> . . . if I can say it all in one sentence, moreover in a sentence befitting to me as a Jew: Honor your master Jesus Christ not only in words and songs, but above all through your deeds.[58]

In his correspondence with the Berlin Goethe League in reference to this statement, Einstein reiterated his emphasis on mentalities developed in peacetime and touched again on insights acquired from Tolstoy:

> If I am to express my opinion on war in a truthful manner, I must in some form say that the glorification of war in peacetime as well as the fostering in general of those thought-feeling complexes that prepare for war in peacetime, must be energetically counteracted by all real friends of human progress. This includes, in my view, *everything* that falls under the term "patriotism." In this respect I must agree unconditionally with Tolstoy.[59]

Einstein ended the first full calendar year of the War commenting on his fledgling pacifist engagement in communications with several key people in his personal life. In late November and early December 1915, he mentioned his involvement in the Central Organization for a Durable Peace to at least four people among his Swiss friends and family: Zangger, his university friend Michele Besso, his older son, Hans Albert, and his estranged wife, Mileva Einstein-Marić. To Zangger he wrote:

> I have become a member of the Great Council of the Dutch Anti-War Council. The times show that everyone must do his bit toward the organization of the whole. Unfortunately so many intelligent people are acting counter to this.[60]

To Besso he wrote:

> Anyway I must go at Easter to Berne, where a meeting of the Anti-War Council is taking place, into whose Great Council they have elected me.[61]

To Hans Albert and to Mileva he also mentioned his plan to attend the meeting in Berne.[62] And to Zangger a couple of weeks later he reiterated:

> At Easter I have to be present at a meeting in Berne of the Anti-War Council, into whose international council I let myself be elected. In these times everyone must do whatever one can for the community as a whole, even if it is only minimal and awkward.[63]

Thus, in the first seventeen months of the War, Einstein responded with horror at it and at the attitudes that sustained it, regarding it as a deplorable madness, particularly in its fracturing of international scientific collaboration. At the same time, he was beginning to develop his lifelong faith in international organization as war's preventative. With the "Appeal to the Europeans," he began a lifelong practice of making ethically based public stands, often—as in this case—in comparative isolation. In joining the New Fatherland League, he had become part of the era's most prominent pacifist organization. Nine months after the "Appeal to the Europeans" was disseminated, he signed the Delbrück-Dernburg petition; three months after that, he provided the Berlin Goethe League with his own summary statement, "My Opinion on the War." He also developed friendships with prominent pacifists such as Romain Rolland, paid close attention to which scientific colleagues shared his views and which did not, analyzed the motives and mentalities of the War's supporters, and made his convictions about the War clear to his closest friends and family. All of this he did prior to the War's bloodiest battles and most horrific sufferings of hunger and disease.

2 Disgust with War, Pessimism, and International Organization

1916–1917

In early 1916, developments in the War included the ending of the battles of Gallipoli after massive casualties, the first bombings of civilian targets on both sides, and the start of the battle of Verdun, which would last through the end of that year.[1] On February 7, 1916, the New Fatherland League (Bund Neues Vaterland) was outlawed in Germany for the duration of the War.[2] In mid-April, Einstein expressed to Elsa his continuing concern about Georg Friedrich Nicolai, the University of Berlin medical school professor and author of the "Appeal to the Europeans" who was attempting pacifist consciousness-raising from inside the Germany army.[3] In late April, Einstein expressed his delight to Dutch physicist Paul Ehrenfest at conciliatory communications between his professional mentors, German and Dutch physicists Max Planck and Hendrik A. Lorentz, aimed at mitigating the harm caused by the "Manifesto of the 93."[4]

The summer of 1916 was a period of fiercely intense combat in World War I on the seas and on the eastern front, in addition to the ongoing battle of Verdun on the western front.[5] On July 1, 1916, the battle of the Somme began in France, during which more than one million soldiers would die.[6] On the German home front, civilians were facing worsening economic hardship and developing pessimism about the War, largely because of millions of casualties in dead and wounded, and due to soldiers' letters home.[7] That summer, Einstein repeatedly communicated to Zangger his views on the War and its hardships as well as his alienation from those who supported it. Yet he also still shared the widespread belief that it had to end soon. In mid-July 1916, while discussing difficulties in relations with his sons, he expressed his sensitivity to the greater tragedies in other families:

> How much better off I am than countless others, who have lost their children in the War! Planck also lost a son that way, the other one has been stuck in French captivity for almost 2 years. . . . I will not come to Switzerland for this vacation, for one because I trust that peace is coming soon, so next spring I could travel there without the chicaneries . . . I shut my eyes as much as possible to the insane hustle and bustle in the world at large, having completely lost my social feeling. How can anyone participate in such a social monstrosity if one is a decent person? A fleeting glance at the newspaper is enough to make one disgusted with our contemporaries. Only in individual people can one find consolation.[8]

A little later in July, Einstein suggested to Zangger a notion that became a repeated theme for him by the end of the War, the idea that hardship had a salutary effect on the character of the German people, whereas victory would do them harm:

> The War is having the most auspiciously educational effect on the people here, or better stated, the food shortage is. If it continues consistently like this, as it seems that it will, folks will become even more entirely sympathetic. No person and no nation can endure great external success undamaged.[9]

Einstein was not greatly bothered when the rationing of clothing began in August. But he continued to feel baffled and dismayed by people's mindless support for the War, detailing their wartime mentalities to Zangger:

> The regulations of August 1st don't bother me much. I won't have to walk around naked, since I am fairly well supplied with clothes and my aesthetic demands in this regard are minimal. My insight into people does not teach me very much that's of interest. Accepting events as if

> they were hail sent by God, a powerless yearning for peace, a lack of understanding of the causal relations and the psychology of others all the way up to the elite intellectual (and loudmouth) heroes, melancholy conviction in one's own integrity and others' depravity, in short, stupidity and still more stupidity. Even the financial robber barons . . . are no worse than the others. They are acting—like the others—consistently with the training that they received, and consider themselves good, useful, and irreplaceable. Everywhere an unhealthy overrating of power and money and an indescribable desolation of the soul.[10]

In comments to Zangger in late September, Einstein touched upon his appreciation for the true spirit of Christianity, another occasionally recurring theme for him:

> I'm now gradually starting to understand what a treasure people had in the Christian ideal of humility. Who will replace the idol of power with a merciful God again?[11]

In the autumn of 1916, aerial bombing continued, Germany gained access to oil fields in Romania, and the multifront dimension of the World War was reflected by developments in Poland, on the Somme, in Italy, and in the Middle East.[12] Writing to Elsa in Berlin while visiting Holland in September and October 1916, Einstein discussed efforts to obtain food supplies unavailable in wartime Germany due to rationing.[13] During that visit, in early October, he probably attended a meeting in The Hague of the Anti-War Council of the Central Organization for Durable Peace.[14]

In mid-October, Einstein accepted an invitation from Werner Weisbach, an art history instructor at the University of Berlin, to join his Association of the Like-Minded (Vereinigung Gleichgesinnter), a discussion group in existence since June 1915 focused on ethical issues pertaining to power politics and common concerns of the European states. Einstein had been nominated for membership in this group by fellow New Fatherland League member Georg Count von Arco, and the group also included fellow "Appeal to the Europeans" cosigner Wilhelm Foerster. Einstein's response to the invitation used nineteenth-century German chancellor

Otto von Bismarck and University of Berlin historian Heinrich von Treitschke as personifications of the spirit of militarism, and it reiterated his conviction that a German victory would foster evils while a Germany defeat would be healthy. He also offered an early expression of his lifelong commitment to an organization of states:

> The intention of your new organization seems to me entirely above criticism. I am convinced that the malady of our times is that moral ideals have almost lost their power. Bismarck-Treitschke personify in quick summary the history of the ailment. And now we are experiencing the crisis:
>
> If Bismarck-Treitschke emerge with the glory of a foreign victory, then the world will be morally contaminated for an incalculable period of time. Humanity will then have to endure an endless chain of betrayals and atrocious acts of violence.
>
> However, if Bismarck-Treitschke are not successful, people will lose faith in the barren ideal of power and willingly and sincerely extend to the states the principles of law. Then our passionately longed-for goal of an organization of states (at least of Europe and America) ruling out war will be achieved in a short time.
>
> This conviction I support publicly, even though I am very aware that personal advocacy of this conviction can only have a weak influence on the masses (including professors!). Victories or defeats abroad are decisive, distressing though it is; let's not deceive ourselves about this. So, include me on your list, so that I at least have the consolation: dixi et salvavi animam meam.[15]

In October 1916, Einstein became involved in discussions of a proposal initiated by Jean Massart, a professor of botany at the Free University of Brussels, also advocated by Lorentz, to establish a German-Belgian commission to investigate atrocities committed by German troops in Belgium and in Lille, France, during the early part of the War. After the War, this initiative developed into a commission of German citizens for this purpose, of which Einstein was a member (see chapters 4 and 5). In mid-October, Einstein mentioned to Paul Ehrenfest and his wife Tatiana plans to speak with Elisabeth Rotten, a Swiss contact person for the New

Fatherland League, who also later participated in this commission.[16] Later in October, he described to Ehrenfest his efforts to spread the idea among academic colleagues in Germany—efforts that had to be cautious given the deeply embedded nationalism and political passivity of the times:

> Recently, while sitting innocently together with [Max] Planck and [Heinrich] Rubens, I planted Lorentz's ideas in their heads, at which their previously normal faces fell noticeably. . . . I shall speak with [Wilhelm von] Waldeyer and [Hermann] Diels as well. If possible, I'm not going to visit them separately, but rather shall try to speak to them at the Academy, so that they do not get the impression that a planned assault is being made. When matters such as these are involved, these otherwise truly highly worthy men show an inhibition, somewhat like someone given a command in a hypnotic state. In our case, this hypnosis happens during childhood.[17]

In November 1916, Einstein continued to advocate the proposed commission in Germany and to communicate with his Dutch colleagues Ehrenfest and Lorentz about his efforts. To Ehrenfest he wrote of his perceived success with Wilhelm von Waldeyer-Hartz, who advocated a delay (which became the eventual future timing of the commission):

> I have spoken now with Waldeyer (Secretary of the Academy) about Lorentz's idea. . . . He received the suggestion most warmly. He considers implementation, though, only possible after the conclusion of peace. Then, however, he intends to support it. He wants to write to Lorentz immediately himself. I was delighted with the reaction of the man.[18]

And to Lorentz he described his efforts with various German colleagues and the "cautious disapproval" with which those efforts were often greeted:

> I am still feeling fully revitalized from the trip to Holland. Not only meeting with highly esteemed men with similarly oriented commitments made this experience so liberating, but also in particular the concurrence in opinion on non-scholarly matters. I have now spoken with

> Planck, Rubens, Waldeyer, and [Walther] Nernst about the matter close to your heart, and can tell you, to my delight, that I had better success with it than I had expected. I encountered with Planck and Rubens, though, a kind of cautious disapproval, which, however, should not be ascribed to ill will but rather to a kind of timidity toward actions with the faintest political flavor. For you know yourself that the former is an extraordinarily conscientious and truthful person. . . . I spoke with Waldeyer recently after the Academy meeting. He . . . spontaneously promised to write to you himself about the matter. He considers implementation to be possible only after the War, but wants to give it his complete and dedicated support. He considers the proposal fair and legitimate and hopes, from the elimination of all confusion, to achieve an improvement in relationships between scholars of the opposing sides. His honest attitude, not stained by any utilitarian considerations whatsoever, encouraged me a great deal. Nernst also welcomes the proposal. . . . He would consider it proper if such a proposal came from locally known men or corporate bodies of neutral foreign countries. Then, no one would overlook the matter. Prudence leads him to agreement, rather than the direct wish to be just. But this too is better than the absence of prudence and a sense of justice, as is usually encountered.[19]

To Ehrenfest he expressed both empathy for and condemnation of these German academic colleagues, again echoing his disgust at the war-enthusiasm:

> Seeing these people from close up dissolves any animosity. Lack of insight, sincere good will, but narrow-minded worship of false gods who send ruin.[20]

In December 1916, Einstein again mentioned to Ehrenfest his continuing efforts on behalf of creating this commission. He also alluded to interactions with Swiss fellow New Fatherland League member Elisabeth Rotten, who was involved in organizational work providing aid to prisoners of war.[21]

In October 1916, writing to Zangger in Switzerland, Einstein reflected on the development of his thought through reading of those whom he opposed, such as Treitschke, but also those with whom he partly agreed on pacifism, though not on all matters, such as Tolstoy:

> In Holland I spent fine days with people of rare intellectual profundity and cultivation who have not swallowed any ramrods and are not suffering from any mass psychosis. I was reading Treitschke with Ehrenfest. You should also do that. This is the name one could apply as the label for the upper class here. Otherwise, right now I am reading My Confessions by Tolstoy, a booklet whose train of thought is somewhat disjointed, but with whose content I feel great affinity. He recognized after a hard struggle that a vibrant and thoughtful person cannot live without clinging onto something beyond the personal. But I cannot agree with him that this supra-personal thing must be the childish belief in God.[22]

Einstein also expressed his continuing admiration and concern for Romain Rolland after it was announced in November 1916 that the exiled French pacifist had been awarded the 1915 Nobel Prize for Literature. Perhaps Einstein identified with his beleaguered loneliness:

> I want very much to write to R. Rolland, once you have sent me his address. The recognition of his artistic achievements by the Nobel Committee is at least a small bandage to his wounds, but only just a small one. But perhaps great things will come to pass, so that he becomes reconciled with destiny; I place my trust in that.[23]

In late November 1916, Einstein provided this optimistic assessment of the duration of the War to his son Hans Albert:

> In the coming year there will surely be peace, so that we will be able to meet more easily.[24]

In early 1917, despite attempts by US president Woodrow Wilson to mediate peace, Germany announced that it would begin to conduct unrestricted submarine warfare. That decision, and the exposure of the Zimmerman Telegram, in which Germany proposed to the Mexican government that it ally itself with the Central Powers if the United States declared war, increased tensions between Germany and the United States. Meanwhile, the "turnip winter" (*Kohlrübenwinter*) brought new hunger in Germany.[25] As 1917 began, Einstein faced ongoing struggles with wartime international money transfers (to support his first wife and two sons in Switzerland). He continued to feel alienated from the attitudes of most Germans and hoped to meet with Romain Rolland. He wrote to Zangger:

> That's quite a calamity about the money. I just found out from my bank that they still could not issue my check to you in the amount of 1,500 francs, because they have not yet been able to obtain the consent to send out the money. This could in the future lead to very significant difficulties. . . . I am living here somewhat like a drop of oil on water, as you can easily imagine, knowing me. The discrepancy even in the unexpressed philosophies of life separates people. . . . When I come to Switzerland at Easter or perhaps somewhat later . . . Then I would also like, if at all possible, to look up Romain Rolland, if he doesn't feel visits by strangers are tiresome.[26]

In January 1917, Einstein continued to express to Zangger his sense that the War was a manifestation of insanity, along with his feeling of its tragic irony:

> The war madness has now reached its boiling point; Switzerland too seems to be on the verge of it, if it's true that it had to sell off a large number of cows to Germany in exchange for coal. . . . Isn't it unimaginable that I, of all people, should have received a Kaiser Wilhelm Institute? [Einstein's position in Berlin was head of the KWI.] Fate sometimes has a sense of humor, although its wrathful qualities work considerably more powerfully. The modern human spared God the task of showering ash and brimstone; he is taking care of that himself.[27]

In early 1917, Einstein continued to stay in touch with Georg Friedrich Nicolai, one of the most prominent war resisters in Germany at that time, although he declined to contribute to Nicolai's latest publication idea.[28] That February, Einstein experienced the first serious symptoms of a gastric ailment that would bother him for most of his life.[29]

Also in early 1917, Einstein showed his compassion and loyalty when Victor Adler, a founder of the Austrian Social Democratic Party, visited Berlin and left his calling card at Einstein's residence.[30] Adler was requesting to speak with Einstein about his son, Friedrich, secretary of the Austrian Social Democratic Party, a fellow physicist whom Einstein had known for twenty years since they were students together. In 1910, Friedrich Adler had advocated, against some xenophobia, for Einstein's ultimately successful appointment to a theoretical physics chair at the German University of Prague.[31] They had also been close neighbors in Zurich when Einstein was teaching there in 1911. The Adlers were seeking Einstein's help because four months earlier, on October 21, 1916, Friedrich Adler had assassinated Karl von Stürgkh, the prime minister of Austria-Hungary.[32] In February 1917, Einstein wrote to Adler's wife offering to help and enclosing an article on a scientific topic for her husband to read while in prison:

> Your misfortune and that of your husband, whom I have always admired, has touched my heart deeply . . . He is one of the most superb and purest men I have known. I cannot judge his deed, because I cannot grasp its motive; I do not believe him capable of rash acts; he is much too conscientious for that. If you think that I can do something for him or for you, then remember me and write me. I am sending you the little essay on [Ernst] Mach by the same post.[33]

Shortly after that, Einstein wrote a strong letter of advocacy for Adler to Emperor Franz Josef of Austria-Hungary. Its contents were consistent with the usual pacifist stance on the death penalty, though here he focused on the circumstances of one individual:

> I am taking the liberty, Your Majesty, to submit a request. The political murder committed by Fritz Adler has most deeply shaken the spirit of every right feeling person. I do not want to gloss over such a horrible act with a single word. Considering the psychological state of the perpetrator, however, it seems to me to be more a matter of a tragic accident than of a crime.
>
> Few would know Mr. Adler as well as I.

> I have known him since he studied theoretical physics together with me 20 years ago in Zurich. A few years ago he was still my closest colleague as lecturer in this field at the University of Zurich; we were also housemates at that time. I came to know Adler in those years as a man of honorable character, of almost unexampled selflessness.
>
> I have encountered very few who have been as absolutely reliable and honest as he, who devote their energy to such an extent to impersonal matters while surmounting their own wishes.
>
> Under these circumstances I fulfill an incontrovertible duty when I hereby, with all my heart, request Your Majesty to make use of the right to pardon, in the event that Adler is sentenced to death. A valuable life could thus be preserved."[34]

Einstein also tried to enlist the aid of Zangger and other Zurich-based friends in this campaign:

> . . . my old friend Fritz Adler. You surely know what the man has perpetrated. . . . But as a person he is a rare guy of extraordinary selflessness, which actually got him into this mess. My feeling for him has again grown so strong that I definitely would like to do something for him. . . . there must be support for him in Zurich, the location of his former activity. I request now that you please arrange in my name a prompt action at the Zurich Physical Society and possibly also at the Zurich Scientific Society, so that a pardon plea is submitted to the responsible authority. . . . [Michele] Besso and [Walter] Dällenbach will handle all the time-consuming aspects for you. However, you must be the "director" so that something comes of this. . . . Dear Zangger, set up the Adler campaign soon, or else it could be too late![35]

While in prison, Friedrich Adler reimmersed himself in physics and corresponded with Einstein about science, in the process indirectly expressing his gratitude for Einstein's help.[36] In April, Einstein offered to testify as a character witness at Adler's trial, and urged him to accept:

> When your case is deliberated in court, I would like to be called there as a witness; you should apply for this. Do not think this is pointless; for not

> only circumstances directly connected to the event are clarified through witness accounts, but also testimony that sheds light on the character of the perpetrator is presented.[37]

In March 1917, Einstein corresponded with Walther Rathenau, the prominent Jewish industrialist who had handled raw materials allocation for the German war ministry at the beginning of the War before resigning in 1915. In discussing Rathenau's recently published book, *Von kommenden Dingen*, Einstein described his own vision of a legitimate state as small and existing solely for the purpose of providing social services, while expressing his opposition to economic elites controlling military hardware:

> I was delighted with your book and have already read through it carefully and completely. . . . I saw with astonishment and joy how extensive an agreement in outlook on life unites us . . . If I must remain somewhere in opposition, it is in the far-reaching function that you assign to the state. In my view, the representatives of economic interests should have no access to military weapons. . . . To me the state seems justified only as a sponsor of institutions for the common good, such as hospitals, universities, police, etc. This is why I do not understand why states larger than the province of Brandenburg could be desirable. Only in small districts, in my opinion, can a people's state have lasting existence. In this regard Switzerland is a model to me . . . I know very well, however, that the world will not take shape according to my wishes![38]

Wartime conditions compromised the security of food packages that Zangger was sending from Switzerland to Einstein in Berlin, as Einstein noted when he responded to Zangger's pessimism about the future economy:

> . . . one of the 3 packages, into which you have put so much effort, has not arrived. The post office here has refused to reimburse, as I recently read in the newspaper. In the face of hunger the best upbringing is powerless. . . . You write very gloomily about the economic future of Switzerland. I shudder at the thought that it could really be so bad. If so,

> then Europe is truly to be pitied. You are well acquainted with my views on political matters, of course. They have only intensified over time.[39]

The War impacted international scientific collaboration in more concrete ways, as conveyed to Einstein by Dutch astronomer Willem de Sitter. Regarding two of his papers on Einstein's theory of relativity in relation to astronomy, which had been published recently in England, de Sitter regretted that he could not send an offprint because he had not received any, suggesting that they were "probably torpedoed."[40]

In March 1917, in a historic cataclysm touched off by the War, the Russian Revolution began. Meanwhile, fighting also continued in the Middle East.[41] In April, the United States declared war on Germany, and the Allies launched several military offenses on the western front, with heavy losses.[42]

Also in March, Lorentz continued discussing with Einstein the initiative to establish a citizens' commission to investigate German war crimes, indicating that he had received "very friendly letters from Planck and Waldeyer," and expressing great optimism about the plan.[43] Einstein's response reflected his continuing political isolation in the academy. He was perplexed at the disjuncture between people's personal and political ethics, and he expressed his feelings about this through an analogy to epidemics and by alluding to other historic situations of mass frenzy:

> I am pleased that Planck and Waldeyer have written you. The latter still hails from the good old pre-Treitschke days. . . . The younger ones, however, are considerably worse than these two. I am convinced that a kind of mental epidemic is involved. For otherwise I cannot grasp why people who are thoroughly decent in their personal conduct take on such an entirely different perspective with respect to public affairs. I can only draw comparisons with the times of martyrdom, crusades, and witchcraft trials. Only very rare independent characters can elude the pressure of prevailing opinion. At the Academy there seems to be no such person.[44]

During April 1917, Einstein continued to try to help Friedrich Adler, writing to Zangger:

> With A[dler] I have taken other measures. I wrote him that I should be called as a witness at the trial. I have a few things to present that will not fail to make an impression on the judges. If the Zurich Physical Society approaches this matter with so little passion, as seems to be the case, then they should just be sent back home. (Dopes!)[45]

Several days later he added:

> Regarding A[dler] I hesitated over whether the petition makes any sense. But now I am in favor of it after all, as I think that it can do no harm. I am writing simultaneously to Besso and sending a telegram to [Emil] Beck. Scientifically I don't think much of A., but got to know him as a remarkably selfless and decent person of rare goodness. He is at the same time almost pathologically stubborn. The denial of egotism in his case veers sharply into its pathological opposite (suffering with ecstasy).[46]

Einstein encouraged his university friend Besso to push the Zurich Physical Society to support Adler:

> I myself was vacillating over whether a petition regarding A[dler] should be made or not, since I doubted its usefulness and because sympathy for A. is supposedly very great in his hometown . . . However, the thing cannot do any harm, and that is why it should be done. . . . I . . . will happily confess before the Society and, if need be, also in the petition, to being the instigator. . . . Overall I give you the authority to do what seems right to you in my name.
>
> It should be emphasized that A. during his years of teaching showed himself to be a selfless, composed, diligent, kind-hearted, and conscientious person, who enjoyed the respect of all.[47]

In early May, Einstein expressed to Zangger his joy that the Zurich Physical Society had passed a motion to testify on behalf of Adler, who was a former member. He added a comment on the realities of wartime censorship:

> I was very glad that the A[dler] matter has now been taken up with such passion. Even if the thing only tips the balance slightly, it really is

> most gratifying that our Zurich people can be counted on when there is something right for them to do. . . . Don't seal your letters; they are always opened by the military. Perhaps this is the reason why some letters have not arrived.[48]

In May, Einstein offered a newspaper interview in support of Adler.[49] And to Besso, while following up on the Adler case, he mentioned deaths from the War and reiterated his alienation from militarist passions:

> Thanks from the bottom of my heart for taking care of the A[dler] matter, also to the Physical Society. The general opinion is that no serious danger threatens the man. . . . [Berlin physical chemist Walther] Nernst lost both of his sons in the War . . . is old Jehova still alive? The psyche of these people is strange. I have forgotten how to hate.[50] (latter ellipsis in original)

May and June 1917 saw a widespread mutiny in the French army, a continuation of the German bombing of England, and the start of conscription in the United States.[51] To Ehrenfest in June 1917, Einstein mentioned Adler's case as well as a slow shift that he believed he saw developing in people's attitudes toward the War. He reflected on the War's brutality, repeated his bafflement at the disjuncture between people's public and private principles, and envisioned the incomprehension of future generations:

> I am maintaining a rather lively correspondence with Adler . . . I hope optimistically that nothing happens to him. He has the sympathy of all sophisticated people, here too more than one would expect. Overall, a transformation is taking place, slowly but surely, toward a straightforward and natural attitude. But a dreadful amount of fertilizer is needed before the little plant can grow!
>
> . . . Have you seen Nernst? He was in Holland, in order to find out whether his son, who crashed as a pilot, is dead. His other son died in action right at the beginning of the War. Old Jehovah still lives. Unfortunately, He also smites the innocent, and He smites the guilty appallingly with blindness, so that they cannot feel guilty. Where does He get ahold of the right to punish and crush? And also of violence? I have become

> much more tolerant, without altering my views in principle in the slightest. I see that often people who are politically the most power-hungry and extreme, as private individuals could not kill a fly. A delusion of epidemic proportions has arisen, which, after it has produced endless suffering, will vanish again, so that the generation after next can marvel at it as something completely monstrous and incomprehensible.[52]

That summer, Einstein continued to struggle with difficulties obtaining the required wartime travel permits to be able to visit his sons in Switzerland.[53]

Sometime between July 1917 and March 1918, Einstein wrote the following response to a questionnaire from the Central Organization for a Durable Peace exploring the prospects for a future League of Nations. It is a disjointed statement in the absence of the questions being asked, but it shows that Einstein was grappling with definitions of nations, notions of self-determination, and questions of the appropriate roles and functions of a League of Nations, which he saw as particularly valuable in the economic realm:

> In response to the questionnaire on the self-determination of nations.
>
> I find a distinction necessary between
>
> a) "People" [*Volk*] = inhabitants of an economic-geographic region that is treated as unified (indivisible).
>
> b) "Nationality" [*Nationalität*] = an aggregation of people who form a more or less tight-knit community of interests through a common language and possibly religion and race.
>
> Addendum to a) The possibility should be just briefly mentioned that there are doubts over which regions should be treated as unified (in relation to state citizenship). I believe that this point will cause no great difficulties, at least in central and western Europe.
>
> ---
>
> **Re I and II**
>
> A "people" has the right to choose its state affiliation through secret direct ballot of all adult inhabitants. All of these votes should have

equal weight. Ceding territory without determining the matter by vote I regard as wrong.

Who may vote is determined by democratically elected commissions of trustworthy men on the principle that an alteration of the result through staged emigration or immigration must be prevented. Neutral commissions must ensure that voting proceeds without influence and that there is no fraud in counting ballots.

Re III

The right of self-determination of "peoples" should continue to exist (referendum, application to the League of Nations, supervision by the League of Nations).

Re IV

1. Yes, insofar as the nationality actually claims its given rights (vote before every appeal to the League of Nations).

2. All rights that you have named, with the addendum that through their incorporation no avoidable hardships should arise for members of national minorities.

Re V

A consideration of the economic interests of the remainder of the state I regard as justified only when a division of all positive and negative property has to take place from legal standpoints, while both parties have the right to appeal for a decision by the League of Nations.

General points about the League of Nations:

In my view the League of Nations can only obtain and retain its paramount significance if its sphere of control is extended to the regulation of economic trade between states.

Allocation and sale of raw materials, or at least supervision of the conditions under which their distribution takes place (customs duties, export licenses, etc.), establishment of minimum wages and other

> protective measures for the economically weak, alongside supervision of the signatory states in these matters.[54]

In July 1917, fighting continued in the Middle East, and on the western front Germans started using mustard gas.[55] In Germany, Matthias Erzberger played a prominent role in the passage of a Reichstag resolution calling for peace without annexations.[56] Also in July, Yugoslavia was founded, and in Russia soldiers and sailors were increasingly resisting the War.[57] In the United States in August, despite some resistance, patriotic pro-War sentiment was growing.[58] That summer, while Einstein was in Switzerland, away from his cousin and future second wife Elsa in Berlin, he often alluded to politics in his numerous postcards to her. He expressed pleasure at the public appeal of June 30 by ten prominent individuals—including Berlin history professor Adolf von Harnack and chemistry professor Emil Fischer—that called for electoral reform in Prussia.[59] In mid-July, following the resignation of Chancellor Bethmann-Hollweg and the passing of the Reichstag resolution renouncing territorial acquisition by force, Einstein's comments to Elsa reflected hope that the new German government might support a peace initiative:

> I read the newspaper articles in the Frankfurter with sincere joy; it does seem that day is dawning.[60]

As Germany cut back on raw materials exports to Switzerland, he touched on continuing wartime hardships:

> I suspect, by the way, that during the winter we'll all be in bed quite a lot (heating substitute). Here [in Switzerland] it will be the same.[61]

Besides fuel shortages and difficulties obtaining travel permits, wartime food rationing in Germany was a fact of daily life, as he discussed with Elsa:

> Your touching telegram protective against hunger arrived yesterday . . . I am eating with all the conscientiousness of which a schoolmaster is capable.[62]

A couple of days later, he added:

> I would gladly wait for you somewhere in Thuringia, but can't get anything to eat there, because I of course have no ration coupons and also have no registration of address there that would entitle me to obtain coupons. . . . Health very satisfactory. Weight 72.1 kg (compared to 70).[63]

Einstein also mentioned to Elsa that he had been contacted during that summer by Paul Seippel, a Swiss professor of French literature and biographer of Romain Rolland, possibly on the topic of supporting Georg Friedrich Nicolai in the military disciplinary proceedings that led to his court martial.[64]

One of the most horrific battles of the War, now known as Passchendaele or the third battle of Ypres, was fought on the western front from August to October 1917. Just as the "turnip winter" of 1916–17 came to signify the brutal harshness of the War to German civilians, and as the mutiny of French soldiers in May and June 1917 came to reflect the death of idealistic innocence in the French army, Passchendaele would be a turning point in British and Canadian consciousness of the horrors of this War.[65]

In August 1917, Einstein presented to Zangger his detailed vision for a postwar order. This may have been sparked by the questionnaire he had answered for the Central Organization for a Durable Peace, as well as by his interactions with various pacifists, including "Appeal to the Europeans" cosigner Wilhelm Foerster, his son Friedrich Wilhelm Foerster, and Romain Rolland. In this era before the League of Nations and the United Nations, Einstein's plan included many pragmatic strategies and principles that would be publicly debated when those two organizations were being created and that would always remain a part of his world view. These included an international court of arbitration, a reduction of standing armies, limits on compulsory military service, democratically elected governments, improved economics through more open international trade, and mutual guarantees of territorial integrity:

> [Paul] Seippel's letter unfortunately arrived in my hands late . . . Seippel found out about my view on the issue of peace from [Pierre]

Weiss, with whom I have been speaking in the greatest detail; it satisfies me very much and confirms to me that my impression agrees with that of Foerster.[66] Recently I have been thinking a lot about the political situation and arrived at a more hopeful conception, a path that I have in mind, which I would like to offer here for your comments:

Now, immediately during the War, from as many of the Entente and perhaps neutral states as possible, a pacifist organization should be founded according to the following principles:

1) A court of arbitration for the settlement of cases of conflict between these treaty states.

2) A mutual institution to decide to what extent these states apply and may apply the principle of universal compulsory military service. Mutual foreign deployment of troops. Reduction of the standing army in accordance with the possibilities afforded by the foreign relations of the treaty states. ~~Mutual military assistance~~

3) Principle of most-favored customs policy among the treaty states, linked with the intention gradually to dismantle the customs barriers between them.

4) Any state that fulfills the following conditions can become a member of the organization:

a) A parliament elected according to democratic principles.

b) Ministers governed by the parliamentary majority (which ministers of course must completely control the executive).

5) Military pacts with states not belonging to the organization are not permitted, on pain of loss of membership in the organization.

6) The organization guarantees each treaty state its territorial holdings against outside aggression.

The advantage that I see in this proposal is essentially that even a federation comprising nowhere near all states can be very valuable in that, at the price of the renunciation of expansion, it secures the existing holdings of its members and brings them an ever greater reduction of the military burdens as more states join the organization. If the Entente were to bring about an organization of this type, which has the United States, England, France, and Russia as members, it could without concern make pacts with Germany, which would be economically forced to seek immediately to join the organization without anyone being able to say that the "national honor" had been wounded.—If you yourself find

> this matter sensible, I request that you support it. I myself know only Weiss on the Entente side, with whom I will speak as soon as I come to Zurich, which should be soon.[67] (strikethrough in original)

Also in August, Einstein corresponded with Rolland, who clearly was one of his pacifist kindred spirits. Rolland expressed a teleological faith in progress: ". . . if I suffer in my heart at seeing the West, and above all my France, bleed itself white, I remain an optimist about things overall and the future progress of humanity. . . . I believe (for the near future) in a richer and greater civilization than that of today. . . ." He also mentioned that he had written two works inspired by the present[68] and that he had "read with passionate interest" Nicolai's newly published book, *The Biology of War*,[69] which he described as "admirable," declaring that he found it "beautiful, in this terrible époque, to discover some great free and serene souls like his. It is sufficient recompense to me for the immense stupidity that is the new universal Flood."[70]

Einstein's reply to Rolland, demonstrating his ongoing attention to public affairs, included a hardheaded assessment of German culture and the German mentality. It included his concept of a religion of power and his continuing conviction that military successes had a deleterious effect on Germans. He also reiterated the principles, proposals, and rationales he had delineated to Zangger a day earlier, including his faith in facts—in this case meaning economic pressure by an international organization—as the only way to influence the German people:

> The evil experiences of the actions of human beings to which we have been subjected in the interim, however, have not made me more pessimistic than I was in reality two years ago. I even find that the surge of imperialistic attitudes, which dominates the influential sectors in Germany, has subsided somewhat. Yet I still find that it would be extremely dangerous to form a pact with Germany as it is today.
>
> Through the military success of 1870 and through the successes in the fields of trade and industry, this country has taken up a kind of religion of might [*Machtreligion*], which found adequate, by no means exaggerated expression in Treitschke. This religion dominates almost all intellectuals; it has almost completely driven out the ideals of the

Goethe-Schiller era. I know people in Germany who in their private lives are guided by an almost unbounded altruism, but who awaited the declaration of an unlimited submarine war with the greatest impatience. I am firmly convinced that this aberration of minds can only be swayed by hard facts. These people must be shown that it is necessary to show consideration for non-Germans as equally worthy, that one is obliged to earn the trust of foreign countries in order to be able to exist, that goals are not achieved through violence and disloyalty. Fighting against the goal only with intellectual weapons seems hopeless to me; people like Nicolai with honest conviction are labeled as "utopians." Only facts can dissuade the majority of the misguided from their delusion that we live for the state and that its intrinsic aim is the greatest possible power at any price.

The best way to lead out of these miserable circumstances seems to me to be the following. America, England, France, and Russia conclude a military arbitration treaty for all time with agreements on mutual assistance and on minimum and maximum military preparedness. This treaty would have to contain the determination of most-favored nation tariffs. This association should allow any state that has a democratically elected parliament, and whose ministers are drawn from the majority within this parliament, to join. This brief sketch will have to suffice.

If Germany, which is dependent on sales of its industrial products, were to find itself confronted with these as permanent conditions, the view would soon prevail that renunciation of the present course is imperative. However, as long as German statesmen can hope that a shift in the distribution of power sooner or later will occur, no sincere change in course will be considered. You see the proof that everything is still the same as it was, in the nature and manner in which the change in chancellorship recently was staged.

With my wish that you find comfort in your gifted artistic creativity in these gloomy times.[71]

Rolland in his response sympathized with Einstein's suffering and acknowledged Einstein's authority in describing Germany. But he argued that what Einstein described was not unique to Germany: "Evil spreads like a splotch of oil. All the nations are interdependent, even though they fight one another; and we have not yet found the way to stop moral epidemics . . . at borders." He characterized "the current War" as "a fight against the Hydra of Lerna. For each head chopped off, two other ones spring back to life." He saw any possible "salvation" as coming only from "social forces," but also accepted that, if it did not come, "it will not have been the first time that a powerful civilization has collapsed." He philosophized, "Life will know very well how to blossom again from the ruins." Rolland here mentioned English mathematician and pacifist Bertrand Russell, with whom Einstein would correspond until the last week of his life, and also alluded to "the new American opposition." He accepted the extreme minority status of pacifism: "I am perfectly convinced that there shall never be more than a handful of us in the world." Yet he took comfort in the thought that "the spirit is never conquered . . . It is ahead of its times."[72]

From August through October 1917, Einstein discussed with both Zangger and his old friend Besso the impact of the wartime weakness of the German currency on his life: it would be much easier to support his first wife and two sons if they moved to Germany instead of staying in Switzerland. He also alluded to the export restrictions imposed by the German government.[73] However, responding optimistically to what he perceived as trends, in late September Einstein commented to the Swiss mathematician Edouard Guillaume:

> Politically, a healthier wind is starting to blow; all shall be well again.[74]

On October 14, 1917, Einstein and forty-six other Berlin academics signed and published in the newspaper *Frankfurter Zeitung* a manifesto addressed to the German chancellor calling for a "peace of reconciliation" (*Verständigungsfrieden*), consistent with the majority vote in the Reichstag three months earlier that renounced all territorial acquisitions by force.[75] In mid-October, Einstein expressed to Werner Weisbach his continuing support for Weisbach's Association of the Like-Minded, which he had joined a year earlier. He reiterated his concept of a "religion of might" (*Machtreligion*) and his sense that Germans would be changed through hard experience, and he alluded to the need for mutual support among pacifists:

> Very belatedly I extend to you my agreement with the decisions of the "Association of the Like-Minded" that were reached on May 30, 1917.
>
> . . . I want to take on the task of winning over colleagues to our organization. The experiences of the last few weeks have unfortunately shown again how deeply anchored the religion of might is among German academics. Hard experience certainly will work incomparably more on our behalf than our conscious advocacy can; yet this little must also take place, if only to spare the enlightened of the discouraging feeling of isolation.[76]

On October 26, Einstein attended a meeting of Association of the Like-Minded at the home of Werner Weisbach in Berlin; on December 4, he attended another.

In October 1917, the War's worldwide character continued with the death of Mata Hari (a Dutchwoman, long resident in Indonesia, executed in France as a spy) and the entry of Brazil into the War.[77] In November, in the second stage of the Russian Revolution, the Bolsheviks took power and began arranging a separate peace with Germany.[78] In late 1917 and early 1918, several eastern European countries declared their independence from Russia.[79] Fighting continued in the Middle East, and tanks were introduced on the western front while trench warfare continued.[80] In December 1917, Einstein was in contact with at least two fellow physicists who were contributing to the military applications of physics: Hans Thirring of the University of Vienna was constructing photoelectric devices for military application, and Gustav Mie of the University of Halle was working on underwater telegraphy between U-boats.[81] During the War, Einstein himself served as a scientific consultant to the Mercur Aircraft Company and designed an airfoil and a gyroscope for naval vessels. But he distinguished himself from those colleagues in expressing outrage at the War's effects and disgust with German militarism, writing to Zangger in December:

> Did you know that [Marian von] Smoluchowski died of dysentery? It's a shame about him. Will there indeed soon be peace? Nobody knows. . . . How can it be possible that this culture-loving time is so horribly amoral?

> I am becoming increasingly inclined to rate everything as trivial compared to loving one's neighbor and benevolence. Wouldn't it be good for the world if degenerate Europe were to destroy itself completely? When I hear the new repugnant term "toughening up" [*Ertüchtigung*], my stomach turns. All of our exalted technological progress, civilization overall, is comparable to an axe in the hand of a pathological criminal. I believe, in all seriousness, that the Chinese should be held in higher esteem than we are, and hope that their healthy proliferation will outlast the extinction of our "toughened" comrades.[82]

The same letter described the concrete impact of wartime food shortages and rationing on his daily life. He assured Zangger of his current good health but noted that he relied on "family packages" from Switzerland. He was living in Germany "as a single Swiss," and thus was only allowed "bachelor's packages," much of which his digestive problems prevented him from eating.[83]

Zangger in his response expressed similar sentiments and the elation of connecting with a kindred spirit. He referred to Einstein's letter as "a great joy among many oppressive things—with your declaration of benevolence as the foremost principle in life." He found "the problem of human decency, of trust in loyalty and faith" more difficult, observing that "Europe must be especially poorly grounded." Nonetheless, clearly deriving sustenance from Einstein's friendship, he declared, "Your grumbling is always like an elevating liberation for me."[84] Zangger also expressed his concern, shared with Einstein, over the fate of the pacifist medical doctor Georg Friedrich Nicolai, author of the 1914 "Appeal to the Europeans," as Nicolai's conflicts with the German military authorities were intensifying.[85]

Also in mid-December, Einstein told Lorentz of his oppressed spirits due to the War:

> I am continually very depressed about the unending tragedy that we must experience. Even the usual flight into physics does not always help.[86]

Coal exports from Germany to Switzerland had been drastically cut back in 1917. In late December of that year, Besso wrote to Einstein from Switzerland about the hardship and the need to toughen up in the future.[87] At the end of December 1917, Planck commented to Einstein on coal shortages in Berlin.[88] Around December 25, Einstein became bedridden with an abdominal ulcer, a condition that lasted several months.

Thus, at the end of 1917, after three and a half years of the War, Einstein was weary and depressed but also unwavering in his convictions. During the prior two years, 1916 and 1917, the carnage and death rates from the War had been historically unprecedented: hundreds of thousands had been killed at Verdun and again at the Somme, and tens of thousands at Passchendaele and in other horrific battles; poison gas and aerial bombing were being used as weapons for the first time; and naval battles were raging along with Germany's decision in early 1917 to pursue unrestricted submarine warfare. Einstein continued to feel alienated and baffled by people's unceasing support for the War despite all of this, but he maintained his ongoing belief that an organization of states could prevent war. In late 1916 and during 1917, he helped develop the idea of a commission to investigate German war crimes. Along with other residents of Germany, he encountered the hardships that the War had imposed, although he also focused on their potential virtue as an instrument for educating German citizens. Difficulties transferring money to Switzerland and supporting his family there from a German salary were also War-induced adversities, and his gastric ailment had been aggravated by wartime conditions, which included the loss of food packages mailed from Switzerland. Yet he carried on, trying to help Friedrich Adler and communicating with Walther Rathenau, Romain Rolland, Heinrich Zangger, and other friends over issues raised by the War. Twice in 1917—once for an organization and once for Zangger—Einstein wrote point-by-point blueprints of his vision of a postwar order. Rolland thought he perceived a loss of optimism in Einstein; nonetheless, in late 1917, Einstein and forty-six Berlin scholars had cosigned a public manifesto advocating a "peace of reconciliation." Einstein was horrified and grieved at the War's collective and individual violence and tragedy and was drawing painful conclusions about human nature, yet he continued to contribute steadily in small ways toward a world at peace and structures for maintaining it.

3 Hope and Optimism beyond the Turbulence of Defeat and Revolution

1918

On January 8, 1918, in his "Fourteen Points" speech, US president Woodrow Wilson announced his plan for the aftermath of World War I, a plan aimed at maintaining future world peace. It envisioned the following: open diplomacy; free navigation and trade; self-determination for colonial populations and subject peoples of the Austro-Hungarian and Ottoman empires; independence for Poland; German forfeiture of territory taken from Russia, Belgium, and France, including Alsace-Lorraine, which had been taken in 1871; and the creation of an association of nations to guarantee political independence and territorial integrity to nations large and small.[1] (Text of the "Fourteen Points" is presented in appendix C.) There were many obvious similarities and common points of reference between Wilson's speech and the two versions of Einstein's point-by-point plan for a postwar order (which he wrote in 1917; see chapter 2).[2] In the months following the speech, Wilson's "Fourteen Points" raised high hopes among Germans and among European pacifists, including Einstein. Wilson's principles and a new postwar order were widely discussed, and by the end of 1918 Einstein would be expressing great optimism about the new Weimar Republic (hopes that would later be sharply disappointed).

At the start of 1918, though, the War was still raging, despite indications—such as strikes on the German home front—that it could not continue indefinitely.[3] In January 1918, fuel shortages were impacting

Einstein and his family and friends in both Germany and Switzerland. Einstein, sick in bed in Berlin since December 24, indicated in his letter to Besso two weeks later:

> I am resting in bed for another 4 weeks or longer, which is at the same time also preferable due to the inadequate heating.[4]

Fuel shortages also affected his son Hans Albert's latest project in Switzerland that January: he tried to make his train electric "due to the coal shortage."[5]

War-induced travel restrictions also continued. In late January, Einstein and Berlin KWI physical chemist Fritz Haber tried to help Gunnar Nordström, a Finnish physicist, obtain permission to travel through Germany from the Netherlands back to Finland.[6] Nordström ultimately concluded that this trip was impossible at present and would have to be delayed until the following summer.[7]

Intellectual interactions between Einstein and fellow physicists were also obstructed concretely by the military's near-monopoly of rail transport. For instance, in February 1918, Gustav Mie, physics professor at Halle-Wittenberg, complained to Einstein that he hoped to speak with him once the railways had become more accessible to civilians again.[8]

Notwithstanding these hardships, Einstein maintained his political engagement. In mid-January, he mentioned to Berlin astronomer Erwin Freundlich that he was still active in the Great Council of the Central Organization for a Durable Peace.[9] On January 29, Einstein's name appeared as the ninth on a Berlin police blacklist of thirty-one pacifists and Social Democrats in the Berlin area.[10] In early February 1918, he corresponded with University of Berlin philosophy professor Ernst Troeltsch about the Popular Alliance for Freedom and Fatherland (Volksbund für Freiheit und Vaterland), an organization comprised of members of the Catholic Center Party, the Progressive People's Party, and the Social Democratic Party, which opposed territorial annexations in the peace with Russia. Troeltsch had given a speech to the organization a month earlier in which he supported electoral reform in Germany, opposed annexations, and favored the fostering of a world economy and an organization of states united by treaty. Troeltsch thanked Einstein for his friendly letter in response to his speech and for informing him of the Association of the Like-Minded (Vereinigung Gleichgesinnter); he also expressed his concern about the incitement then being engaged in by radical rightists, as well as his belief that liberal intellectuals must take action.[11]

In the first few months of 1918, with Russia collapsing into civil war and the German military advancing rapidly in the East, many small nations declared their independence from Russia. In March 1918, the Germans signed the Treaty of Brest-Litovsk with the Russians, which officially ended the latter's participation in World War I.[12] Between February and April of that year, the Second Battle of the Somme—a new German offensive during which large quantities of poison gas were used—caused a half million casualties in killed and wounded on both sides.[13] In February, Einstein responded to a package he had received from University of Vienna professor of zoology Karl Camillo Schneider, who claimed to be a pacifist yet argued in favor of force and against democracy and who wrote articles favoring a German victory with annexations. Einstein's response to Schneider's overture utilized the diplomatic device of "no accounting for taste":

> You gave me great pleasure by sending me some of your booklets. I admire the versatility of your interests and knowledge, as well as your fine and amusing writing style. I have read almost all of them. . . . Only your strong-arm attitude that sails under the Germanic flag goes very much against the grain for me. I prefer to hold with my countryman Jesus Christ, whom you and your ideological comrades consider definitively obsolete. Suffering is truly preferable to me than to exert force. History may teach where we will end up with this attitude that you and so many contemporaries in this country extol; who can tell? But beyond all setting of goals: De gustibus non est disputandum; my taste is otherwise.[14]

Einstein's conciliatory inclinations extended beyond the political relations of nations into his personal and professional life. After Dutch physicist Paul Ehrenfest in March 1918 declined the invitation of German mathematician David Hilbert and physicist Max Planck to attend an annual lecture series, citing his outrage at the ongoing "rape" (*Vergewaltigung*) of Belgium, in April Einstein acted as peacemaker between his fellow scholars and friends, apologizing to Hilbert and making excuses on behalf of Ehrenfest.[15]

In April 1918, Berlin geophysicist Wilhelm Schweydar, who was working in Romania and who shared some of Einstein's views as well as his sense of isolation, wrote reinforcing Einstein's observation about the disjuncture between people's private and public ethics. He also broadened Einstein's understanding of the War by offering a different geographic and situational perspective:

> You are right, intelligence without good character turns people into beasts. This simple truth was clear to me long before the War, and I did not understand why people whom I had highly esteemed before the War were seeking means that would be suitable for annihilating millions of people in one stroke. If I had not had Ad[olf] Schmidt with me at Potsdam, I would have felt completely isolated. Here in the occupied region one also sees more clearly the other side of what people at home so loudly extolled to me as patriotism and what I once in anger designated the most dangerous of all virtues.[16]

During April 1918, Einstein discussed food rationing and shortages in both Germany and Switzerland with his Swiss friend Zangger.[17] To his son Hans Albert, also in Switzerland, he continued to express a sense of endurance and the focus on basic survival:

> It's a shame that your projects are suffering so much under the wartime deprivation, but slowly, slowly it will pass. Still, the main thing is that there is still enough to fill your hungry mouths, even if a substantial amount of time must be spent in getting everything.[18]

In April and May 1918, Einstein solicited the participation of University of Göttingen professor of mathematics David Hilbert in a plan by Georg Friedrich Nicolai, a collection of essays by prominent intellectuals from various countries aimed at international reconciliation. Condemning the destructive impact of the October 1914 German nationalist manifesto

"To the Civilized World" (or "Manifesto of the 93"), he envisioned the proposed collection as offering renewed support to the prewar international community of scholars and to isolated individuals of ethical intent:

> Countless times in these miserable years of widespread nationalistic blindness, men of science and the arts have issued statements to the public that already inflicted incalculable damage to the feeling of solidarity that was developing with such promise before the War among those who devote themselves to higher and freer purposes. The outcry of narrow-minded preachers and of servants of the dreary power principle is rising so loudly and public opinion is being misled to such an extent by decisive silencing of the press that those with better intentions, feeling wretchedly isolated, do not dare to raise their voices. The danger is increasing daily that even those who until now have been holding on with all of their strength to the ethical ideals of a happier phase of human development are gradually despairing and falling victim intellectually to the widespread breakdown. This serious situation places before those who have attained an elevated prestige among scholars throughout the civilized world, through fortunate intellectual achievements, a task that they must not evade: they should make a public declaration that could serve as support and consolation for those who in their solitude have not yet lost their belief in moral progress.
>
> I am envisioning the matter this way. Each of us provides an avowal, in the form of a short essay of up to about 10 printed pages that is designed to be effective in the above-indicated sense. These essays appear together in a little volume on the book market as evidence of international sentiment, actually probably at first in neutral countries abroad. In order to stress the international character, we could try to acquire articles by men from countries presently at war with one another and from neutral nations.
>
> This letter is being sent to you and also to a very few highly respected men. I ask you please to tell me frankly whether you approve of this undertaking and whether you would be inclined to offer an article.[19]

Hilbert's reply may have contributed to Einstein's growing political savvy. It reflected external pressures and self-censorship operating among German intellectuals during the War as a response to virulent attacks by the radical rightist press on liberal or moderate public figures and initiatives. Hilbert advised against Einstein's "well-meaning and appealing undertaking," which he predicted "would be tantamount to self-denunciations." In Hilbert's view, "even your name would afford you no protection—since the very word 'international' has the effect of a red cloth to a bull for our colleagues." He conceded, "I can perfectly understand the desire, which I feel just as you do, to do something," but he argued that such an initiative would "damage the cause." He advised "waiting until the hurricane of madness has spent itself and reason has the opportunity of returning."[20]

Another recipient of Einstein's proposal was University of Berlin philosophy professor Ernst Troeltsch, who, in his pragmatic response, said he saw the proposal as questionable at a time of intense and irrational feelings under wartime conditions. In the past, Troeltsch had taken internationalist stands publicly, for example, "in an article in the *Frankfurter Zeitung* specifically directed against hatred between nations." He believed, however, that "the War has grown beyond the dimensions of other wars and threatens European civilization at its very roots." In his view, "the realities . . . are running their course and can no longer be restrained and overcome by reason." At a time when irrational feelings were dominant, he explained, "an appeal to reason does not save us, because reason is powerless everywhere." Troeltsch also saw direct danger to pacifists: "We must unfortunately . . . allow the disaster to take its course, and save our own skins in the midst of the tremendous dangers. For I still regard the War as extremely dangerous for us. . . . I am too much of a realist not to sense the messiness of the hurricane and the crisis resulting from it for us."[21]

In May 1918, Einstein continued to discuss Nicolai's idea of publishing a collection of great works expressing the true principles of European culture, both with Nicolai, then serving military detention after a court-martial, and with Elsa's daughter, his future stepdaughter Ilse Einstein, who was working as his part-time secretary and informal liaison with Nicolai.[22] Always ready to incorporate the thoughts and information of valued friends and colleagues into his perspectives, Einstein changed his mind about advocating this collection, probably in part due to the input of Troeltsch and Hilbert. While Einstein was withdrawing his support for the idea, partly by invoking his Swiss citizenship, Nicolai urged him to continue with the project, emphasizing Einstein's European identity. Nicolai argued that "deliverance from the coming cultural collapse will only be possible if the European idea . . . can gain widespread acceptance . . . This idea of a worldwide cultural organization must inspire the masses." He saw the current time as crucial, maintaining that "especially because the danger at the present moment is so great . . . I believe that the good sense of humankind will instinctively understand this and not tolerate the desecration of this beautiful Earth." His concern was for the future: "For times to come, for the future Europe . . . something should be done."[23]

But Nicolai failed to sway Einstein. In his final comment to Hilbert on the matter, Einstein cited insufficient support, especially among those from whom he would have expected it most. He also upheld the ideal of a primary identity as a person of culture instead of a national identity:

> I thank you belatedly for the extremely warm response that you gave to my last letter with the circular. Since then, I have become sufficiently convinced that the psychological basis for my envisioned endeavor is lacking, in that just those people on whom I would have depended do not feel predominantly international sentiments. The whole endeavor should not be aimed at any *political* goal. Those people of whom I would have expected that they would hold our common cultural achievements far above the current division should have extended a hand to one another. It thus would have been, in a way, the antithesis to the infamous Manifesto of the 93. This says: I am foremost a German, only secondarily a person of culture; we, however, should have said: I am foremost a person of culture, only secondarily a German or a French person (insofar as it can be made compatible with the former!). But I gave up the endeavor after the responses that I received. If you undertake anything along these lines later on, naturally you can count on my participation.[24]

April to July 1918 was another period of strong German offenses with poison gas on the western front. Both sides engaged in aerial bombing campaigns.[25] In the east, the Germans occupied more territory and signed the Treaty of Bucharest with Romania. In Russia, the Bolsheviks killed the

Tsar and his family, and more countries declared their independence.[26] On June 7, 1918, after University of Breslau mathematics professor Adolf Kneser characterized Einstein's scientific work in German nationalist terms, Einstein responded diplomatically but firmly in declaring his non-national identity:

> Thank you for sending me your original and inspired lectures. In them, however, one thing pained me, which I do not want to pass over in silence. It hurts me when my name and my work are abused for chauvinistic propaganda, as has been happening more frequently in recent times. This is out of place even objectively. I am by ancestry a Jew, by citizenship Swiss, and by fundamental attitude a human being and only a human being, without special inclination toward any state or national entity whatsoever.[27]

On June 12, 1918, after being separated throughout the War, Einstein and his first wife Mileva Einstein-Marić signed a divorce agreement, although it would not be finalized until 1919.[28]

In early July 1918, Einstein sent his good friend Michele Besso in Zurich a letter of introduction for Otto Buek, one of the four cosigners with Einstein of the "Appeal to the Europeans" in October 1914.[29] Later that month, he followed up with Besso regarding Buek, also recommending to him a new book, *Pure Reason and State Reason* (*Reine Vernunft und Staatsvernunft*) by Rudolf Goldscheid, a Viennese sociologist and fellow member of the New Fatherland League.[30] In July and August, Einstein corresponded with Friedrich Adler in prison, primarily about Adler's scientific work. Adler had been convicted in May 1917 for assassinating the Austrian prime minister and had been sentenced to death; however, the emperor commuted his sentence to eighteen years' imprisonment. Einstein also recommended Goldscheid's book to Adler.[31]

On the western front, the tide turned in August and September, and the Allies, now buttressed by American troops, began pushing forward, while using poison gas in their attacks.[32] By August 1918, the "Spanish flu" pandemic was raging, and Allied troops were invading eastern Russia and Siberia in an anti-Bolshevik campaign.[33] In August 1918, the influenza epidemic was a compelling topic of discussion between Einstein and both Zangger and the engineer Walter Dällenbach. Einstein wrote from his seaside vacation to Dällenbach:

> There is no sign of influenza here at the Baltic Sea. But in a few days I am going home again to Berlin.[34]

Zangger wrote to Einstein that he had lost four pupils to the influenza.[35]

In early September 1918, Einstein corresponded with the revolutionary pacifist and secular Jewish publicist Kurt Hiller about Hiller's League for the Goal (Bund zum Ziel), which in November 1918 was reconstituted as the Political Council of Intellectual Workers (Politischer Rat geistiger Arbeiter). Hiller invited Einstein to a meeting on September 21, where the group planned to organize a politically progressive intellectual elite with the intention of taking decisive public action. Hiller believed the "psychological moment" was at hand because "the Parliaments are breaking down" and "salvation cannot come from the official appointees, but rather only from an association of intellectually oriented." He enclosed the text of a letter from a military officer, commander of an engineer battalion on the western front, who made a proposal along the same lines, and he referred to a similar suggestion in an article by Otto Flake in the June issue of *Friedenswarte* (*Watch-Tower for Peace*, the main pacifist journal of the German-speaking lands at that time). According to Hiller, the League for the Goal, which had been dormant since its Westend Resolutions of August 1917, was now ready to take its first action. He assured Einstein, "You will meet exclusively people who . . . agree with the basic perspective, whose reliability is beyond any doubt, and who are young in spirit." Yet he valued any input from Einstein: "Should you be prevented from appearing, you would be of service to the gathering through a short written description of your perspective.[36] In his response, Einstein used his Swiss citizenship to decline participation in the meeting; he also told Hiller forthrightly that he found Hiller's proposals too vague to be realistic. He conveyed his views on the source of the mass mentality supporting the War and his sad conclusion about the disjuncture between people's private and public ethics. He considered the model of western democracy to be the best pragmatic course for Germany, and he reiterated his long-standing identity as a pacifist:

> I read your brochure as well as your note of September 7 with great interest and do not want to fail to thank you and express myself on the matter.
>
> I would have liked to put in an appearance at your meeting. However, it is not appropriate for me as a Swiss to involve myself in local political affairs.
>
> Your brochure is brilliantly written, your proposal however seems to me in its present indefinite form to contain little that falls within the realm of the possible. If you call together those local men who have distinguished themselves through significant intellectual achievements in their various fields, you will assemble a collection of power or "pragmatic" politicians who—if they are truthful—will agree on the tenet that private morals have no significance in the relations between states and nations, that only the right of the stronger is decisive, that without war humanity would degenerate. Concessions to pacifism are only to be taken as pretty gestures for hoodwinking. . . . etc. etc.
>
> These people are simply the children of their time, their wishes (which are determined by their views) the product of a development of the mass psyche, which is stamped with the name of Bismarck.
>
> When a great scholar from a neutral country abroad was very eloquently presented with the foregoing political gospel, he said: "Perhaps the world is so, but then I do not want to live in it!" I am firmly convinced that the "world full of enemies" is forged out of this reaction.
>
> I see rapid and radical democratization according to the model of the western powers as the only salvation for Germany. For only such a constitution—great as its shortcomings otherwise may be—seems to guarantee a sufficiently far-reaching decentralization of the drive for power that a repetition of the conduct of 1914 appears excluded. Nobody abroad trusts the current regime any more.
>
> This in a few words is my conviction. As one who was already a pacifist before the War, I have the right to express it now.[37] (ellipsis in original)

In September and October 1918, the War was moving rapidly toward a conclusion, with German officials urgently requesting a nonpunitive peace and the Allies repeatedly rejecting this.[38] The same months saw high death rates from influenza among soldiers and civilians in countries throughout the world.[39] Germany was experiencing extreme inflation, and more eastern European countries were being formed.[40] In late September 1918, Einstein's main wartime pacifist affiliation, the New Fatherland League, which had been disbanded in February 1916, was reconstituted.[41] On October 19, the league adopted a declaration that called for an investigation into the question of war guilt, demanded the introduction of civil liberties, and advocated "a complete restructuring of the German constitution and civil service along democratic lines . . . and done by convening a National Assembly with legislative powers, with equal, secret, and direct right to vote, including for women and soldiers."[42]

That fall, Einstein sensed that the War was about to end. He was still hindered by wartime inconveniences, and the influenza epidemic was an ongoing presence. To Ehrenfest in Holland, he expressed revulsion at the difficulties of wartime travel, along with relief that the War showed indications of ending soon:

> You are such a kind person that I have no alternative but to go visit you, despite the horrendous traveling conditions . . . Send me an official invitation very soon and take steps from there very quickly so that my entry is permitted immediately. The official way through the local embassy takes a minimum of 3 weeks . . . I am indescribably happy at the prospect of seeing you again and of breathing the Leyden air, now that the pressure of the hideous War no longer weighs upon everything.[43]

To his son Hans Albert in Switzerland, he expressed worries about the influenza and, again, enthusiasm about the impending end of the War:

> The cursed flu is on the loose again, here too it's frighteningly rampant. . . . I hope Tete [Eduard, his younger son] does not catch the flu. How satisfied are you all with his health? . . . Now there will be peace, I hope. Then there will be no more Wall of China between us and letters will arrive quickly again.[44]

In late October 1918, Hans Mühsam, a physician at the Jewish Hospital in Berlin, then serving at a military base hospital in Liège, Belgium, wrote to Einstein agreeing on the tragic absurdity of the War: "Isn't it despicable that thousands of young human lives depend on the haggling of those in power, on their willingness to respond more or less quickly, and on their double-dealing intrigues?" However, his views on social equality, grounded in quasi-scientific social Darwinism, were quite distinct from Einstein's: "If justice means equal rights for all, this . . . [would] lead to conflict and confusion. The natural law of organized nature demands privilege for the strong." He then applied this belief to the League of Nations, which he believed should be based on "scientifically secure foundations," declaring that if it were to "encompass all peoples or nations of the Earth," then "the League of Nations should restrict itself to the protection of the weak, but . . . only . . . guarantee them an economic existence." He then applied this notion back to individuals, arguing, "I can just as little accept equal social rights for all within a nation . . . efforts at leveling differences go against nature. Protection of the weak must also at the individual level be limited to the guarantee of his physical subsistence." Essentially, for Mühsam, nature was "aristocratic, not democratic." But he hastened to add, "I consider our autocratically feudal system [based on heredity] . . . just as nonsensical as the equalization sought by the Social Democrats . . . Everyone must first personally produce proof of his superiority." On the prevention of war: "I believe that the world will not progress in steady healthy development until the social structure of the states and the coexistence of the nations are placed on scientifically secure foundations. Wars will not be preventable until then either. . . ." Mühsam exemplified the socially hierarchical world view common in their milieu, in an era when the destruction of monarchies and the Bolshevik challenge made the definition of justice a widespread concern. Meanwhile, in Mühsam's immediate situation in Liège at War's end, "the Belgians are confronting us daily more and more impudently and provocatively."[45]

Einstein's most prominent colleague among physicists in Berlin, Max Planck, had in 1914 been one of the ninety-three figures who signed the German nationalist manifesto "To the Civilized World." Since then, though, he had moderated his stance toward the War. Nonetheless, he declined to sign a declaration sent by Einstein in late October, which may have been a new statement of founding principles by the New Fatherland League, whose prohibition had been lifted that month. Planck worried that such a statement might motivate the Allies to be harsher against Germany. He was placing his trust in the government of the new chancellor, Max von Baden, who on October 4 had sued for peace based on Woodrow Wilson's "Fourteen Points." Referring to the "public declaration" that Einstein had presented to him, Planck declared, "I am in full agreement with its content, at least its main point." But he saw the issue as "whether its publication will have the effect that the signers wish and expect." He explained that he could imagine "that such a publication could serve our opponents as a stimulus to stretch their demands against us even higher," and he felt certain "that this step will elicit an agitated reaction on the other side, which could again spoil the program of our current government, which in my opinion is presently following a promising course."[46] Thus, Planck's declining to sign the declaration highlighted Einstein's isolation among his academic colleagues as an advocate for peace.

Events moved swiftly in October and November 1918, with the end of the War, the German Revolution, the kaiser's abdication, and the founding of the Weimar Republic.[47] Casualties were unprecedented in World War I. It would be estimated that approximately ten million soldiers had been killed in the War, more than seventeen million were wounded, and ten million civilians had died due to the War.[48]

On November 9, 1918, the day of the German Revolution, as recalled by fellow physicist Max Born, Einstein gave a speech to the Students' Council in the Reichstag. On that day he also met briefly with Friedrich Ebert, leader of the Social Democratic Party, to whom the prior chancellor had transferred his position that day. Based on his notes for a speech he gave four days later for the New Fatherland League, Einstein probably touched on the following themes in his speech to the students (and possibly in his conversation with Ebert): that their shared goal of democracy would only be achieved if the beliefs, healthy judgment, and will of the people was implemented through the people voting in elections; that rule by class had been defeated, but that class tyranny by the right must not be replaced with class tyranny by the left; and that they must not let vengeful feelings lead them into violence, that violence nurtures only bitterness, hatred, and reaction.[49] In the early, heady days of the Weimar Republic, Einstein was celebrated as an influential political figure, and he in turn felt great optimism about the dramatic changes underway, despite ongoing hardships.

As the War reached its conclusion, influenza and economic hardship remained challenges. On about November 9, Einstein expressed these concerns to his first wife in Zurich:

> All will go well if my income is not considerably reduced by the consequences of the War. . . . Here too the influenza is very potent and virulent; I have been spared thus far, however.[50]

Zangger, who had gone with Einstein's university friend Michele Besso to Davos, Switzerland, to recover from the influenza, graphically described the epidemic among the medical community in Zurich: "Now multiple deaths have broken out in the student and doctor communities; yesterday we buried the 4th member of the university hospitals since my becoming dean." The decision fell to him as dean: "Should I have the clinics closed as in the summer . . . and dismiss the students . . . ? If it goes on like this, we shall have in a single year a death rate of 10% of the population."[51]

Along with the influenza, political upheaval, socialism, and changes of leadership held the attention of Einstein and his close friends and family. In the early stages of Swiss strikes of November 9–14, 1918, which led into a Swiss general strike, Besso wrote of the turbulence in Switzerland, expressing his worries that the people were not prepared for all of the implications of changes in the political order: "Conditions are critical even in Zurich these days. We can probably expect a frenzied state of disruption everywhere . . . The mental preparation is totally lacking. People were little acquainted with the functions that the former social structures fulfilled, when they were cultivating the emotional conditions for the new creation . . . of the oncoming new order."[52]

Notwithstanding the upheaval in Germany, however, Einstein expressed confidence in the political changes and new leadership, writing to his mother:

> Don't worry. Up to now, everything has gone smoothly, even impressively. The current leadership seems to have really grown to meet the task. I am very pleased with the way things have developed. Now it will start being really good for me here. The collapse has done wonders.
>
> All of us are doing well. We are healthy, and Haberland Street is peeping half curiously, half anxiously into the world. I'll be writing you more often now, so that you don't worry. Among the academics, I am some kind of high-placed socialist [*Obersozi*].[53]

To his sister and brother-in-law, he rejoiced with even more unadulterated enthusiasm at the German Revolution and Germans' apparent receptivity to it, alongside lingering concerns about health:

> The great event has taken place! I had feared some sort of breakdown of order. But up to now the movement has been taking a truly impressive course, the greatest external experience conceivable. And the funniest thing of all is: people are adapting to it remarkably easily. That I could live to experience this!! No collapse is so great that one would not gladly accept it in order to receive such magnificent compensation. Here with us, militarism and the stupor of monarchical appointees have been thoroughly obliterated. . . . All of us have been doing well up to now, also health-wise.[54]

In the midst of the German Revolution, Einstein wrote to Leo Arons, who two decades earlier had lost his academic position as a physicist at the University of Berlin due to his support for the Social Democratic Party. Einstein expressed his support for Arons and his alienation from German professors due to his disgust with the 1914 manifesto "To the Civilized World." Nonetheless, with typical self-confidence he stood firm in his own convictions, declining to support an initiative by Arons to get academics to support a socialist but not Bolshevik communist government:

> Since I know that you, as a courageous champion of free speech, certainly cannot feel offended by the following, I may be completely frank.

> Professors during this War showed evidence that from them we can learn nothing on political matters, that on the contrary it is urgently needed that they learn one thing, namely:
>
> To shut up! [*Maul halten!*]
>
> That is why I cannot endorse your proposal.
>
> In the hope of soon making your personal acquaintance . . .[55]

In the early days of the German Revolution, Einstein expressed the usual pacifist commitment to social justice by supporting the new German government; he also tried to help guide it in a democratic direction. On November 13, in a speech to a New Fatherland League gathering, he supported the values of parliamentary democracies, warning that tyranny by the left could be as bad as tyranny by the right. He encouraged support for the leaders of the new government but added that it was critical for a constitutional assembly to convene as soon as possible.[56] According to his handwritten draft of the speech, he stated:

> Comrades!
>
> Please allow a longtime supporter of democracy, who is not a recent convert, a few words:
>
> Our common goal is democracy, that is, the rule of the people. This is only possible if the individual holds two things sacred, namely:
>
> faith in the healthy judgment and healthy will of the people,
>
> willing subordination to the will of the people that is expressed through the electoral process, even when this popular will is in contradiction with one's own personal will or judgment.
>
> How do we attain this goal? What has already been achieved? What still has to happen?
>
> The old rule by class has been eliminated. It fell due to its own sins and through the liberating action of the soldiers. The soldiers' council that was swiftly elected by them, in conjunction with the workers' council, must for the time being be regarded as organs of the popular will. Thus, we owe these authorities our unconditional obedience in this critical hour, and must support them with all our powers, whether or not we approve of their decisions in detail.

> On the other hand, all true democrats must be vigilant lest the old class tyranny of the right is replaced by a class tyranny of the left. Do not let yourselves be tempted by feelings of vengeance to the disastrous view that violence must be fought with violence, that a temporary dictatorship of the proletariat is necessary in order to hammer freedom into the heads of our nation's people. Force generates only bitterness, hatred, and reaction.
>
> We must therefore unconditionally demand of the present dictatorial authorities, whose directives we must willingly follow, that, irrespective of party interests, they immediately prepare for the election of a legislative assembly, so that all fears of a new tyranny be dispelled as soon as possible. Only after this assembly is convened, and its task is satisfactorily completed, can the German people state with satisfaction that they have achieved freedom for themselves.
>
> Our present Social Democratic leaders deserve our unqualified recognition. With proud awareness of the appeal of the program that they represent, they have already decided in favor of convening a legislative assembly. Thus they have shown that they hold high the democratic ideal. May they succeed in leading us out of the grave difficulties in which we have been mired by the sins and half measures of their predecessors.[57]

Einstein reiterated his liberal political identity on November 14 to the Nobel laureate in chemistry Svante Arrhenius in Stockholm, expressing both his aversion to the War and his genuine hopes at the start of the Weimar Republic:

> For me as an old, convinced democrat and republican and a person almost fanatically focused on rights, joy is drowning out all other apprehensive responses, especially since I am Swiss. Hopefully, Europe will soon recover somewhat again from the terrible blow, and hopefully the more refined feelings and interests will soon make themselves felt again, so that life regains a happier character. Then the economic pressure will not make any decent person overly sad. It is quite remarkable how flexibly most people here have now already adjusted to such entirely new conditions; it is the most surprising experience of all of the surprises.[58]

On November 15, as part of the executive committee of the New Fatherland League, Einstein cosigned a telegram to fellow league member and prominent pacifist Ludwig Quidde, a former member of the Progressive People's Party in the Bavarian parliament and currently an official in the Bavarian Provisional National Council. The telegram requested Quidde's consent to their nominating him to the league's executive committee and summarized the league's current goals:

> New Fatherland League strives toward realization of socialism through democracy. Aims to propagate this idea among all circles of the population. Next goal is convocation of a constitutional national assembly.[59]

Quidde declined due to reservations at such a vague articulation of goals, expressing thoughts that may have contributed to Einstein's selective political engagement: "'The realization of socialism through democracy' is much too ambiguous a formulation for me to be able to commit to an association whose entire program is summarized in these words." Quidde advocated a balancing of individual freedoms and socialist goals: "To the extent that I do want the realization of socialist ideas, the safeguarding of personal freedom, that is the protection of individual claims, is to me of equal importance." Also, referring to "the newly founded Democratic Party, whose appeal for support is also signed among others by Professor Albert Einstein," Quidde worried about duplication of the new Democratic Party's program: "I do not understand what the New Fatherland League needs to do outside of this party, [which might cause] unnecessary and perhaps damaging fragmentation."[60]

As alluded to by Quidde, also in mid-November, Einstein showed his support for the liberalization of German politics by signing a public appeal encouraging people to join the German Democratic Party (Deutsche Demokratische Partei).[61]

In October and November 1918, the War having brought about the collapse of the Austro-Hungarian and Ottoman empires (as well as Russian and German empires), new states were forming in central and eastern Europe.[62] In Germany during November, the newly formed temporary government moved quickly to draft a constitution and convene a national assembly to ratify it.[63] Meanwhile, industrialists and unions reached an agreement that met several workers' demands, and the newly founded German Communist Party (KPD) pushed for a fully socialist revolution.[64]

In late November 1918, Einstein's brother-in-law Paul Winteler wrote from Lucerne, Switzerland, about the dramatic developments in Germany and Switzerland at the War's end. Winteler offered an eyewitness account of the general strike in Switzerland. He also raised concerns about revolutions leading to a new absolutism. He wondered about the new German Republic—whether the "child" would resemble "the eastern Bolshevik father or the western revolutionary mother." In Switzerland, he related, "Our general strike petered out after a few days, despite participation by the railway workers for a week." His concern was that the strike committee's nine demands, "whose legitimacy . . . was . . . universally recognized or . . . could not seriously be disputed: recall of the Federal Council, replacement of the Federal Assembly, 8-hour workday, old-age pension and disability insurance, voting rights for women," and which were "regarded by all as good or worthy of discussion," had not been "introduced by popular initiative and the exercise of voting rights, but . . . forcibly imposed through a general strike." He worried that if the strike committee took over from the Swiss Federal Council and Federal Assembly, they would not be "responsible to anyone, nor constrained by any constitutional or legal regulations in the implementation of the demands"; this would lead to "absolutism of the purest form," thereby "degrading the voter to a subject." He reasoned that "the leap from the 18th century into communism, without adoption of the democratic achievements of the French revolution, only leads from absolutism to absolutism. . . ." In his view, to be a republican and a democrat, "this higher way of thinking needs the longest and most persistent education and self-education."[65]

In a letter from Zurich, Einstein's fourteen-year-old son Hans Albert expressed ongoing concern about his father's health. He then described the events of the Swiss general strike with youthful alacrity and in more concrete, detailed terms: On Saturday, streetcar operators "abandoned their duties and went home." Then, "building construction sites" were abandoned and "all workers left their bosses in the lurch." Then events took a serious turn: "A lot of military were immediately brought to Zurich, probably close to 10,000 men." By Monday, a general strike had been decided upon for Tuesday. "Then suddenly on Wednesday evening . . . a paper was circulated, as there had been no newspapers previously either. They were sold by students and printed by employers." Students were helping keep the postal service going, and "a few trains were also running, operated by engineers and the military." In what could not have been a comfort to his distant father, Hans Albert concluded, "Apart from twice, when the military had to fire shots in order to disperse the crowds, the strike in Zurich went completely peacefully. And on Friday everything was back to normal." Einstein's son, who would become an engineer, also noted, "One amusing thing that happened during the strike was that by the 3rd day the tram rails were already completely rusty."[66]

Meanwhile, economic hardship confronted even academicians in their formal interactions, at the mundane level of paper shortages, as Einstein explained to the Zurich mathematician Hermann Weyl:

> Yesterday I tried to submit your paper to the Academy. It could not be done, however, because—something I had forgotten—a prior decision had been made that papers by nonmembers that are longer than 8 printed pages may no longer be included in the minutes under any circumstances. I therefore believe it would be best if I set your manuscripts aside until you have dealt with it otherwise. You can inform me by letter or telegram what I should do with the manuscript. The above decision arises from poverty—paper shortage and unaffordable printing costs; poverty is in general the specter that grins at us everywhere.[67]

In December 1918, US president Woodrow Wilson traveled to Europe to participate in the Paris Peace Conference, but his influence there was far less than the inspirational impact of his "Fourteen Points" speech might have predicted.[68] Late December 1918 in Germany was a time of political violence in the streets and of ideological conflict between the more moderate Social Democratic Party and the more leftist parties that might have been its political allies.[69] In early December 1918, Einstein described to his longtime friend Michele Besso in Zurich the beginnings of early Weimar's inflation, other troubling aspects of the German Revolution, and the contrast between those rejoicing versus those grieving. Yet he was still highly optimistic for the new republic:

> A major thing has really been achieved. The military religion has disappeared. I believe it will no longer return. Admittedly, nothing has taken its place. While southern Germany promises to develop more after the Swiss model, here the Russian example is alarmingly prevalent. Runaway slaves without a true public spirit and without overall perspective. The government, which of course is dependent on the masses, is struggling tirelessly but with minimal success against the economic collapse that is relentlessly brought about by the cursed circle: forced wage increases—accelerated bill printing—currency devaluation—forced wage increases . . . In spite of everything, most people feel relieved by the drastic cure, even those who know very well that they will have to part once and for all from their money. The Academy meetings are amusing now; the old folks are for the most part completely disoriented and dazed. They perceive the new era as a sad carnival and are mourning over the old economy, whose disappearance means such liberation to the likes of us. When I see Treitschke on the university campus with his ostentatiously arrogant posture and his solid belly of bronze, he seems to me like a mammoth from the Bismarckian ice age. Long ago it was . . . that is, about four weeks, when he was still a god beside the higher god Bismarck, who now suddenly appears to many in the same light in which he always appeared to us. If only people survive the drastic cure; some symptoms are a bit too suspicious. But I will not be dissuaded from my optimism. I am enjoying the reputation of an impeccable socialist; as a consequence, yesterday's heroes are fawning on me, under the impression that I could check their fall into the void. Strange world![70] (both ellipses in original)

Also in early December, Einstein agreed to help some New Fatherland League colleagues who were now part of the new German government, such as Helmut von Gerlach and Eduard Bernstein, by speaking on conditions in Germany at the Paris Peace Conference (although this trip ultimately did not come to pass). He described to Ehrenfest his intention to inform the Allies of the food shortages and hunger in Germany:

> In the next few days I am going to be traveling via Switzerland to Paris, in order to ask the Entente to save the famished population here from starvation. After so much falsehood, it is difficult to help give credence to the bitter truth. But I think people will believe me if I give my word of honor. Incidentally, I must say that the people here endured the collapse with composure and calmness, after the truth about the causes of the War had somewhat dawned on them. . . . The flu is less virulent here than elsewhere. We have been spared thus far.[71]

Einstein's scientific colleagues regarded him as optimistic about the new political developments, but this reinforced his isolation from many of them. For example, the Munich physicist Arnold Sommerfeld commented bitterly to Einstein in early December, "I hear . . . that you believe in the new era and want to participate in it. . . . I find everything unspeakably dreadful and idiotic. Our enemies are the greatest liars and scoundrels, we the greatest imbeciles . . . Money rules the world."[72]

Einstein's response was cautious, informed, and nuanced as usual, but he did indeed show an optimistic faith in current developments, in the feasibility of an international organization and its effectiveness in preventing war, and in an improved quality of life in Germany in the near future:

> It is true that I have hopes for this era, despite the many ugly things that it is bringing forth in the particulars. I see the political and economic organization of our planet advancing. If England and America are sensible enough to agree, wars of a certain magnitude will not be able to occur at all any more. Also, the military economy that is so offensive to me will virtually disappear. If the transition period now becomes rather oppressive for us in particular, in my opinion it is—frankly speaking—not entirely undeserved. However, I am of the firm conviction that culture-loving Germans will soon again be able to be as proud as ever of their fatherland—with more reason than before 1914. I do not believe that the present disorganization will leave lasting damage.[73]

In response to his older son's description of the Swiss general strike in Zurich, and in much the same spirit, Einstein wrote back to both of his sons, Hans Albert and Eduard (Tete), narrating the series of dramatic events in and around Berlin. However, he concluded with an incisive reiteration, on a very personal level, of a common pacifist concern that children should not acclimate to warfare as a form of normalcy through playing at it:

> Your strike was just a weak reflection of the revolution here. This one started in Kiel. The sailors there refused to fight on, ousted their officers, and then took control of the public buildings and the administration of the city. After that they sent delegations to the soldiers in other cities, who followed their example. The workers everywhere joined in, and everywhere workers' and soldiers' councils were formed, which took the place of the previous government. Finally the coup was carried out in Berlin, the Republic was proclaimed, parliament was dismissed, the kaiser and crown prince were forced to flee, the generals and officers were removed, and a socialist government was installed, which is accountable to the representatives of the workers' and soldiers' councils, called the Executive Council. A normal National Assembly that is elected by all men and women of Germany is supposed to convene shortly.
>
> In the meantime, however, quite wild things are going on here. Many are trying to take advantage of the confusion. The funds and other properties of the state are being handled carelessly. Paper money is being printed in vast quantities, so money is constantly dropping in value. Workers are demanding enormous wages, although there is nothing to do in the factories. All factories are on the verge of going bankrupt.
>
> Today the soldiers marched in. That was a cheerful flurry. Many boys like you drove and rode around in the streets with the soldiers and everyone was walking around in order to welcome the men who had to endure such very long, hard ordeals and dangers. But Tete should not engage in vile soldier games. I don't like that in the least.[74]

In mid-December, having failed in all of his attempts to persuade their mother to move to Germany, Einstein stressed to her the hardship imposed by the worsening exchange rate:

> It's truly frightening that the current exchange rate is now 8,000 francs to 14,000 marks. Who knows whether I might not one day be faced with

> the necessity of leaving here because I couldn't raise the money for you otherwise.[75]

Also in December, in a statement titled "To the Society 'A Guaranteed Subsistence for All'" (An den Verein "Allgemeine Nährpflicht"), Einstein expressed the pacifist concern for social justice, his own moderate Social Democratic perspective, and familiarity with the ideas of a contemporary social reformer. Offering his own independent thoughts in response to their proposals, Einstein agreed in principle with some aspects while disagreeing on the pragmatic level with others:

> The guarantee of a minimum level of existence for every citizen proposed by Popper [Josef Popper-Lynkeus] I also regard as the most important social goal for which we must strive. But I can by no means agree with the idea of a "subsistence army," because I am convinced that such a thing would produce the most inefficient results. Everybody should have the right to get the absolutely essential things one needs to live from public soup kitchens, in case of disability or unemployment. I am convinced that the striving for an improved existence through wage labor, combined with the contempt people have toward nonworkers, are sufficiently strong motivational forces to guarantee the healthy development of economic life. Experience seems to show that dire need and bitter troubles have a paralyzing rather than work-promoting effect.[76]

Thus, 1918 was a year of dramatic changes in Germany, with the final stages and conclusion of the War and founding of the Weimar Republic. Einstein maintained his political engagement throughout the year, through all the turbulence and uncertainty about the future. At the start of the year, while he and the rest of the German population were suffering from food and fuel shortages and strikes were breaking out in German cities, he continued interacting with people and organizations that favored peace and opposed annexations. In the late spring, he explored the prospect of a collection of essays advocating international reconciliation—an initiative that evolved into discussions about prevailing attitudes among the German people and the political viability of such an endeavor, during which Einstein demonstrated his interweaving of pragmatism and informed idealism. Starting in the late summer, he shared with most people an alert apprehension regarding the influenza epidemic. In the early fall, as on other occasions, he expressed to the radical pacifist Kurt Hiller his faith in Western-style democracy, probably partly inspired by the principles articulated in Woodrow Wilson's "Fourteen Points" speech. At the point of the armistice in November, Einstein gave a speech to students in the Reichstag building, met with the republic's future first chancellor Friedrich Ebert, and gave a speech for the New Fatherland League, which had been revived in September. Despite tremendous political upheaval, both in Germany where he lived and in Switzerland where his sons lived, Einstein felt great optimism about Germany's future, albeit tempered by awareness of the German Revolution's imperfections and of a possible tyranny by the left, which he viewed as potentially as dangerous as tyranny by the right.

4 Uprisings, Paris Peace Conference, Treaty of Versailles, and the Publication of *Lille*

1919

In 1919, as Einstein became a more prominent figure due to his scientific achievements, he increasingly utilized his name recognition to take public stands on political and humanitarian issues. After Germany's defeat in the World War and the fall of its empire, at a time of hunger, strikes, revolution, and counterrevolution, he based his hopes for Germany's new democratic republic on education of the German public. The volatile first months of the Weimar Republic were marked by the murders of the revolutionary leaders Karl Liebknecht, Rosa Luxemburg, and Kurt Eisner, and the rise and destruction of the Bavarian Soviet Republic; all the while, Einstein either initiated or took part in a variety of proclamations and actions on behalf of international reconciliation, the fledgling German Republic, some socialist and Zionist goals, and humanitarian concerns; these included lifting the blockade of Russia and liberating political prisoners. During 1919, he cultivated his relationships with pacifist and liberal political and cultural figures such as Helene Stöcker, Georg Count von Arco, Paul Natorp, Elisabeth Rotten, Otto Lehmann-Rußbüldt, Prussian minister of education Konrad Haenisch, Hellmut von Gerlach, Georg Friedrich Nicolai, Hans Delbrück, and Alexander Eliasberg.

Dramatic and violent events took place in Berlin in January 1919 involving the Spartacus uprising, violent clashes between rightists and workers, and the Weimar period's first wave of political murders of prominent leftists.[1] Hunger as a consequence of the continuing blockade of Germany was interwoven, internationally, with immediate postwar disasters, the Paris Peace Conference, and the founding of the League of Nations.[2] Meanwhile, a draft of the new German constitution was published and various German elections and government conferences were held.[3] While Einstein was in Switzerland in January 1919, he followed events in Berlin with interest, as he indicated to fellow Berlin physicist Max Born and his wife Hedwig on January 15.[4] Four days later, he expressed to the Borns his continuing affinity for the struggling German people and his ongoing confidence in the new German Revolution:

> Sparkling scenery and well-fed citizens who have nothing to fear. That is what it looks like here. But God knows, I like people with cares, for whom the next morning threatens uncertainty. How will everything turn out? One can barely take one's mind off the changed and still changing Berlin. I believe it will turn out well, in my sense of the word, when calm is restored again. The young people who have experienced all of this will not become bourgeois so soon.[5]

Notwithstanding Einstein's comments about well-fed citizens, Switzerland was experiencing a postwar shortage of heating fuel, which caused the University of Zurich to close down for a "heating vacation" (*Heizverien*) of three weeks, thereby delaying the start of Einstein's lecture series there.[6] University authorities also felt obliged to place guards outside lecture halls, blocking those who wished to enter primarily to warm themselves, to enable people registered for the courses to find sufficient seating.[7]

Sometime after it was drafted on January 23, 1919, Einstein signed the "Declaration on the Liebknecht-Luxemburg Case" ("Erklärung in Sachen Liebknecht-Luxemburg"), produced by the League for the Promotion of Humanity (Liga zur Beförderung der Humanität). He was one of 327 signatories who included Paul Natorp, Anita Augspurg, Walther Schücking, Alfred Hermann Fried, Maximilian Harden, Ludwig Quidde, Minna Cauer, Walther Rathenau, Hans Paasche, and Käthe Kollwitz. The declaration passionately invoked the pursuit of reconciliation, rather than the brutal murder of political opponents, to save Germany from annihilation and offer hope for the future:

> The undersigned, who represent the most varied perspectives and Parties, protest the spirit of brutal violence that currently once again rules in almost all strata in Berlin.
>
> Defenseless prisoners have been shot without justice or legal proceedings . . . Apart from these nameless people, the Spartacus leaders Karl Liebknecht, Rosa Luxemburg, v. Lojewski, Jordan, and Marx were killed as prisoners.
>
> In this case, strict, impartial justice must be demanded. Only absolute openness can prevent Liebknecht's death from becoming a similar catastrophe for Germany as the sinking of the Lusitania was at that time, an event which also received loud approval from almost the entire nation—at least, aroused no protest.
>
> Have we still not yet seen that only lawfulness can save us.
>
> . . . even the most common criminal may not be subjected to capriciousness, but rather is protected by law. Yet Liebknecht and Rosa Luxemburg were no common criminals! . . .
>
> So we challenge the government to protect the life and freedom of citizens and the most basic human rights with all of its power.
>
> But we also request that our fellow citizens bear in mind the internal domestic struggle for the new principle of reconciliation and understanding, which is directed against the resolution of conflicts by the sword, and therefore also must take primacy as most advantageous in domestic strife, if we don't want to arrive at the incalculable horror of annihilation.
>
> The crime of the murder of political opponents, which we have experienced in recent days, intensely spotlights the situation in which we find ourselves: that every step further on this path leads to the precipice. Only when we can show that we are filled not with blind hatred and petty self-interest, but rather with sincere love for all, only then will we be able to overcome the fanaticism of hostile opinions, only then will we finally succeed in building out of this world still full of blood and terror a world of trust and love, in which it is worth living.[8]

In February 1919, amidst a growing rightist movement in Germany, a provisional government was established at a meeting in Weimar.[9]

Meanwhile, warfare continued east of Germany through late 1920.[10] On February 8, 1919, in his ongoing efforts to help steer the German Revolution in a positive direction, Einstein cosigned the "Appeal to the Free Youth of all Classes and Nations" ("Aufruf an die freie Jugend aller Stände und Völker"), along with other members of the Council of Friends of the International Youth League (Freudesrat des internationalen Jugendbundes) who included Käthe Kollwitz, Elisabeth Rotten, Walther Schücking, and Max Hodann. The appeal urged young people to create a new world culture of goodwill in public life. It extolled the scientist's highest ideal of reason as the sole source of legitimate authority, as opposed to party doctrines and slogans:

> All states are tottering. All historical structures are breaking apart. All party dogmas are being blown up. Only one rock remains, on which a new national life and life of the nations can be justly constructed: reason.
>
> Moral reason is the creative revolutionary and legislator in every human being, not empty, irresponsible understanding or self-interest. Reason alone befits sovereignty. Every other claim of constitutional power must be derived from it, otherwise this entire claim to power is null and void.
>
> Reason is universal and cannot be satisfied with a single form. It delves deep under the surface of political life, in order to eradicate fundamental evil from its roots. Thus the unsatisfactory nature of all newly founded entities that take up some particular interest, party doctrine, or universal slogan while extolling the highest principle. . . . Reason seeks absolute freedom in order to create justice according to its eternal law. Solely on the basis of the constitution of reason, which is independent of all outward violence, is a truly complete sweeping reformation of societal life at all possible.
>
> On this unshakeable basis, in the middle of the World War in the heart of Germany, the cornerstone of a new international party, an international of reason, a party of reason, was laid.
>
> This cornerstone is the International Youth League [Internationale Jugendbund]. Despite pressure from the police and censorship, it succeeded in creating strong working groups in Berlin, Munich, Frankfurt a. Main, Magdeburg, Kassel, and Göttingen. . . . Thus the International

Youth League shall become the powerfully driving seed of the party of reason.

Because among youth reason lives still unbroken. Therefore youth must be, and alone can be, the source of lasting renewal. Youth only needs to develop a lasting form of selection and of inner consolidation of its noble powers, in order to achieve the guarantee of a better future for the nation's life. The International Youth League offers this lasting form to the young generation. . . .

The demobilization has taken place. We proclaim a new mobilization: that of good will in public life. We call upon the young generation to help bring a new world culture to victory. . . . Declare finally closed the war of the true enemy of all peoples, that of the domination of despotism and of the accidental in all of their forms! . . . Do not believe in the saving power of universal slogans . . . Do not make as a final goal of politics what in the best case is one of its means! It is not done with mere democracy and the mere Republic. Therefore free yourselves from the spell of every phrase and make yourselves serious about Wilson's words: "Abolition of every arbitrary power of the state"![11]

On February 10, 1919, Einstein was among several intellectuals invited by Harry Count Kessler and a few colleagues to join an independent "Fabian Society." The group was intended to engage in objective research and discussion on political, economic, and ethical matters, according to the published diaries of Kessler.[12]

On February 14, Einstein's divorce from Mileva Einstein-Marić was finalized. The divorce decree was delivered to Einstein in Berlin in April.[13] In late February, Einstein responded favorably to Zangger's request for information on several prominent Germans who might help stabilize the political situation in Germany, as part of a plan advocated by Wilhelm Muehlon, formerly of the German Foreign Office, who was engaged in food distribution in postwar Germany.[14]

During February 1919, further details in the structure of the Weimar Republic were ironed out, while political violence continued, including assassinations and street battles between left and right.[15] Meanwhile, with communism on the rise in Russia and Hungary, fascism was on the rise in Italy.[16] In March, Germany and the Allies reached an agreement on the

German merchant marine and the feeding of Germany.[17] In late March, Arnold Sommerfeld complained to Einstein about being prevented from attending German Physical Society meetings in Berlin, almost certainly because strikes were impeding railroad travel in Germany.[18] At the end of March, Einstein was still hoping that the Weimar Republic would take decisive action to live up to its expressed ideals. In a letter to Swiss professor of mechanical engineering Aurel Stodola, he used rather pungent imagery when describing postwar political conditions in Germany:

> The main thing is, now, that you overcame the unavoidable flu and feel even better than before. . . . Here things are lively again. The country is like someone with a badly upset stomach, who, however, has not yet vomited enough. When Scheidemann, Noske, etc., are out as well, the patient will gradually get better. He can finally recover once he can make a firm resolution to change his lifestyle. Your optimism gives me joy and I share it. The preparation of the sluggish masses for a socialist consciousness is making clear progress, even if I often can only detect it in the hypocrisy of our splendid newspaper Berliner Tageblatt.[19]

In April 1919, the Allies invited German officials to Versailles to receive but not negotiate the peace treaty.[20] Internationally, German troops engaged in one legally permitted action in Latvia, and British troops, in an event of interest to pacifists who were following Mohandas Gandhi's nonviolent national liberation movement in India, weakened their legitimacy in India in the wake of the Amritsar massacre.[21] Einstein was communicating with fellow members of the New Fatherland League, including Georg Count von Arco and the socialist feminist Helene Stöcker. The latter urged him to join her and others in signing an appeal to the government, which she enclosed. In light of the uprisings of January and March and the serious dangers of class conflict, that appeal made eleven demands of the German government; these included barring the military from politics, expediting the trials of the murderers of Liebknecht and Luxemburg, establishing a commission to investigate and assess war guilt, and replacing compromised members of the German peace delegation.[22] Arco wrote inviting Einstein to speak before a meeting of members of the

Berlin branch of the German League of Monists (Deutscher Monistenbund). About a month later, its members would cosign an appeal against civil war circulated by the New Fatherland League.[23]

Also in April, in a letter to his former Zurich neighbor and ex-wife's divorce attorney, Emil Zürcher, Einstein made it clear that he was following the actions of the Russian Bolsheviks. He showed astute judgment in pointing out that the Allies would be ill-advised to crush Germany economically:

> Reliable statements about Russia that I have been able to gather have multiplied so much that further doubt is unfortunately no longer possible. The country is being plundered by countless minor potentates = leaders of robber bands, which are systematically shooting all people who do not belong to the lowest class. These bands are recruited for the most part from the Chinese. Fine prospects for us, too! If the Entente were well oriented and farsighted, it would make efforts to maintain Germany in a state of somewhat healthy life. France should receive damages for the destruction, but should not want to take German land. Now would be the time to speak out, if only one were to be heard.[24]

Sometime prior to April 26, 1919, Einstein joined a commission of private citizens with the purpose to investigate charges that the German army had committed atrocities in Belgium and northern France during the War. This commission, which he and some of his colleagues had planned during the War, intended to publish its findings in Germany. Its hope was to inform Germans of the conduct of their army, thereby helping them to understand the bitterness of their wartime opponents, thus furthering international reconciliation. In the fall of 1919, the commission published its first version of a booklet entitled *Lille*, named after a city in northern France that was prominent in these allegations of human rights violations. This first version reprinted original German wartime documents—none, though, that addressed French charges contained in documents attached to a wartime protest statement by the intellectuals of Lille, although it did include this protest statement. The French documents described arrests of civilians, deportations, a forced march of ten thousand women, and other similar wartime atrocities.[25] This first version was deemed unacceptable by Einstein and others when it came out in the fall of 1919, and efforts were made to suppress and recall the copies that had been distributed. A revised version was prepared during 1919 and published at the start of 1920, as discussed in more depth later in this chapter.[26]

Shortly after joining the commission, Einstein described to Lorentz its purpose, expressing his belief that revealing the truth was the only way to expunge ongoing postwar hatreds. He invited Lorentz to join the commission so as to expand its multinational representation:

> You will certainly still remember the sad conversations about the war atrocities that we had together 3 years ago in Holland. The memory that you took the most lively interest in these terrible things encourages me to present the following to you.
>
> A commission has been formed here by a few private individuals, which has set for itself the task of scrupulously examining the accusations about Germany's conduct of the War that have been publicized abroad and found to be justified, by means of local and, when possible, also foreign official documents, and of publishing the results here. I myself am on the commission, along with five other people, for whose absolutely honorable and earnest convictions and sense of justice I can vouch.
>
> In order to acquire materials and, where possible, documentary evidence to serve as a basis (grounds) for our claim, we must expand our commission membership to members who live in neutral foreign countries. We would be happy if you, esteemed colleague, would join the commission, because your great experience in human affairs and your name would be exceptionally useful to our effort. If, however, you do not wish to do this for any reason, I request that you please suggest to me a person in Holland who in your view would be willing and able to join our cause in the manner described.
>
> I believe that a slow disintegration of the terrible hatred will only be possible when the truth is completely uncovered. There are many thoroughly honorable people here who consider all these serious accusations, to the extent that they have penetrated to them in muted form, to be tendentious lies; to enlighten such people is our goal.[27]

Sometime before April 23, Einstein agreed to deliver a speech at a meeting of the League for the Promotion of Humanity (Liga zur

Beförderung der Humanität), to be held at the Berlin Opera.[28] By April 30, the organizers of this event had listed him as a possible speaker.[29]

During May 1919, the terms of the treaty received at Versailles provoked outrage in Germany.[30] The German delegation met in Germany and drafted a counterproposal, which was presented to the Allies and the German press in late May.[31] The Paris Peace Conference would also have War-related repercussions for Japan and China. The Ottoman Empire's defeat in the War would lead to its dismantling as well.[32]

In early May 1919, Lorentz wrote to Einstein regarding the private war crimes commission. He declined to join it, though he fervently hoped that Einstein would be able to implement the plan and that it would "contribute toward gradually alleviating the current vehement bitterness"; he added, however, that "at a time when Germany was still victorious, it could have been more successful." He also raised concerns about the difficulties that would face a private commission that lacked the legal authority to access all available documentation. Such a commission would find it difficult to attain a "certainty that is convincing to all," for it would "have to limit itself to collecting already published material and testimonies by credible persons." Lorentz did, however, offer extensive help with disseminating the word to western European contacts, and with "procuring the necessary data . . . from countries that suffered from the War." And he assured Einstein that he would "vouch for your absolute integrity and the probity of your convictions."[33]

Related to this commission's work, and consistent with Lorentz's concerns about bitterness in western Europe, sometime before May 6, 1919, Einstein promised to sign the international declaration written by Romain Rolland, "An Appeal: Proud Declaration of Intellectuals" ("Un Appel: Fière Déclaration d'Intellectuels"), provided that prominent French and British intellectuals also signed.[34] This appeal by intellectuals who had maintained their neutrality during the War was motivated by the need to propagate justice and international reconciliation; it called for an investigation of the question of War guilt conducted in such a manner that thoughtful people of all countries would see its handling as an act of justice, not of power or violence.[35] The appeal was published in late June 1919 in the French journal *L'Humanité*, with Einstein among the signatories.[36]

On May 7, Einstein signed a public appeal drafted by the New Fatherland League. The thirty-four cosigners included the leaders of the German branch of the Women's International League for Peace and Freedom (Internationale Frauenliga für Frieden und Freiheit), Anita Augspurg and Lida Gustava Heymann, as well as prominent pacifists Kurt Hiller, Käthe Kollwitz, Helene Stöcker, and Hans Wehberg, and was endorsed by several other organizations. This appeal, a prelude to pacifist women's demonstrations planned for May 27 in eight major cities in Germany, called for domestic as well as international peace:

> Four years of war have proven that turning to weapons brings no clarification to the solution of economic and political conflicts in modern life, but rather, quite the opposite, brings about ever worsening confusion, most prominently the destruction of irreplaceable human life and great cultural treasures.
>
> This experience applies to both international and national life. We demand and need pacifism also in domestic politics!
>
> We therefore challenge all people of good will to join in a united front with all of their powers to take back the red and white guards, so that the civil war that is driving Germany toward total destruction stops. The soldiers do not exist in order to resolve political questions, but rather solely to serve as police officers. The government and the terrorists of the right and left must be pressured by public opinion to pursue the path of negotiations.
>
> Away with violence!
>
> We do not want to bleed to death in a civil war!
>
> We demand negotiations!
>
> New Fatherland League.[37]

On May 9, Einstein wrote to Lorentz's son-in-law, the Dutch physicist and mathematician Wander de Haas, expressing a fervent commitment to the war crimes commission. He explained why Lorentz's preference to start the commission's work before Germany had been defeated would not have been possible. He also emphasized that, due to current political agitation, the commission would not be requesting documents from the German government until after the peace treaty had been finalized. (He did not realize that the unrest would only worsen after the treaty had been signed.) He reiterated the commission's goal, which was to counter a vengeance mentality among Germans, and he noted again, with compassion, the political chasm between himself and his most significant colleague in Berlin, Max Planck:

> My collaborators and I were very pleased that Mr. Lorentz so willingly wants to help us. During the War years, such an undertaking would have been not only dangerous but also simply impossible, since every last piece of documentary material would have been inaccessible. Now our enterprise should be almost exclusively dedicated to the enlightenment of local public opinion. We want to seek to obtain the evidence through our neutral associates and try to verify it against documents here. The results are intended to be publicized. We originally wanted to approach the government with our request only after we had conferred with our associates from neutral foreign countries, in order to come forward with greater weight and to have a greater prospect that the local documents would be made available to us. The present moment is not favorable, partly because of the agitation of political passions, and partly because our efforts could be falsely interpreted as long as the peace [treaty] conditions are not established. We must avoid the appearance on either side that we are representing any sort of political goal in the usual sense of the word. That is why I did not send a telegram but rather chose the slow route by letter.
>
> With the enormous abundance of material, it cannot be our goal to follow up on the crimes of any individuals. We must rather, in my view, restrict ourselves to whatever indicates criminal orders and intentions from the top. We are requesting material selected according to this standpoint from our neutral friends and will then attempt, on the basis of local documents, to find evidence that would be convincing to people here. Thus we hope to contribute to an improvement of the mentality of the latter, and particularly to counteract the development of the idea of revenge [*Revanche-Idee*]. My name may be mentioned everywhere in this connection. I am doing nothing that I cannot justify before anyone. Unfortunately I cannot get anywhere with [Max] Planck. I never speak with him about things that even remotely touch upon politics. I would only cause him pain, without our understanding each other.

> These people are as though they were under hypnosis; Planck is already so shattered by all the experiences that he absolutely must be spared. He is otherwise so upright and selfless that one must excuse him in this respect. As soon as the preliminary peace [treaty] has been concluded, we shall turn to the government, and I will then immediately report to Mr. Lorentz about it all.
>
> Meanwhile, I thank him from my heart for his willingness to promote this good cause.[38]

On May 11, Einstein wrote to Paul Natorp, a professor of philosophy at the University of Marburg, indicating his support for two appeals that Natorp and several associates were then circulating in draft form. These two appeals addressed intellectuals and working-class leaders, urging them to bridge the gap between each other and cooperate in building socialism, to trust intellectuals as leaders in developing socialism, and to oppose violence or a dictatorship of the proletariat.[39] Einstein expressed enthusiasm for these sorts of attempts at cooperation:

> I have studied your new draft and can now thoroughly approve of it. I heartily hope that now a fine little band of like-minded people can be found to cooperate in earnest. Most people, even the most intelligent, regrettably are exclusively guardians of their own—special or class interests. I am very curious to hear from you about the success of this action.[40]

When the harsh conditions in the preliminary version of the Versailles treaty were revealed on May 7, the German government declared a "week of mourning" (*Trauerwoche*) from May 10 to 16, 1919.[41] In a letter to his mother on May 14, Einstein mentioned the dismay in Germany over the Versailles treaty and alluded to the ongoing food shortages:

> We are currently having an official week of mourning in honor of the projected peace. . . . In August let's travel back to Berlin together, dear Mama. By that time, the food situation will be better here as well.[42]

Also in mid-May, consistent with typical pacifist opposition to all state-sponsored violence, including its individual application in the death penalty, Einstein joined with sixteen political, religious, and academic figures in signing a telegram urging the Munich military tribunal not to implement the death penalty against Eugen Leviné. The telegram reflected a savvy assessment of the current political situation. Leviné had led the short-lived Bavarian Soviet Republic from April 13 to May 3 and was charged with high treason.[43] Fellow signatories included Protestant and Catholic clerics Karl A. Aner, August Bleier, and Hans K. Francke, University of Hamburg philosophy professor Ernst Cassirer, Foreign Office Secretary of State Karl Kautsky, Independent Social Democratic Party (USPD) leaders and Reichstag members Hugo Haase and Hugo Simon, 1914 "Appeal to the Europeans" author Georg Friedrich Nicolai, and six of Einstein's fellow members of the war crimes commission, Georg Count von Arco, Hellmut von Gerlach, Maximilian Harden, Max Hodann, Elisabeth Rotten, and Helene Stöcker. The telegram warned of the detrimental effects of mob mentalities on legal processes and urged careful adherence to law and cautious implementation of the judgment:

> We undersigned members of the most varied political parties regard military tribunals as a great danger to the interests of the domestic peace of our nation. We fervently insist that the judgment that will be made against communist leader Leviné by the military tribunal in Munich, despite the agitated political atmosphere in which the process is taking place, will not depart from the laws of justice and humanity. In any case, we raise our voices against the judgment being immediately executed. More than ever is it necessary to avoid anything that can whip up anew the passions of our people.[44]

These political developments coincided with a major turning point in Einstein's professional life, and a significant step in his personal life. On May 29, 1919, a total solar eclipse was observed by two British expeditions to the southern hemisphere, Arthur S. Eddington in Principe, an island near Africa, and Andrew Crommelin in Sobral Ceara, Brazil. The results of their observations verified Einstein's theories and launched him from prominence in the physics profession to iconic world fame.[45] On the personal side, on June 2 in Berlin, a few months after his divorce from his first wife, Einstein married his cousin Elsa Einstein.[46]

At the beginning of June 1919, Einstein described to Zangger in Switzerland his impressions of a deflated and ambivalent postwar political atmosphere in Germany, one that mixed relief and rage:

> Here the political wave has subsided. People have no more energy for great emotions, but rather are more or less passive. The end of the War is felt among the people as a liberation, the reappearance of vegetables as a relief. Among entrepreneurs a lack of initiative due to insecurity about the future, among the masses a weak sense of self-esteem combined with a disinclination to work. The Bolshevist wave seems to be more or less neutralized by the peace terms; but one can easily be mistaken and must humbly await what will come. In any case, I am not pessimistic, but rather foresee a not exactly splendid but a tolerable future approaching. Truly free conditions cannot evolve here, where the people know nothing other than subservience and—rage.[47]

Three days later, Einstein wrote to Max Born with a somewhat prescient assessment that the Treaty of Versailles would never be fully implemented. He was, however, still relatively optimistic about postwar prospects in Germany:

> I don't view the political situation as pessimistically as you do. The [peace treaty] conditions are harsh, but will never be realized. They will serve to satisfy more the eye than the stomach of the "enemy." Ludendorff was undoubtedly much worse than the Parisians. The French act like that out of fear, but Ludendorff has Napoleonic rapacity. The hardships from French mistakes will be mellowed by a never-failing sloppiness, just as in my former fatherland Austria. Ultimately, the threat of Germany and also the unity of the opponents will dissipate like a puff of smoke, in which a definite hysteria is prevailing, though. May a hardcore x-brother [*X-Bruder*] and determinist with tearful eyes say that he has lost faith in people? It is precisely the impulsive behavior of the people of today in political affairs that tends to keep the belief in determinism very much alive. I am convinced that what is coming in

> the next few years will be far less difficult than the experiences of the past few years.[48]

The postwar political upheaval affected scientific collaborations even domestically, as Leonhard Grebe, a fellow physicist at the University of Bonn, showed on June 6 when he wrote Einstein that he could not travel from Bonn to Berlin for a planned scientific consultation because of the political situation.[49] On that date, workers' councils in Berlin had called a one-day strike to protest the execution of Eugen Leviné by the Bavarian government.[50] Meanwhile, inflation was hurting Germany's foreign exchange rates, compounding Einstein's difficulties in sending financial support to his first wife and sons in Switzerland, as he wrote to his sons in mid-June.[51]

During June 1919, the Allies conveyed their peace terms to Austria. For various reasons, these were more lenient than those imposed on Germany.[52] With Germany, the Allies made a few concessions, but they intensified the war-guilt clause, leading to several stages of heated debate and resignations in the German National Assembly. Ultimately, though, after a few gestures of defiance, Germany signed.[53] In mid-June 1919, Zangger noted proto-*Anschluß* sentiment (with territory attaching in the opposite direction) when he speculated to Einstein about the idea of Bavaria seceding and uniting with Austria.[54] In reply, while he viewed the Treaty of Versailles as less harsh than most Germans regarded it, Einstein focused on the current class conflict and the obtuseness of elites in not recognizing what he saw as the need to implement at least modest socialist and democratic goals:

> In politics it seems to me to be a similar situation here to 1914. Again murkiness and no equilibrium can be found. I do not believe that the attempt to smother in blood the yearnings of the many will be successful. The peace [treaty] now seems to have turned out to be fairly lenient, as I always expected. It is just too impractical and arduous to transform the world into a prison. The potential undoing resides in the dull resentment of the masses, who feel that they cannot trust the "bigwigs." Until it is given the feeling of power and responsibility, the society cannot

> become healthy. When the right of individuals to control businesses and then profits at their own discretion stops, the worker will no longer be exploited. It is difficult to grasp what an idiotic and uncomprehending attitude almost all "intellectuals" have toward these things, moreover not only those supposedly interested through property. I believe that in the fighting out of this conflict Switzerland will no longer play the role of happy island.[55]

On June 26, 1919, Paul Natorp sent Einstein a revised version of the appeal, "To the Intellectuals of the Proletariat," a preliminary version of which Einstein had approved in early May. In his cover letter, Natorp expressed passionate disagreement with Georges Sorel's *Réflexions sur la Violence* regarding whether the working class struggle justified violence. Natorp saw Sorel as believing in "the sanctifying power of war . . . in the violent war [*Gewaltkrieges*] of the general strike." Natorp argued that "justified defiance should attain its goals . . . through . . . tenacious adherence to *justice*"; this was the sole guarantee that freedom would be attained. Natorp saw "more serious danger" in "the path of violence." Therefore, he recommended standing with the "intellectuals of the proletariat."[56]

Also on June 26, *L'Humanité* published Romain Rolland's "An Appeal: Proud Declaration of Intellectuals" ("Un Appel: Fière Déclaration d'Intellectuels"), which urged investigation of War guilt in a fair manner. Einstein was among its signatories. He had signed the German version, "For the Independence of the Mind!" ("Für die Unabhängigkeit des Geistes!"), in early May.[57] The German version was signed by German pacifists as wide-ranging as Paul Natorp, who advocated that Germany sign the Treaty of Versailles, to Nobel Peace Prize laureate Alfred Hermann Fried, who opposed Germany signing, as well as iconic cultural figures including Hermann Hesse, Heinrich Mann, and Käthe Kollwitz.[58]

These public appeals generated various expressions of support in Einstein's broader circles. At the beginning of July 1919, University of Berlin geophysicist Adolf Schmidt responded enthusiastically to Rolland's appeal. Schmidt proposed that the appeal be circulated more widely, suggesting that it would find thousands of signatures among German intellectuals, and that this could "thereby contribute to the circulation of these perspectives and to the struggle against the attempts at hatred and revenge that are hostile to civilization." He urged Einstein to form a committee for this purpose.[59]

Several days later, University of Breslau historian Robert Holtzmann, having seen the declaration of intellectual workers in the newspaper, wrote to Einstein that the appeal had found much support among his local "Cultural-Political Study Group" (Council of Intellectual Workers; Kulturepolitische Arbeitsgemeinschaft [Rat geistiger Arbeiter]) due to its "exceptionally appealing content." On the group's behalf, Holtzmann expressed a similar desire to join a campaign for additional signatures.[60]

In mid-July, Einstein's wife Elsa wrote to him in Switzerland, mentioning the continuing postwar volatility and disruption in Berlin, with strikes and no trains running for eleven days.[61]

In late July, Adriaan D. Fokker, a young physicist who had studied under Lorentz, also responded to the Natorp appeal. Writing to Einstein from a sanatorium in Switzerland, Fokker discussed the nationalism and vengeful spirit among the Germans there, as well as the difficulties in advocating reconciliation because of the Allies' behavior in Paris and because people "invariably had heard other facts." Fokker complained that, "with screeching and protesting and the slogan that this forced peace [*Gewaltfrieden*] *must* lead to a new War, they are immediately starting an asinine warmongering all over again." He found it courageous that Jan Christiaan Smuts of South Africa, at the Paris Peace Conference, had advocated reconciliation with Germany.[62] In his reply, Einstein told Fokker that German attitudes and chauvinism did not surprise him. Yet he continued to express optimism about the future effect of the League of Nations as well as faith that Woodrow Wilson's principles would prevail:

> You have the opportunity there to get acquainted with the German political mentality. Only too easily do people create for themselves an overly simple image of foreigners. These people still have no clear idea that they were the blind tools of an arrogant, unscrupulous minority. Therefore, to them indignation over the "forced peace" is no slogan or hypocrisy, but rather truthful experience. But the truth is gradually seeping through, due to the efforts of individual personalities to wash

> themselves clean—at the expense of others. However, I am not so pessimistic about the future, but rather believe in a gradual success of the League of Nations. Even if great hardships and injustices in particular were to take place, the constancy in the development of the idea of rapprochement would awaken trust. It is especially gratifying that America, which is not burdened with the fatal traditions of Europe, has the leadership.[63]

In late July 1919, Lorentz wrote to Einstein about his continuing efforts on behalf of the private war crimes commission. Lorentz indicated that, while visiting Belgium and France, he had discussed the matter with Marcel Brillouin, Paul Heger, and Adriaan Noyons, who had "pledged to me their cooperation in the procurement of the desired documentary evidence." In particular, Noyons had "very extensive and . . . conclusive material at his disposal" about the destruction of Louvain (Löwens) and was willing to travel to Holland with his documents to meet with Einstein at Lorentz's home. Lorentz also discussed the French and Belgian resentment of Germans and the related delay among scholars in welcoming German scientists back into international conferences: "there is generally . . . a feeling of bitterness against Germany; whenever this is expressed, the deplorable Manifesto of the 93 yet again comes to the fore." However, he emphasized, they also understood that there were people in Germany like Einstein. Still, regarding professional meetings, "it is clear that for the time being no Germans will be invited . . . but there is no discussion of a formal exclusion; the door will be held open for them . . . in the future."[64]

At the end of July, Einstein began to express concerns that would begin to weaken his initial optimism about Germany's political prospects. In a hedged prognosis for Germany written to Auguste Hochberger, a friend of his mother's, Einstein reiterated his faith in hardship as educational but worried about the trend toward vengeful feelings:

> Once the social difficulties are somehow overcome satisfactorily, then we will be able to delight in this land in which we live. It will prove once again that failure and poverty are the best teachers and purifiers. The process is painful, certainly, and people are only too inclined to place the blame on the external enemy. I hope, however, that the opinion

around here will become more just, as the old sins are gradually brought into the light of day. In that respect, unfortunately, much too little has happened so far.[65]

Also in late July, Einstein mentioned to his wife his ongoing interactions with both Lorentz and Swiss philosophical-political writer and fellow New Fatherland League member Elisabeth Rotten with regard to their collaboration on the war crimes commission.[66] Einstein wrote to Lorentz thanking him for his help. He acknowledged the impact of the "Manifesto of the 93" on feelings among the French and Belgians, and he restated that the commission's larger purpose was to counter Germans' desire for vengeance by helping them understand their neighbors' feelings:

> It is extraordinarily valuable to our cause that you are providing your support to us to such an extensive degree. We also must thankfully welcome and acknowledge the fact that the French and Belgian scholars whom you mentioned want to help us. Right up to my departure on June 28, we were not yet able to gain access to the documents that are piled up there; there is open and especially covert resistance. Of course, here I find out nothing about what has taken place since my departure, so I can only respond properly to your friendly letter when I come back home (mid-August).
>
> I understand the bitterness due to the famous Manifesto of the 93, which, by the way, was drafted by the writer [Hermann] Sudermann. Once, when I happened to sit next to him and expressed myself in no uncertain terms about that manifesto, he openly and without much shame admitted to being its author. He is, incidentally, a decent person of rare good will, as long as one does not wave the red flag "politics" before his eyes.—As I said, I understand that bitterness, but it does have a slight tinge of the comical for me, if also not nearly so bad as the manifesto itself with its manly defenders. When a group of people is infected with collective insanity, one should defuse it; but hate and bitterness cannot consume great and perceptive people for the long term, unless they themselves are infected by it. We must indeed bear in mind that the moral qualities of people do not differ very much at all, on average, from country to country. . . . German scholars may be excluded from international social exchanges of scholars for a number of years; this

> may be a lesson in humility for them, which will not do much harm at all—hopefully even be beneficial. The matter in no way has much significance.
>
> But what seems important to me is that people in Germany learn to understand the attitude of the "enemy," so that there is no room for the abhorrent idea of revenge to take hold, from which new misery could later grow. Our modest goal is to work in this direction.[67]

Severe coal shortages were anticipated for the coming winter, largely because the Treaty of Versailles required that millions of tons of coal be sent from Germany to the Allied countries as part of reparations but also because of strikes by transport workers. Coal and gas rationing had already begun.[68] Einstein mentioned this postwar hardship to his mother:

> We must count on spending the winter here without heat. Just thinking about it almost turns one into a chunk of ice.[69]

After the adoption of the Weimar constitution, the German National Assembly's first attempts in August 1919 to investigate the War were undermined by rightist and antisemitic backlash.[70] Meanwhile, in defiance of Allied directives, German troops in eastern Europe and the Baltics stayed on past their mandate.[71]

In mid-August 1919, in his response to Robert Holtzmann's letter encouraging efforts to attract more signatories for the declaration of intellectual workers, Einstein offered an assessment of political conditions in Germany that was less sanguine than his earlier prognoses. He displayed a sharp awareness of vengeful attitudes among the German population, doubts that the declaration would attract many more signatories, and an astute understanding of the ways in which political processes could backfire:

> After my return . . . , I found your letter, whose supportive content pleased me extraordinarily. I must admit that I am not one of the originators of the appeal, but rather merely added my signature upon request. Therefore I do not know whether a collection of signatures is desired. From what I have recently encountered of the opinions among

the educated in our country and elsewhere, it seems to me, unfortunately, that such an undertaking at the present moment promises little success. I believe that our most important goal must be to prevent the spread of the idea of revenge among us.

I fear that, through a declaration signed by a larger number of kindred spirits, we would provoke a counter-declaration of considerably greater "impact on the masses" and thereby bring more damage than help to the good cause.[72]

Einstein reiterated similar reservations in his response to Adolf Schmidt, in which he also described the Lille commission and his concern over the rise of vengeful attitudes:

On my return . . . , I found your friendly note. . . . Above all it greatly pleases me to have you in this matter again as a kindred spirit—as already many times before.

Regarding the appeal, I am not among its originators. Because I am acquainted with the greatly embittered opinion among literary icons in various countries, I know very well that such attempts at reconciliation currently promise only the most minimal success.

I must frankly admit to you that a collection of signatures in the sense that you suggest seems to me to have just as few prospects, partly because of the painful gaps that it would reveal, and partly because of the rebuttal that would probably arise. For the time being, it should be valuable, first of all, to cultivate individual international personal liaisons and thus quietly and slowly pave the way to better relations. Similarly, I consider it beneficial to publicize knowledge about the heavy sins of the German army leadership in Belgium and France, in order to clear the way for a better understanding of the others' state of mind. I am a member of a small private commission that is currently preparing publications in this area. For the present, I consider it most important to prevent the germination of the idea of revenge, so that the current wretched state of affairs is not reinforced by such brutal means.[73]

Three days later, Einstein again conveyed a moderated political optimism to his ailing mother's friend Auguste Hochberger, reiterating his old trust in tough conditions as a mechanism for moral teaching:

> You reminded me of my prophecies and views. In those days it was easier than today. All the same, my outlook is not so very somber, but rather I believe that in a few years bearable conditions will arise among us. Next winter we will be shivering, but at least not starving. It is in England's interest that we not be dragged down too terribly low; it alone knows how tautly the bow may be drawn before it breaks. For our country's inner development this time will be beneficial, because it is politically educational and sweeps away the luxury and pampering, also because it augments the moral weight of southern Germany, whose population thinks, feels, and acts more humanely than that of northern Germany.[74]

At the end of August 1919, Einstein expressed to Hedwig Born his trust in the League of Nations' capability to adapt and improve over time. On the eradication of anarchy in international relations, he alluded briefly to what would become his lifelong faith in a supranational organization with executive powers:

> In politics I side more with your husband than with you. I believe in the League of Nations' ability to develop and also believe that the hardships connected with its creation will fade away with time. The conflicts of interest among the Allies are already so great now that some things are becoming more moderate—the constitutional incident regarding Austria; the Allies' intervention in Silesia. The greatest danger for future development, in my opinion, is that the Americans could withdraw; hopefully, Wilson will be able to prevent this. I do not believe that humanity as such can change in essence, but I do believe that it is possible and even necessary to put an end to anarchy in international relations, even though the sacrifice of autonomy for individual states will be significant.[75]

In September 1919, the University of Berlin astronomer Wilhelm Foerster, one of Einstein's three fellow signatories of the 1914 "Appeal to the Europeans" and father of the pacifist philosopher Friedrich Wilhelm Foerster, wrote Einstein with the same suggestion that Adolf Schmidt had made, urging him to circulate Rolland's appeal "For the Independence of the Mind!" among other potentially interested individuals, emphasizing that, in the appeal, "France is reaching out a hand of reconciliation to us."[76]

In early September, Einstein described various postwar hardships to his mother:

> We, though, are doing well currently. Only the hardships that arise from the general economic situation are making themselves felt. We have to give up a room (rent it out). Starting tomorrow, the elevator won't be working any more, so each outing will involve a climbing expedition, and, in addition, a lot of shivering faces us this coming winter. However, for big money, we can at least get something to eat.[77]

In September 1919, the Allies and Austria concluded the Treaty of Saint-Germain.[78] In October, an attack by German troops in Latvia prompted stronger measures on the part of the Allies, as a result of which German troops were compelled to withdraw from the Baltics.[79] Meanwhile, US president Woodrow Wilson, who had propagated many seminal concepts in international affairs, suffered a debilitating stroke.[80] On September 12, in explaining to Ehrenfest why he was declining the offer of a position in Leyden, Holland, to stay in Berlin, Einstein aligned himself with the new political direction in Germany. But he was aware that this direction might change:

> It would be doubly mean of me if I, right at this moment of the fulfillment of my political hopes, perhaps in part for the sake of material advantage, without necessity turned my back on people who have shown me love and friendship and to whom my departure at this time of supposed humiliation would be doubly painful. . . .
>
> So you see how it stands for me. I can only go away from here if there is a turn of events that makes it impossible for me to continue to stay. Such a change could occur. If it doesn't take place, my departure would be equivalent to a contemptible breach of faith toward Planck, and moreover also disloyal. I would have to reproach myself later.[81]

Later in September, Einstein wrote to his other significant mentor in physics, Lorentz, regarding the war crimes commission's first version of

Lille and the reasons for its withdrawal. He also discussed the rejection of German scholars by scholars abroad, encouraging sympathy for Germany's limited understandings during the War, and his desire to moderate the current trend toward vengeful thinking in German public opinion. He saw value in deflating the German hunger for power but had some fears regarding potential trends in Germany:

> I wanted to wait with my letter until I could send you our first publication (about the events in Lille). It was completed during my stay in Switzerland. When I received it in printed form a few days ago, I had a real shock. The foreword was tactlessly written; the local German records hardly referred to the factual findings in question, so the brochure could have done more harm than good. So there was no alternative but to withdraw the whole (luckily small) print run, in order to re-print it after supplementing and improvement. There is still so much material to be processed that, for the time being, it does not yet seem appropriate to travel to Holland for perusal of the records that were kindly made available to us. . . .
>
> Regarding the judgment of scholars here by their foreign colleagues, it does seem to me to be too harsh, despite the abominable things that were done. The majority, as a result of years of coercive manipulation of public opinion, still have absolutely no clear judgment on the events. Those on the outside have no concept of how difficult it is to elude mass suggestion. Also, the naked facts have not been properly made known here, or not at all. That is why such information is so direly needed here.
>
> It is already unbelievable that the inhabitants of an entire large country should be presumed to be morally inferior! The Manifesto of the 93, foolish as it was, was neither written nor signed with an awareness of its injustice. People whose passions are whipped up to the greatest bitterness, through skillful manipulation by every method, are not capable of any objective scrutiny of matters of fact. If we can now even belatedly awaken the recognition that the gravest injustice has been committed—what should these people now do to cleanse themselves? Since they have now sunk into the greatest powerlessness, every explanation must awaken the mistrust that it is dictated by characterless opportunism. That is exactly what is feared by the decent people, who

until now believed that they had justice on their side. At least those who are dispassionate should consider this. Incidentally, there is no harm if these people are made to feel their dependence through boycotts by foreigners. In this way the last remnants of megalomania and the hunger for power that the economic boom brought with it earlier will be eliminated.

I am convinced that it is precisely these conditions that allow "Prussians" to thrive; and my fear is that the same might now grow again somewhere else under the entirely altered conditions. Hopefully this fear is not justified! . . .

I hope I can soon send you our first publication.[82]

In late September 1919, Einstein was notified that the results of the British scientific expeditions on May 29, 1919, to observe the solar eclipse in the southern hemisphere had confirmed his theory of relativity. He had been long been prominent in the physics community; once the results were announced, he would become world-famous among the general public.[83] This led people increasingly to solicit his support for various political causes and appeals, in response to which he was judiciously selective. It also prompted Einstein himself to utilize his renown to further his ethical convictions.

In early October 1919, at the conference of the Federation of the German Academies (Kartell der deutschen Akademien), Einstein advocated freedom of action for individuals in international professional relationships, unrestricted by group resolutions.[84] Also in early October, Einstein suggested to Elisabeth Rotten, fellow New Fatherland League member and collaborator on the war crimes commission, the idea of simultaneously publicizing information about war crimes carried out by wartime opponents on both sides, as a strategy to undermine the spirit of militarism:

A Dr. Joseph . . . has made available to me material about the prisoner camps in Dahomey . . . This material could be of interest to us if we want to expand our publications to cover the nature of imprisonment. We could then perhaps overtly contrast the allegations against both hostile parties, if broad similarities can be found. This could possibly be the most effective way to work against the militaristic spirit.[85]

Also in October 1919, Berlin geophysicist Wilhelm Schweydar alerted Einstein to his anxiety—which he correctly judged Einstein would share—about the potential militarization of the Geodetic Institute in Potsdam through a governmental administrative consolidation that would place a general at the head of the institute. Schweydar feared that the institute might lose its independence; if that happened, it would make a bad impression on foreign scientists. So he asked Einstein to lend his "powerful hand" in preventing this from happening.[86] Concerned as he was about militarism, infringements on academic freedom, and damage to international scientific collaboration, Einstein shared Schweydar's misgivings over this prospect, as evidenced in his letter of a few days later to Carl H. Becker, an undersecretary of state in the Ministry of Education:

> I take the liberty . . . to point out . . . a threat of severe damage to German science. . . . The Geodetic Institute in Potsdam, until now one of the most reputable scientific institutes throughout the world in this field, is in danger of coming under the leadership of a general who is only remotely concerned with scientific projects. If this decision takes place, it would seriously damage our scientific interests and even more severely damage the reputation of German science abroad.[87]

Letters to family and friends in October 1919 alluded to postwar material and economic hardships. Yet again, he asked his ex-wife in Switzerland to move with their sons to Germany because of the German inflation and exchange rates:

> The exchange rate conditions have now become so difficult . . . that it's becoming impossible for me to support you in Zurich. I cannot raise the 8,000 francs that I contractually guaranteed you any more. It is therefore necessary that you move into German territory, in fact as soon as possible. . . . In Constance, for example, it is at least as nice as in Zurich. I will help you find an apartment . . . The schools in Baden are superior and their atmosphere is excellent. I will give you enough so that you can live as well as or better than in Zurich.[88]

Yet again, his ex-wife declined, as she always would. To his mother, he mentioned this plan, as well as his anxiety over a possible financial collapse in Germany:

> I have now decided to relocate Miza and the boys to Germany, probably to the Baden area. I can't withstand the currency exchange any longer. . . . I am in suspense about the famous national bankruptcy, or rather what form it's going to take.[89]

To Max Born, Einstein referred indirectly to the coal shortages of the summer that had led to rationing in anticipation of winter, and the depressing effect on everyone of shivering in winter.[90]

In mid-October 1919, expressing a typical pacifist concern for social welfare, Einstein wrote to one of the New Fatherland League's co-founders, the publicist Otto Lehmann-Rußbüldt, urging efforts to stop the blockade of Russia:

> It has been brought to my attention by politically neutral parties that the population of Soviet Russia is being most severely afflicted by epidemics. The greatest part of the guilt for these catastrophic health conditions lies with the blockade imposed on Russia, which also applies to medicines, etc. In my view, it is our duty to use our influence at home, and also abroad through our foreign contacts, to get the blockade relaxed, at least to the extent that the unfortunate civilian population will be somewhat spared. . . . I urgently request that you consult with our friends as quickly as possible about what can be done from our side. I myself will take steps in Holland and in Switzerland.[91]

Einstein had also raised the same concern with Zangger in Switzerland. His friend agreed to help but saw limited value in sending medicines. Zangger saw nutrition as the primary factor, pointing to the link between hunger and epidemics. In Russia, he saw "the accretion of power by oppressors who sacrifice all others for power" as "the tragedy of our times." From the Ukrainian refugees he encountered every day, he had learned of "the brutality, the almost sadistic atrocity" then reigning in

Russia. As for fair, peaceful distribution of medicines, "the increasing bestialization makes it improbable that medical considerations would govern the distribution; even supplies would become the object of the most cruel speculation." The only ray of light that Zangger could glean from the situation was that the blockade was saving people in Europe from infected carriers.[92]

On November 6, 1919, at a meeting in England of the Royal Astronomical Society, the results of the solar eclipse expeditions of the previous May were announced. They confirmed Einstein's theory of relativity, which elevated him to world renown. From that point on, he would be able to use the weight of his fame to speak out more with regard to his political-ethical convictions and advocate for causes he supported; he would also encounter a tremendous increase in solicitations for that support.[93]

In early November, still financially stressed by Germany's weak exchange rate, Einstein was contemplating bringing his terminally ill mother from Switzerland to live in his building in Berlin. Eventually, this was done.[94] Ten days later, he again pleaded with his ex-wife to move with their two sons to Germany so that he could better support them.[95]

In November 1919, the German National Assembly's Committee on Investigation took testimony from various rightist-nationalists and militarists. It was during these hearings that General Paul von Hindenburg launched the "stab-in-the-back legend" (*Dolchstoßlegende*), which blamed German civilians rather than the military for losing the War.[96] Also in November, the United States Senate voted against signing the Versailles treaty.[97] As discussed above, sometime prior to November 1919, the faulty first version of the booklet, *Lille*—the findings of the private commission to investigate the German army's actions in the War—had been published while Einstein was away in Switzerland. The first version of *Lille* was unclear about the commission's objectives (which were to raise awareness among Germans of their actions during the War, thereby dampening vengeful feelings and fostering international reconciliation); also, it failed to include the documents that were most critical to verifying German atrocities in the War. Einstein was astutely aware of the harmful impact this first version of the booklet could have on German public opinion. He wrote to the secretary-general of the Central Organization for a Durable Peace, Benjamin de Jong van Beek en Donk, in The Hague, Netherlands, to explain why the first version of *Lille* was being withdrawn and revised:

> I recently found out through Prof. H. A. Lorentz on the occasion of a visit to Holland that you have in your possession a brochure on the Lille events that is signed by me among others. Because I am for the most part not in agreement with the content of this brochure and consider a circulation of it as undesirable and damaging, I place importance on informing you of how it came into existence.
>
> The signatories and a few other people got together in order to work toward understanding and reconciliation here through publication of War incidents. While Dr. [Elisabeth] Rotten and I were in Switzerland, some members of the commission with good intentions but unfortunately little expertise put together this brochure and—without showing it to us beforehand—printed it and already sent out a few copies. I can no doubt spare myself a critique of the content; the deficiencies of the brochure are all too obvious. As soon as we got wind of the matter, we suppressed further distribution with the intention to publish a better brochure about the Lille incidents as soon as possible.
>
> I urgently request that you keep to yourself or destroy the copy that is in your possession and make your best effort to prevent this embarrassing business from causing any damage. In the hope that we are soon in the position to patch up this ugly matter.[98]

The first version of *Lille* tended to defend German actions in the War. It suggested that if American, English, or French troops had occupied Germany during the War, they likewise would not have acted as "pure angels" (*reine Engel*). It posited that individuals were not guilty of war crimes; rather, the psychosis of war was responsible. No belligerents were culpable; instead, "the War itself holds the guilt, it has sinned against all of us" (*der Krieg selbst trägt die Schuld, er hat an uns allen gefrevelt*), and all of the wartime belligerent states should establish similar commissions to investigate their own actions in the War. This first version's analysis of German documents focused solely on health and economic measures taken by the German army, such as suppressing prostitution, although it did include a protest statement by the intellectuals of Lille, France, describing a forced march of ten thousand women and similar violations.[99]

A second version of the booklet *Lille,* published in early 1920, clearly stated that its goal was to educate the German people about war crimes committed in their name. It would then be possible for them to condemn those actions and thereby redeem themselves as sharing the same moral standards as other peoples. Through rejection of such violence, Germans could atone and rejoin the community of nations, which by extension might end the intellectual blockade of Germany. Documents in the revised version more specifically addressed forced labor and deportations, house searches of scholars, thefts from scientific institutes, the use of children and the elderly in the line of fire, and the taking of hostages.[100] However, Einstein, perhaps skittish because of the too-hasty publication of the first version of the booklet, elected not to have his name included among the authors of the second version, although he had certainly contributed to the impetus behind its creation.

In mid-November, Einstein wrote a eulogy for the physicist Leo Arons, whose academic position at the University of Berlin had been terminated around 1900 due to his support for the Social Democratic Party. Einstein praised Arons's courage, independence of mind, and commitment to social justice:

> A man has departed from us whose simple greatness will always be remembered by all who knew his work and his life: the physicist Leo Arons. His social compassion and his passion for justice led him to the circle of socialists, and motivated him to advocate his socialist convictions in public, notwithstanding the severe restrictions and animosities that he had to suffer in a reactionary-led state. He was one of those figures, so rare among our circle of academics, who represented not only intellectual autonomy but also independence of character, unaffected by the prejudices of his class and with the selfless courage to make sacrifices. He acted on what he saw as self-evident, with simplicity and without the grand gestures of the martyr. . . .
>
> [In the next four paragraphs, Einstein describes Arons's scientific work.]

> These lines can only convey a vague idea of the tireless achievements of the man. . . . as a human being, he constantly strived to promote what, according to his convictions, the times and justice demanded.[101]

Also in mid-November, Adriaan Fokker wrote to Einstein again from the Swiss sanatorium with information about militarist activities: "I was told by what seems to me a well-informed source that the U-boat officers from Germany since the armistice have been hiring themselves out on a massive scale to the Japanese navy. That is indeed a genuine business in killing, in which any excuse of patriotic passion ceases to apply." However, Fokker was finding solace in "the Indian point of view" in Rabindranath Tagore's *Nationalism,* from which he "got an idea of how much we are chained to old and probably wrong traditions."[102]

Later in November, Adolf Friedrich Lindemann, an astronomer from England, wrote to Einstein about the spate of English newspaper coverage of the recent verification of his theory of relativity. Lindemann's letter reflected the positive impression Einstein had made abroad by not signing the "lamentable" manifesto "To the Civilized World."[103]

Also in late November, Robert W. Lawson, a physicist at the University of Sheffield in England who had spent the War in civilian confinement in Vienna, wrote to Einstein requesting an article for a British journal: "I know that an article from your hand would be greatly treasured . . . It would also bring us one giant step closer to a desirable collaboration among scientists of various nationalities." For Lawson, "it is my most cherished wish to work toward healing . . . the bitter wounds that this War inflicted . . . and toward re-creating the possibility of continued collaboration." Lawson's letter reflected a sensitivity among internationally oriented people to the hardships in Central Europe. He had learned from friends "that living conditions in Berlin are almost intolerable . . . [and] . . . in Vienna the food supply and heating problems are even more desperate than at the time when I left Vienna." Lawson found this a "bitter tragedy" and sincerely hoped "that life on the continent will soon be substantially eased."[104]

A couple of days later, Lawson noted that Einstein's fame had led to the publication of an article by Einstein in the *Times* of London.

Paradoxically, the *Times* was owned by Alfred C. W. Harmsworth, Lord Northcliffe, who, as part of the Lloyd George government during the War, had advocated harsh peace terms for Germany.[105] Although the article focused primarily on his science, Einstein's concluding quip, in response to the *Times*'s description of him, displayed his well-developed sense of irony about nations and their nationalisms:

> A final comment. The description of me and my circumstances in the Times shows an amusing feat of imagination on the part of the writer. By an application of the theory of relativity to the taste of readers, to-day in Germany I am called a German man of science, and in England I am represented as a Swiss Jew. If I come to be regarded as a bête noire, the descriptions will be reversed, and I shall become a Swiss Jew for the Germans and a German man of science for the English![106]

Einstein described this article and commented similarly to Ehrenfest in early December 1919.[107] Also in early December, he expressed to Fokker his sharp disappointment and discouragement at the rise of militarist attitudes in Germany. Yet he continued to have faith in the good will of internationally minded intellectuals past and present:

> My political optimism has also suffered a blow. It seems as if everyone who has the opportunity to do so is developing into a "Prussian." I'm gradually concluding that large-scale processes are so hopeless that I'm starting to feel the political disinterestedness and international feeling of the intellectuals of earlier centuries as the only kind possible and worth striving for. . . . Our English colleagues not only achieved excellent work, but also behaved superbly in the human sense. Today I even received a very kind letter from [Arthur S.] Eddington, in which he also mentions [Erwin] Freundlich. So far I have only thanked them in a small article in the Times, because under the current conditions I did not know whether a personal communication would be possible or desired. Now, however, I see that these men really rise above the situation.[108]

In turning down a request for an article made by Vienna's daily newspaper *Neue Freie Presse*, Einstein indicated that international reconciliation had been a key motivation for writing his article for the *Times*: official relations between scholars of the Central Powers and those of the Allies had been suspended since the War, and he had regarded an article in an English newspaper as the only way to thank English astronomers for their expenditures and extraordinary service in proving his theory.[109]

In late November and December 1919, a few residual War-related matters were settled: the "Spanish" influenza epidemic was officially declared over, the Allies and Bulgaria signed the Treaty of Neuilly, and German troops withdrew from the Baltics.[110] Meanwhile, the German government further consolidated itself legally and defied the Versailles treaty with regard to the extradition of accused German war criminals.[111]

At the beginning of December 1919, Einstein wrote to Moritz Schlick, a philosopher at the University of Rostock, who had hosted him during a recent visit. He alluded to economic hardships in Germany and widespread predictions of an impending economic collapse:

> I know that those days of heated rooms and sumptuous feasting were not a mere friendly gesture but rather a powerful achievement, even a heroic act of self-denial. For now you are again sitting by the only heat-producing stove, and the lady of the house is keeping a watchful eye on the bread for the next few days . . . Yesterday evening I attended a discussion by experts about the economic situation, in which a gradual but inevitable total collapse of the economy was predicted. The difference from Austria would be merely temporal. I am only reporting; I myself do not have the necessary overview for making predictions, the others also probably not. At any rate, all shared the same pessimism. The tone was: a year ago, we really could have done this or that, but now, though, . . . [112] (latter ellipsis in original)

Several days later, Einstein communicated to Prussian minister of education Konrad Haenisch both a sensitivity to the widespread economic deprivation in Germany and a personal hardship that even his fame was not sufficient to avert:

> A few days ago I received a report of the National Budget Committee of the Prussian Constitutional Congress, according to which it is planned

> to place 150,000 marks from the State Treasury at my disposal to support research in the field of the theory of general relativity. However great my joy and feeling of gratitude for this truly generous gesture, I nonetheless cannot suppress painful reservations. In this time of extreme poverty, won't such a decision justifiably trigger bitter feelings among the public? I believe that we can effectively promote research in the field of the theory of general relativity even without the use of special state funds, if the country's observatories and astronomers would merely place a portion of their apparatus and labor in service to this cause. . . .
>
> Finally, I would like to mention a small but nonetheless urgent request that pertains to my own working conditions and for which I implore immediate resolution. By arrangement of the housing office of the city of Schöneberg, a room in our building was granted to me to accommodate my terminally ill mother and her nurse. The building owner is resisting this . . . so that I'm actually facing the necessity of making room in my study for the ill person if the matter is not speedily and favorably resolved. A word from the Ministry of Culture directly to the rental office of the city of Schöneberg . . . would bring an immediate decision to this matter of such importance and urgency to me.[113]

Einstein was ultimately obliged to accommodate his mother in his own study during her final days. Meanwhile, Einstein repeated to Besso his ongoing anxiety about the economic hardship resulting from the devaluation of German currency when trying to support a family in Switzerland on a German salary:

> It would be a huge relief to me if my boys together with Mileva would move to Germany in the course of the next year. . . . Nothing can be achieved with German money in Switzerland, and it is certainly not right to use up the small savings in Swiss money.[114]

In mid-December, Einstein reiterated to Zangger the same postwar themes of personal financial hardship and societal economic crisis:

> The affairs of my Zurich family worry me. . . . Through the catastrophic currency exchange rates, my Zurich family's circumstances, which

> seemed so well arranged through the divorce contract, have now again gone completely awry. Earning in Germany and spending in Switzerland is an impossible combination. . . . I therefore see no possibility of leaving Mileva and the children in Switzerland; in Germany, I can provide for them excellently.
>
> Here all conditions are variable and not for the best; large-scale corruption and impoverishment. Over the long term, however, I believe that the country will develop in a satisfactory direction. The disadvantages of the defeat are felt immediately, the benefits of it only little by little. I am determined to hold out here as long as I am capable of properly providing for both households.[115]

Also in December, Einstein expressed to Max Born his increasing disillusionment both with the reactionary spirit among the German people and with the actions of the Allies in Germany and prospects for the League of Nations:

> I was with Schlick in Rostock for a few days on the occasion of the university's anniversary celebration, and heard there at that event malicious inflammatory political speeches and observed rather delightful small state politics. . . . It was amusing to see how men of the old and new governments were seated in two proscenium boxes, one below the other. Naturally, the newcomers among the academic notables were tormented with jabs of every conceivable type, the ex-duke receiving an interminable ovation. No revolution helps against the ancestral spirit of servitude! . . .
>
> The conduct of the Allies is now starting to become revolting, even to my taste. My hopes for the League of Nations seem unlikely to materialize. At any rate, France also seems to be suffering very much, despite coal imports, as one can see from its new railway travel restrictions. Here furniture and real estate are being bought up by foreigners, to the point that we become an Anglo-American colony. It's a good thing that our brains are not for sale or subject to a German emergency relief tax![116]

His relationship with socialists was another area in which Einstein displayed a nuanced and well-informed independence of thought on political matters. He supported some of the social justice goals, as well as convictions that socialists and pacifists sometimes shared in common, but he did not agree with their sectarian stridency and suppression of individuality. This was evident in his response to Hermann Schüller, a publicist in the socialist youth movement who had solicited Einstein's participation in a school for workers:

> From your program, I infer that there would be little purpose in your visiting me. It seems to involve less a school in which the talented proletarian can acquire knowledge and methods of thinking than much more a propagandistic institute with the intention of planting unproductive hatred.[117]

In mid-December in Berlin, at a New Fatherland League event held in the former Prussian House of Lords—and in the provocative context of the investigations of wartime human rights abuses by the German army in Belgium—Einstein courageously gave the following welcoming address to Paul Colin. Colin was a leader of the Belgian branch of the Clarté movement (with which Einstein was also briefly affiliated in this period) dedicated to building a just postwar political order:

> In the name of the New Fatherland League, which during the terrible War also held high the ideal of humanity and of reconciliation between peoples, I very heartily greet you, as the first French person who, in service to the sacred goal of the reconciliation of peoples, has come to us since the War.
>
> Our duty is serious and the hour is difficult. It is difficult to say whether with you the victory or with us the defeat has more disastrously provoked the nationalistic passion that threatens to perpetuate the state of blood revenge between both neighboring lands. However, the root of the sickness lies not in the present historical moment, but rather in traditions that are passed on from generation to generation in the educated strata of the European states, in defiance of the outwardly emphasized teachings of Christian morality: Rape and oppression bring honor and glory, to endure injustice brings disgrace and humiliation. We want to counter these old evil traditions, which threaten to drive our continent

> completely into the ground, with the passionate belief in the human feeling of solidarity, without which a bearable coexistence of individuals and of states is not possible.
>
> When Treitschke's teaching about states was skillfully presented to my Dutch colleague, he said something like, "It is possible that the world is so, but I would not like to live in such a world."
>
> May the courageous and humble enterprise of our present guest help this spirit to take root on our side and on the other side of the Rhine![118]

Moritz Schlick wrote to Einstein that he had read in the newspaper of Einstein's Clarté meeting and asked how he could establish a local branch of Clarté in Rostock.[119] From Zurich, the Swiss writer and journalist Carl Seelig was sufficiently inspired by Einstein's participation in the Clarté event that he invited him to contribute a written work for a collection he was compiling. The volume would offer twelve contributions from an international group of prominent intellectual and altruistic thinkers, among whom he hoped to include Romain Rolland, Hermann Hesse, Stefan Zweig, Henri Barbusse, Georges Duhamel, Rabindranath Tagore, Carl Hauptmann, Max Picard, and Frederik van Eeden. The goal was to present a varied image of modern intellectual work, reflecting on the past, mirroring the present, and glimpsing into the future.[120] While wishing him success in his project and supporting Clarté, Einstein declined Seelig's request:

> It is true that I welcome the Clarté movement with great joy. I will make an effort to advertise Clarté in Germany, although this is not easy in the currently prevailing atmosphere of unspeakable bitterness. Unfortunately, I am so overextended that I cannot even think of writing the booklet that you desire, as much as I also would like to. . . . I would greatly appreciate the books kindly offered to me, even more to make your personal acquaintance on the occasion of my next stay in Zurich. . . . Best wishes for the success of your project.[121]

In late December, Einstein expressed to Lawson in England his hopes for Clarté as a vehicle for internationalism:

> On the suggestion of the secretary of Clarté, we are right now on the verge of forming a German Clarté group here. I hope that this excellent undertaking will reawaken the international spirit.[122]

Also in late December, in the context of attempts to recruit him to positions in Switzerland and Holland, Einstein reiterated to Zangger his belief that the present hardship was good for the German people, the implication being that it would foster a more republican consciousness:

> My local colleagues and officials credit me highly for not chasing after economic prosperity but rather steadfastly sticking it out here. Ever since things have been going badly for the people, I've been liking them incomparably better. Misfortune suits people insurpassably better than success.[123]

In explaining to Edgar Meyer his decision to decline a renewed offer of a position in Zurich, Einstein alluded to the same affinity that he now felt with the people in Germany:

> Despite the insecure situation, I simply cannot turn my back on my colleagues here in these times. If only Planck alone were here, I would also not do it. I could no longer be at peace with myself. . . . Ever since the exhibition of power here has been gone, the human milieu has become tremendously sympathetic to me, so that I happily accept the superficial difficulties and small deprivations that are a part of life here.[124]

The year 1919 saw political agitation, violence, and economic hardship in Germany. Einstein started the year observing political developments with avid interest, signing declarations protesting the murders of Karl Liebknecht and Rosa Luxemburg and supporting a progressive youth league's interpretation of the German Revolution. By late March, however, he was feeling disappointment, both in the failure of democratic courage among the leaders of Germany's new Weimar Republic and in the vindictive approach being taken by the Allied nations that had won the War. Nonetheless, he continued to act on his ideals. By late April, he was participating in the commission of private citizens investigating allegations of atrocities by the German army in France and Belgium, which eventually published its findings. The intention was to help the German public better understand the resentment of the Allied populations. In May, he signed two more public declarations alongside other prominent pacifists, urging that war guilt be investigated as a matter of justice, not of power, and encouraging the German people to halt the wave of political violence. He also signed a group appeal to commute the death penalty against Eugen Leviné, a leader of the defeated Bavarian Soviet Republic.

The first half of 1919 was a time of personal and professional transitions for Einstein: his divorce from Mileva, his remarriage to Elsa, and the confirmation of his scientific theories by the English solar eclipse expedition. By mid-1919, at a time when most Germans were outraged at the harsh terms of the Versailles treaty, he was confident of what in fact eventually came to pass—that its terms would never be fully implemented. At this point he still felt optimistic about the League of Nations and still believed that hardship would help educate the German people; but at the same time, he was well aware of the dangers posed by Germans' urge for revenge. In October, he took a couple of stands on behalf of academic freedom in Germany. Meanwhile, economic deprivation continued in Germany, in Einstein's case taking the form of difficulty supporting his first wife and sons in Switzerland on a salary paid in the inflated German currency. At the same time, his lifelong propensity to aid others was expressed in his concern over the hunger and epidemics in Russia. By the end of the year, his international reputation had fully flowered, leading to an expanded foreign correspondence, especially with scholars from England. As the year ended, Einstein felt far more pessimistic politically than at the start of the new republic a year earlier, alienated by the reactionary spirit among Germans, and disgusted with the actions of the Allies. Nonetheless he spoke publicly to welcome the French pacifist Paul Colin to a German pacifist event.

1. Albert Einstein in his study, Berlin, 1916. Courtesy of Albert Einstein Archives, The Hebrew University of Jerusalem, Israel.

2. Mileva Einstein-Marić with sons Eduard and Hans Albert, Berlin, 1914. Courtesy of Albert Einstein Archives, The Hebrew University of Jerusalem, Israel.

3. Albert Einstein with Elsa, Ilse, and Margot Einstein, Sellin, island of Rügen, Germany, July–August 1915. Courtesy of Albert Einstein Archives, The Hebrew University of Jerusalem, Israel.

4. *Left to right*: Albert Einstein, Paul Ehrenfest, Paul Langevin, Heike Kamerlingh-Onnes, and Pierre Weiss, Leiden, October 1920. Courtesy of Albert Einstein Archives, The Hebrew University of Jerusalem, Israel.

5. Albert and Elsa Einstein arriving in the United States on board the SS *Rotterdam*, April 2, 1921. Courtesy of Albert Einstein Archives, The Hebrew University of Jerusalem, Israel.

6. Albert Einstein, April 27, 1921. Courtesy of Albert Einstein Archives, The Hebrew University of Jerusalem, Israel.

7. Albert Einstein, May 6, 1921. Courtesy of Albert Einstein Archives, The Hebrew University of Jerusalem, Israel.

8. Albert Einstein, March 30–April 1, 1922. Courtesy of Albert Einstein Archives, The Hebrew University of Jerusalem, Israel.

9. *Back row, left to right*: Albert Einstein, Paul Ehrenfest, and Willem de Sitter; *front row, left to right*: Arthur S. Eddington and Hendrik A. Lorentz, Leiden, September 26, 1923. Courtesy of Albert Einstein Archives, The Hebrew University of Jerusalem, Israel.

10. *Left to right*: Albert Einstein, Hendrik A. Lorentz, and Arthur S. Eddington, Leiden, September 1923. Courtesy of Albert Einstein Archives, The Hebrew University of Jerusalem, Israel.

11. Max Planck presenting Albert Einstein with the Max Planck medal of the German Physical Society, Berlin, June 28, 1929. Courtesy of Albert Einstein Archives, The Hebrew University of Jerusalem, Israel.

12. Heinrich Zangger, ca. 1905–1920. Courtesy of Ursula Cohrs and Zentralbibliothek Zürich, Nachlaß H. Zangger 94.

13. Georg Friedrich Nicolai, ca. 1910–1920. Courtesy of Institut für Zeitgeschichte, Munich.

14. Romain Rolland, ca. 1914–1918. Courtesy of Schweizerische Nationalbibliothek (Swiss National Library), Prints and Drawings Department, Collection of Portrait Photos, Bern.

15. Friedrich Adler, ca. 1920–1930. Courtesy of ETH Bibliothek, Zurich.

16. German soldiers in trench and leaping from trench, Battle of Lys, Belgium, April 9–29, 1918. Courtesy of Mary Evans Picture Library, London.

17. French soldiers in trench with wounded comrade, France, 1915. Courtesy of Mary Evans Picture Library, London.

18. British soldiers with gas masks, in captured German trench with German dead, Flanders, Belgium, 1917. Courtesy of Mary Evans Picture Library, London.

19. Paul von Hindenburg and Erich Ludendorff, between 1910 and 1915. Courtesy of Library of Congress, Prints & Photographs Division, LC-DIG-ggbain-20468.

20. David Lloyd George, Vittorio Orlando, Georges Clemenceau, and Woodrow Wilson, Paris, May 27, 1919. Courtesy of Library of Congress, Prints & Photographs Division, LC-DIG-ggbain-29038.

21. Albert Einstein and members of the Princeton World Federalists, Princeton, December 1947. Courtesy of Albert Einstein Archives, The Hebrew University of Jerusalem, Israel.

22. Albert Einstein, NBC TV speech regarding the H-bomb, October 2, 1950. Courtesy of Albert Einstein Archives, The Hebrew University of Jerusalem, Israel.

5 Economic Crisis, Reactionary Politics, and Ongoing Engagement

1920–1921

At the start of 1920, the German people were feeling and often expressing a collective anger at the harshly punitive provisions of the Treaty of Versailles, which they widely regarded as an illegitimate "dictated peace" (*Friedens-Diktat*). The Versailles treaty had been ratified on January 1, 1920, and had come into effect on January 10.[1] Also on January 10, German troops were compelled to relinquish Upper Silesia to Allied troops, who were mostly French and tended to support the Poles.[2]

On January 6, 1920, in a letter to fellow New Fatherland League and Lille commission member Hellmut von Gerlach, Einstein advocated on behalf of political prisoners. He suggested that Gerlach, in his capacity as undersecretary of state in the German Ministry of the Interior, pressure the government to release political prisoners from the Bavarian Soviet Republic:

> Friends of mine have brought my attention to the fact that the moment has now arrived to put pressure on the government to obtain the immediate release of political prisoners. I am convinced that the current situation is regarded by a large number of influential personalities throughout the country as intolerable, and that, for example, a declaration bearing their signatures that is published by a suitable political group or elected representative . . . could meet with success in the National Assembly. The point in time seems so appropriate for this, because the lifting of the state of siege without the release of political prisoners is

> certainly only a half measure, entirely apart from the fact that imprisonment under the current bad food conditions is especially hard.
>
> We are all agreed that the written statement for which signatures will be solicited must be written by you. Because you too cannot be closed off to this understanding, I am venturing here to request this kindly of you. You will certainly manage to avoid the difficulty of a political incident arising out of the matter, which would be very damaging to the undertaking. Because the moment should be seized, I request you urgently, regardless of your answer, to reply to me immediately.[3]

Organizations soliciting Einstein's support were not always politically ambiguous, and Einstein was comfortable using technicalities to avoid meaningful discussion when refusing them. For example, also in January, the League of German Scholars and Artists (Bund Deutscher Gelehrter und Künstler), a German nationalist and rightist-leaning organization, sought Einstein's endorsement of one of its campaigns, the German League for the Protection of Germans Outside of Germany (Deutsche Schutzbund für die Grenz- und Auslandsdeutschen) and for its Border Fund, whose aim was to fund voters to travel to their places of birth in order to vote.[4] Einstein refrained from engaging in any political discourse with this group, resorting to his Swiss citizenship to decline participation in it:

> I request that you please refrain from including my name in the list of the Committee for Arts and Sciences for the Border Fund. This concerns a matter directly connected with the foreign policy of the German Reich, so it would not be appropriate for me, as a Swiss citizen, to participate actively in it.[5]

Although Einstein occasionally expressed Jewish communal feeling, cultural pride, and advocacy on behalf of Jews facing hardship, a formally religious side to Einstein's Jewish identity was almost nonexistent. (Einstein's Jewish identity will be discussed in chapter 6.) In this period of his life, one of his most significant comments about religion was simply that he saw the world as better off with the presence of religion than without it. In January 1920, the pacifist Georg Graf von Arco, whom Einstein had known as a fellow member of the New Fatherland League since 1915, and fellow Lille commission member, inquired on behalf of the German Monist League (Deutscher Monistenbund) whether Einstein would be willing to head its international commission with "the goal of reviving contacts with like-minded free-thinking efforts abroad."[6] Einstein declined, explaining that, although he shared the German Monist League's internationalism and advocacy for individual freedoms, he did not support what he saw as monism's antagonism toward religion. Despite formal religion's imperfections, Einstein felt that religion offered more good than evil:

> I also see that I would be particularly suitable in the respect that I, as a dedicated internationalist, could be useful to the sustenance of your international association. It is, however, impossible for me to take on such a mission, because I have always had reservations on principle about practical Monism. Specifically, I am of the conviction that Monism's hostile stance toward religious organizations is in principle not justified. The supra-personal content that religion conveys, although primitive in form, is more valuable to humankind, according to my conviction, than Haeckel's materialism. I believe that the elimination of the religious tradition, even today, would mean further spiritual-intellectual [*geistige*] and moral impoverishment, crass and ugly though the attitude and actions of the clergy may be in many respects. The defects of the existing tradition are merely much more frequently noticeable than its modest virtues! Monism's struggle for freedom of the individual against encroachments by religious communities is however also in my view a thoroughly commendable and necessary one.[7]

On January 12, two days after the Treaty of Versailles came into effect, Einstein wrote to Lorentz indicating that the second version of *Lille* had been published. He also bemoaned the apparent failure of educated Germans to be receptive to its message and learn from it. His initial optimism at the start of the Weimar Republic a little over a year earlier had clearly shifted to greater pessimism about prospects for change in people's political attitudes:

> The new publication about Lille is in print and has already been the object of strong animosity. You shall receive a copy very soon. It is sad

> that even highly educated people are completely unable to extricate themselves from the petty nationalistic perspective, not even when it is a matter of right and wrong. The evil from the past lives on and on, with no end in sight.[8]

During January 1920, Robert Lawson asked Einstein about the Clarté movement; he wanted to support anything that might foster "the resumption of international collaboration."[9] With his reply, Einstein sent Lawson a copy of the Clarté statutes.[10]

That January, the Allies also encountered frustrations in attaining their goals. Meanwhile, violent reactionary developments, both political and social, took hold in Germany.[11] In the fall of 1919 and again in January 1920, at the University of Berlin medical school, rightist German nationalist students disrupted the lectures of physician Georg Friedrich Nicolai, author of the 1914 "Appeal to the Europeans" and other pacifist works, calling him a traitor. During the War, Nicolai had attempted to conduct himself as a conscientious objector from within the Germany army, including during his imprisonment for insubordination. Eventually, in June 1918, he had fled Germany for the remainder of the War—the incident on which the students now based their accusations.[12] In solidarity with a fellow pacifist with whom he had been associated since October 1914—and astutely forecasting the hostile response of the faculty senate—Einstein wrote to University of Berlin historian and liberal politician Hans Delbrück asking him to join in a public statement in support of Nicolai:

> For some weeks the pan-German press has been trying to obstruct the academic activity of Prof. Nicolai. The matter is presently being investigated by the university rector and senate according to their rules, thus by an authority whose members by nature are by no means kindly enough disposed to Nicolai's ideas that one can be convinced of their objectivity in this matter.
>
> The press's agitation seems all the more reprehensible, a sample of which is the enclosed article from the Deutsche Tageszeitung.
>
> Because the investigation by the rector and senate may still drag out for a long time, it would be desirable, in order to counter the

suggestive effect of these articles, to publish a short statement by university teachers.[13]

The enclosed draft of the declaration condemned the press campaign and student riots and supported Nicolai's actions and his right not to be harassed:

> In recent days a systematic hate campaign has started up in the newspapers against Professor Nicolai, who is well known domestically and abroad as a pacifist writer and courageous defender of his convictions, after pan-German students had already earlier made his lectures impossible through shameful riotous scenes at the university. The undersigned university teachers consider it their duty to express their deep regret at these incidents, which, according to their conviction, as a symptom of narrow-minded intolerance can only damage the spirit and reputation of the University of Berlin.
>
> We who know Nicolai's work and actions firmly deny that he did even the slightest thing to harm Germany. On the contrary, his actions only helped to increase sympathy for Germany.
>
> But even if one has a different opinion about the effect of Nicolai's actions, one should not therefore attempt to oppose him with blatant untruths and slanders.[14]

At the end of January, Einstein described to Hedwig and Max Born the clashes between leftists and reactionaries in Germany, contextualizing Nicolai's struggles and the Matthias Erzberger slander case as manifestations of this larger conflict. He also touched upon economic deprivation in Germany and accurately predicted that the reparations clauses of the Treaty of Versailles would be unenforceable:

> Politics are actually developing consistently in the Bolshevik direction. It seems to me that the great external successes of the Russians, in connection with the increasingly intolerable situation of the West, but especially our situation, are pushing inexorably in that direction. But until it gets that far, streams of blood must flow, because the reaction is getting increasingly heated. Nicolai is being challenged and abused so much that he no longer can lecture, not even at the Charité. I had to support him publicly yet again . . . France is really playing a rather sad role.

> (At least it is to their credit that they have freed themselves of the tiger.) Even a victory is hard to endure. The Erzberger lawsuit is comical; those with clean hands (and purses) may stone him, if any exist! I confess to you, incidentally, that the Bolsheviks are not so entirely unappealing to me, as funny as their theories are. It certainly would be damned interesting sometime to take a look at the matter from up close. At least the effectiveness of their slogans is great, because the military equipment of the Allies, which wiped out the German army, is melting away there in Russia like snow in the March sun. Those guys have politically talented people at the top. I recently read a brochure by Radek—with the greatest respect, he understands his business!
>
> You think I should speak out in England? . . . But I see that they are deeply in the throes of their own difficulties. What should they do in order to alleviate this poverty? They, like the Americans too, are certainly sending out relief missions. But little can be achieved in the face of this mass destitution. The peace treaty goes much too far, though. But since its execution is certainly impossible, it is better that its demands are objectively unrealizable than just barely intolerable. One must consider that the citizen on the opposing side would have to receive something printed in black and white as thanks for the glorious bravery. Taking action against the treaty would only make sense if one believed in its real significance, which is not the case with me.[15]

In February 1920, Germany suffered some territorial losses in Lithuania and Poland, but the Allies agreed not to force the extradition of accused German war criminals.[16] In early February 1920, Hendrik A. Lorentz conveyed to Einstein his favorable impression of the revised version of the Lille commission's booklet. He considered it objective, and he anticipated that its effects would be positive, for it would allow readers to form their own judgment.[17]

On February 20, 1920, Einstein's mother, Pauline Einstein, died, after residing briefly in his Berlin household. Three days later, she was buried in the Maxstrasse cemetery in Berlin-Schöneberg.[18]

On February 27, Einstein addressed the juncture between politics and academia when he wrote to his university friend Marcel Grossmann, a mathematician, of the Swiss Federal Institute of Technology

(Eidgenössische Technische Hochschule, or ETH), to recommend two academic colleagues who might do better in Switzerland than in Germany owing to their pacifism and internationalism:

> Now I would like to entrust you with the suggestions of two more men, each of whom would be a truly great gain for Switzerland:
>
> 1. M. Schlick, currently a private lecturer of philosophy at Rostock. . . . He is in somewhat precarious circumstances and, as a non-Kantian, modern-thinking person and pacifist, has little prospect of a German philosophy professorship. National considerations would be completely misplaced with this extraordinarily fine and entirely internationally oriented person.
>
> 2. The very well-known and original Viennese biologist Paul Kammerer finds himself in such oppressive circumstances that he is essentially deprived of any possibility of work. . . . Because he is known as an internationalist, he (too) has little prospect of a professorship in Germany.[19]

That same day, Einstein wrote to Zangger, also in Zurich, again recommending Schlick and Kammerer.[20] He then followed up with Schlick himself. In that letter, he also touched on Clarté, showing his cautious way of assessing political groups:

> I have just written to a good friend in Zurich, urging him to support your candidacy there. . . . In addition, I will advocate for you personally in Zurich when I get there again. I often feel ashamed on behalf of others that you still have no proper teaching position. Please also inform me if a position is becoming vacant in Germany, so that we can do something for you here. . . .
>
> I intentionally wrote you nothing more about Clarté, because I do not yet properly see in what kinds of hands the German Clarté group is. Ultimately, it's primarily the people who matter, not the slogan. As soon as I know something reliable and have developed confidence in it myself, I shall write you again about it.[21]

In late February and early March, developments leading into the Kapp Putsch were under way, involving renegade army officers and German nationalists challenging the government and defying its orders.[22] On March 5, 1920, in the midst of these volatile reactionary political maneuverings, the University of Berlin faculty senate voted to expel Nicolai from the faculty, accusing him of having undermined the German war effort, a resolution made public on March 9. The following day, Minister of Education Konrad Haenisch annulled the decision and guaranteed Nicolai's position, declaring that the faculty senate had overstepped its authority in basing its verdict on political considerations.[23] Einstein wrote to Haenisch praising his action, with particular reference to Germany's reputation abroad:

> My joy at your worthy and just decision in the Nicolai affair is so great that I cannot refrain from specially thanking you. You have not only reinstated the dignity of German universities abroad, which had been injured by the senate's decision, but also proven that the government of the new Germany is determined to protect freedom of speech. It is a fine counterpoint to the Arons case, of which you and the new Germany may be proud.[24]

On March 12, Georg Count von Arco, who had just visited Haenisch's Ministry of Culture in person, wrote to Einstein, urging him as well as others who were close to Nicolai to keep a close watch on him and try to prevent him from acting rashly in the present volatile circumstances.[25] That same day, Haenisch wrote Einstein expressing his heartfelt appreciation for Einstein's friendly letter, which he found "particularly encouraging" in the context of "sharp attacks from the right wing."[26] The following day, the day of the Kapp Putsch, Schlick responded to Einstein thanking him for his reference for positions in Zurich, celebrating "Haenisch's support for freedom of instruction in the Nicolai case," and predicting that the joy of the rightists who were fomenting the Kapp Putsch would not last long.[27]

The Kapp Putsch was carried out on March 13, 1920, the day after the Erzberger-Helfferich verdict was announced technically in favor of Erzberger and three days after Haenisch guaranteed Nicolai's position. The regime collapsed after only four days, defeated largely by a general strike.[28] While it lasted, a workers' uprising in the Ruhr was violently suppressed by German army and Freikorps forces, which led to the French occupation of several cities in that region.[29] On March 19, the US Senate failed to achieve the two-thirds majority vote necessary to ratify the Treaty of Versailles; the United States never became a signatory to it.[30] On March 18, 1920, immediately after the Kapp Putsch failed, Einstein conveyed to Lorentz his cautious hope that the booklet *Lille* was starting to exert a positive influence. He also expressed his concern over the one-sided punishment of war crimes, along with his compassion at the postwar hunger in Germany:

> I am glad that you were now able to approve of the booklet about Lille . . . Here the conviction is gradually beginning to penetrate that heavy sins were committed and that people are not blameless for the terrible hatred. It is unfortunate, though, that the punishment of war crimes is not being applied internationally; holding only guilty Germans responsible, although terrible things also happened on the French side to their prisoners, leaves the satisfaction in this act of justice incomplete. The English apparently proceeded very properly. It's a great stroke of luck that our most recent reactionary putsch failed so pathetically; it would have been intolerable to carry on here among those people. However, the poverty of the people here is terrible. Very many are dying from the effects of inadequate nutrition.[31]

On March 26, in a letter to his two sons, Einstein alluded to difficulties arising from the inflation in Germany and to the political volatility and its impact on living conditions during and after the Kapp Putsch:

> In May I'm going to Holland. From there I will be able to send you 1,000 francs, and then the same again in the summer. Sending it from here is out of the question, because our money is worth nothing abroad (also not worth much here) . . . Rather wild things have been going on here again, God knows what more is yet to come. One week we were without light, gas, sometimes even without water. The university was closed.[32]

That same day, he expressed to Zangger disgust with the tragic failure of the Weimar Republic to curb rightist and militarist violence in Germany:

> I received your postcard late, of course, because of the political carnival . . . Here we have confusion, corruption, and dictatorship by the sword. The military is murdering with impunity, that pathetic riffraff. The barbarity is terrible. The government is weak and always leans toward foul compromise.[33]

In the immediate aftermath of the Kapp Putsch in late March and early April, along with changes of government officials and struggles over compliance with the Versailles treaty, there were ongoing battles in Germany between the army and the Freikorps on the one hand and communists and militant workers on the other.[34] Internationally, this was also a period of prisoner-of-war exchanges.[35]

In early April 1920, for the Zurich newspaper *Neue Zürcher Zeitung*,[36] Einstein wrote an appeal promoting processes to help Central European university libraries and educational institutions acquire scholarly literature. His article reflected his internationalism, his compassion regarding the deprivations brought about by the War, and his commitment to the well-being of the scholarly community:

> Truly effective help would be achieved through the organization of an exchange of journals and books, in which the journals given in exchange from the Central European countries would be billed at their actual value, without reference to the currency exchange rates. American institutes have already reached out to local institutes in this way in a truly liberal manner . . . It would be most useful to contact the Prussian Academy of Sciences, which is . . . continuously making efforts for the preservation of threatened scientific life.[37]

Also in early April, Einstein described vividly the postwar hunger and deprivation in Germany to his older son, Hans Albert. Despite the usual currency exchange problems, however, at this point he agreed that it would be best for his first family to stay in Switzerland, due to conditions in Germany:

> Only the fall vacations remain. Then you really should come visit me, if possible with Tete [his younger son, Eduard]. A trip to Switzerland is

> too expensive for me. I'm already incredibly happy when I can raise the money for your support in Swiss currency . . .
>
> It is dreadful how the people here have to suffer. Clothing and shoes are exorbitantly priced, as well as most foods. The death rate is terrible. In the hospitals the sick die of starvation, that is, they perish due to insufficient nutrition. I am happy in any case that you are in Switzerland. Never before has it been as bad as it is at present. Supposedly it's even worse in Austria. God knows how this will end.[38]

That same day, Einstein expressed to an admirer his dislike for the use of his work to propagate nationalism, again clearly affirming his identity as a pacifist:

> I must confess that it always pains me when the few little flowers that have grown in my garden must suffer by serving as decoration to the flag of nationalism. I am a pacifist, you see, and always attempting, in these times of hatred, to foster reconciliation.[39]

Also in first half of April, Einstein cogently summarized to Ehrenfest the conditions of the early Weimar Republic during the Allies' extension of their blockade of Germany past the end of the War. His summary touched on resentment, chauvinism, hunger, infant mortality, and the impotent German government:

> Here calm has superficially returned. But there are gaping differences of enormous severity. Saber-rattling despotism and rage against it. Poverty and hunger are terrible in the city. The infant mortality rate is appalling. Where we are heading politically no one knows. The state has sunk into utmost impotence. At its side the main powers are feuding: weapons, money, extreme socialist unions.[40]

In mid-April, Einstein learned that University of Berlin classical philologist and former rector Ulrich von Wilamowitz-Moellendorff—a German nationalist who made speeches supporting the German military during the War and who had lost his son on the Russian front in 1915—was hesitating to support the Anglo-American literary aid effort for Central

Europe because doing so would place him on the same list with Einstein, whom he regarded as an independent socialist (although Einstein was not a member of the USPD). On April 19, Einstein responded by offering to withdraw his name from the list (using the convenient excuse of his Swiss citizenship) if this might help what he viewed as an important effort toward restoring international scholarly collaboration. Also, with his typical pragmatism and self-confidence, he invited Wilamowitz-Moellendorff to meet with him in person to discuss the matter.[41]

Also on April 19, Einstein commented to Zangger on famine and disease in Germany, the continuing domination of the Weimar Republic by the military, and the Allies' failure to take a wise or effective approach in their dealings with Germany:

> Here the poverty is now very extreme, everywhere discouragement, malnutrition, tuberculosis. And there is still no improvement in sight, no imports, no exports, insufficient food. Politically, after as before, the blind extreme dominates—so it seems, at least. The military is dangerous, after as before, issues orders to the government, and never considers obeying it. It is of course thoroughly monarchist. If the Allies were not . . . France is behaving like the little brother, who under lucky conditions has ripped out a huge tuft of the big brother's hair.[42] (ellipsis in original)

In April and May 1920, the Weimar Republic took further steps to establish voting laws and governmental structures.[43] On May 17, French and Belgian troops withdrew from the cities in the Ruhr region in Germany that they had occupied in the aftermath of the Kapp Putsch.[44] At the beginning of May 1920, University of Berlin botanist Gottlieb Haberlandt encouraged Einstein to sign an appeal by German academics in support of a republican constitution. Responding to a prior question from Einstein, he assured him that other socialists would be signing it, including University of Berlin sociologist and ethnologist Heinrich Cunow. Haberlandt showed that political progressives continued to rely on and value Einstein's courageous integrity when he stated, "Your name would be even more welcome to us, as the number of colleagues who completely agree with the content of the declaration but who don't want to sign for opportunistic reasons is unfortunately greater than we originally expected."[45] Einstein signed the appeal at the end of the month.

On May 10, Schlick wrote about reverberations from University of Rostock faculty who had supported the Kapp Putsch, speaking of a faculty "shaken by convulsions," some of whom had "Kapp-ed" themselves.[46]

In the immediate postwar decade, which was pacifism's twentieth-century zenith, meanings of words like "pacifist" were contested, and conceptualizations of a League of Nations or a World Parliament, which would become core elements of Einstein's world view in future decades, were still being debated and defined. In an example of widely differing understandings of these terms, Carl H. Unthan claimed to be a pacifist and saw his ideal as "a parliament of the nations," but he declared to Einstein that "the only way to peace leads through struggle and war." Unthan boasted that he had struggled energetically against such highly respected pacifist stalwarts as 1911 Nobel Peace Prize winner and publicist Alfred Hermann Fried, Hellmut von Gerlach, and journalist Carl von Ossietzky, all three of them leaders of the German Peace Society (Deutsche Friedensgesellschaft). Unthan categorized them as "vermin" (*Geschmeiß*) because they were attempting "to burden us with the sole guilt for the War" (which was directly inaccurate in the case of Fried).[47]

On May 30, 1920, Einstein was among signatories of the "Declaration of German University Professors for the Republican Constitution" ("Kundgebung deutscher Hochschullehrer für die republikanische Verfassung"), about which he had corresponded with Gottlieb Haberlandt. The declaration was published in major Berlin newspapers on that day. This appeal, in support of the Weimar constitution, condemned violence from both the right and the left and warned of the destructive danger to political freedom caused by defamation of the constitution.[48]

The rightists at the University of Rostock reported by Schlick were not an anomaly. On May 30, Paul Epstein wrote to Einstein about Edgar Meyer's observations at the University of Tübingen, stating that Meyer was appalled at the reactionary spirit at the University of Tübingen.[49]

The summer of 1920 in Germany saw a sharp increase in political murders, political violence, and death threats. In addition to leftists, pacifists were increasingly targeted.[50] In Reichstag elections in June, in a trend that would accelerate throughout the Weimar years, the original Weimar coalition lost its strong majority while votes increased for political parties of the more extreme right and left.[51] Meanwhile, treaties and plebiscites were redrawing national boundaries in Europe, and Germany continued to defy the Allies by blocking aid to Poland.[52]

In this summer of violence and threats against leftists and pacifists, Einstein repeatedly identified himself publicly as an internationalist. On June 6, 1920, when Columbia University awarded him its Barnard Medal for Meritorious Service to Science, Einstein used the occasion to advocate internationalism in scholarly circles:

> Apart from the personal gratification, I believe that your decision may be seen as a herald of a better time, in which the international feeling of solidarity will reunite the scholars of the various countries.[53]

On June 11, Einstein wrote to Arthur S. Eddington, the Cambridge Observatory astrophysicist whose solar eclipse expedition of May 1919 had confirmed Einstein's theory of relativity, and who was of Quaker and hence pacifist background. In declining an invitation to visit England, Einstein again mentioned his international orientation, as well as demonstrating his strategically pragmatic side:

> I had the feeling that a trip to England at this time would seem to our English and German colleagues like an ugly attempt to curry favor; I wanted to avoid this. This is where even the internationally oriented person [*der international Fühlende*] must restrict himself if he does not want to damage the good cause.[54]

On June 15, writing to Lorentz, Einstein again characterized himself as internationally oriented. He particularly emphasized his desire to connect with French people, Germany's wartime enemies:

> With great pleasure I accept the invitation to the Solvay Conference. . . . It is a heartfelt need of mine to see especially our French friends again after such a long, difficult time and to shake their hands. I'm also pleased

> that it is still possible even today to be treated as an internationally oriented person [*internationaler Mensch*], without being locked into one of the two big drawers.[55]

On June 10, in the aftermath of the Kapp Putsch, Einstein was informed that he, along with all civil servants, would have to take an oath of allegiance to the Weimar constitution. Notwithstanding his repeatedly self-avowed international orientation, ever the pragmatist, and generally conciliatory in his personal interactions, Einstein did so on July 1, 1920.[56]

Also at the beginning of July, the French writer and translator Gaston Moch praised Einstein for standing alongside other prominent German pacifists (such as Georg Friedrich Nicolai, Hermann Fernau, Hellmut von Gerlach, and Friedrich W. Foerster) in representing the dignity of the older, pre-Bismarck Germany, in contrast to most Germans.[57] Not one to lose his broader perspective on human beings regardless of nationality, Einstein replied to Moch defending Germans:

> I find your judgment of the Germans somewhat harsh. It seems, indeed, that no people is safe from falling victim to an imperialistic trend, especially when its internal balance is endangered by external success.[58]

In the first half of July, at a conference in Spa, Belgium, the Allies refused to compromise on the military disarmament of Germany, which strengthened profound feelings of resentment in Germany.[59] On July 8, the German League for the League of Nations (Deutsche Liga für Völkerbund), after consulting with Einstein's fellow New Fatherland League member and Lille commission member Elisabeth Rotten, presented Einstein with a plan to place a short article in several German newspapers regarding the crisis of German science and the action by English universities to aid German academic libraries. The league asked Einstein whether he would be willing to author the article.[60] Sensitive to political currents, to the intensity, volatility, and unpredictability of German nationalist feeling, and to the limitations of his political understanding (despite his ongoing efforts to stay informed), Einstein declined. In doing so, however, he expressed his heartfelt affinity with the League of Nations:

> After careful consideration and consultation with scholarly friends, I have decided not to write the article in question. I am sorry in this case not to be able to be of service to this goal that I hold so close to my heart. The political perspectives from inside and outside the country that would have to be taken into account in the crafting of the article are of so complex a nature that I do not feel capable of doing the job properly, especially considering that my insight into the state of affairs is insufficient. Also, the fact that I am Swiss is a reason to refrain.[61]

Einstein was more prepared to speak out regarding the more elemental crisis of physical survival, when he was approached in early July by Elsa Hermann of the German Central Committee for Foreign Relief. Hermann told him that a new round of fundraising was starting in foreign countries for the German children's relief effort. She described the American Quaker relief as currently feeding 632,000 children, and the English Quakers as "especially looking after young people and the needy lower middle class." Other charitable gifts were focused on "needy regions, sectors of the population, and institutions." The need for this aid was especially urgent for "growing youth, expectant and nursing mothers, the poor and the sick." In seeking Einstein's help, she explained that, although official reports and statistics were helpful, "people abroad . . . also want to hear statements from popularly beloved and well-known personalities, which fully acknowledge the size of the relief effort."[62] In response, Einstein wrote the following statement praising the postwar hunger relief efforts in Germany by Anglo-American Quakers (the Religious Society of Friends), a pacifist religion that clearly linked the propagation of peace with that of social justice and meeting human needs:

> In these hard years that have brought such great disappointments to every friend of human dignity and justice, there is no greater consolation for me than the thought of the beneficial and well-calculated effectiveness of the American and English Quakers. I myself have seen that precious human beings have been saved from extreme deprivation by them. Whatever great political disappointments we have experienced and will still have to go through, we cannot abandon hope for a just

> and satisfactory order in the world, when we see how in both of these countries, which fate has led to the highest level of power, such powerful forces are put to work for the well-being of all. No other participant in public life is suited to the same degree to strengthen mutual trust between peoples, and even more should be done to raise a stronger consciousness in the people of the beneficial work of the Quakers.[63]

Shortly thereafter, he wrote a similar statement on the same topic, this time emphasizing Quaker actions as conducive to the development of the League of Nations and to fostering receptivity to the league in Germany (despite Versailles):

> It is admirable what the Quakers have achieved in the alleviation of poverty in Central Europe. However sad the political experiences that we had to go through during the past six years may have been, as long as the world is under the leadership of countries that are willing and able to provide such significant powers and means toward the well-being of people without regard to race or political affiliation, we may in spite of it all believe with good reason that the psychological preconditions for a fruitful development of the League of Nations are at hand. Attention should be drawn here more often than in the past to this work of unselfish philanthropy, so that—despite Versailles—this idea can take root in Germany.[64]

In late July 1920, Einstein expressed to Besso his somewhat abated admiration for US president Woodrow Wilson, the early proponent of the League of Nations, alongside an ironic critique of political leaders and their followers:

> You really are right in your comment about the docility of people, except that there actually is no superior leadership. All are links in a chain, pulled by God knows what, but not by insightful people who are free from selfish motives. Only exceptionally is a stroke of luck realized in an enlightened despot (Marcus Aurelius or someone like Wilson, if good intentions would also be insight and action).[65]

In August 1920, Germany granted amnesty to participants in the Kapp Putsch, and the Bavarian government refused to disarm under the terms of the Treaty of Versailles.[66] Battles continued between Poles and Russians around Warsaw and between Poles and Germans in Upper Silesia.[67] In Turkey, nationalists were taking over from the Ottomans and changing the terms of their treaties with the Allies.[68] Meanwhile, in early September 1920, Einstein praised the Norwegian Students Association for rising above nationalism to propagate internationalism and healing in Europe:

> In grateful remembrance of the magnificent summer days that I was permitted to spend as a guest of the Norwegian students in Kristiania, I hereby send you a warm greeting from my study in Berlin. In you I came to know a student body which, free from stifling nationalist tendencies, holds high the cultural ideal common to all. . . . Alongside the impersonal factual elements that unite discerning people of all countries, for the recovery of international relationships, warm-hearted free human beings of liberal views and good will are needed. You are such people, and it is to be hoped that, as students of a neutral country, you will make valuable contributions to the recovery of the European psyche.[69]

After the Bad Nauheim conference (discussed in chapter 6), where he was clearly supported by legitimate scientists, Einstein expressed optimism about the scientific field to his good friend and fellow physicist Lorentz. However, he felt simultaneous hope and disappointment at developments in the larger political arena, as not only the Weimar Republic but other countries too were wasting a rare historic opportunity to establish international organization:

> The Natural Scientists conference offered nothing new of considerable significance, but it did show that, despite the War and economic crisis, interest in pure science has remained very much alive. Otherwise, a good many things have really changed. From all of these experiences, people's dispositions are so malleable that an eminent statesman could achieve great things; I'm thinking of a united states of Europe. It is unfortunate that in France or England there is no broad-minded,

> visionary leader at the top, who envisions more than the narrow momentary material interests of his own country.[70]

Sometime after late September 1920, Einstein wrote a statement for a volume being produced by the German Social and Scientific Society of New York (Deutscher Gesellig-Wissenschaftlicher Verein von New York) giving his answer to the question of how and what intellectuals of different countries could contribute in daily life to help reconcile the wartime enemy nations and bring about a lasting alliance of humankind.[71] In it, he extolled the creation of the products of the arts and sciences as most conducive to fostering an international spirit. He also advocated, on a pragmatic level, the value of young people studying abroad:

> The most important contribution of intellectuals to the reconciliation of nations and to a lasting alliance of humankind, in my opinion, lies in their scientific and artistic creations, because these elevate the human spirit above personal and national-egotistical goals. The concentration of energies on those questions and goals that are common to all intellectual people intrinsically generates a feeling of comradeship, which always must reunite the true scholars and artists of all countries, although it is unavoidable that the less generous and independent natures among them will occasionally be turned against one another by political and other passions. Intellectuals should never become weary of emphasizing the internationality of the most beautiful possessions of humankind, and their associations should never be party to fostering political passions through public declarations or other public steps.
>
> Finally, I would like to comment that, in my view, it would be of value to the reconciliation of nations if young students and artists were to study abroad in the former enemy countries to a greater extent than previously. Direct experience most effectively counteracts the disastrous ideologies that took root in people's minds under the influence of the World War.[72]

During October and November 1920, the German Communist Party (KPD) expanded its influence, and Prussia, Germany's largest state, consolidated its government.[73] Internationally, the Polish army made territorial gains in the East, leading to a ceasefire.[74] In late October 1920, the Göttingen theoretical physicist Paul Hertz appealed to Einstein to intercede on behalf of József Kelen, a Hungarian engineer who had been the people's commissar of social affairs in the Hungarian Soviet Republic, which had been defeated in April 1920. In this era of political vengeance, Kelen was now on trial for charges similar to treason.[75] In 1917, Einstein had sent a plea on behalf of Friedrich Adler (see chapter 2) during three months when Adler was sentenced to death (the sentence was later commuted), in which he expressed the typical pacifist opposition to state-sponsored violence in the death penalty and to political persecution. Similarly, now Einstein wrote an appeal on Kelen's behalf. He sent it in early November, probably via Hertz to Kelen's wife, and it was published in Vienna on November 19:

> Through various colleagues, I have learned of the dire situation in which the engineer Kelen in Budapest finds himself. I would very much like to undertake a drive together with Viennese professional colleagues on his behalf. . . . I hereby declare myself ready to take a stand in writing everywhere in favor of this valuable man. . . . It is among the most important responsibilities of our day to protect good and valuable people like Kelen against partisan political persecution.[76]

Also in early November, Friedrich Adler asked Einstein to sign an appeal for amnesty for ten Hungarian people's commissars (including Kelen); the same appeal protested against leading figures of the Hungarian Soviet Republic being tried as common criminals. This appeal was being circulated by Otto Bauer and Sigmund Kunfy, editors of the *Arbeiter-Zeitung*, the publication of the Austrian Social Democratic Party. Along with signatories from ten nations, Einstein signed this appeal and protest, which was published in Vienna on December 15, 1920.[77]

In mid-November, Paul Hertz thanked Einstein for his efforts on Kelen's behalf, adding that Georg Count von Arco, Einstein's fellow Lille commission member, had also taken steps to help Kelen and that Theodor von Kármán, a professor of aerodynamics and mechanics at Aachen, had published a supportive article anonymously in the *Frankfurter Zeitung*, which had been passed on to the Hungarian government.[78] Also in mid-November, Kelen's wife, Jolán Kelen-Fried, wrote thanking Einstein for his help and asking him to write directly to the Hungarian prime minister. She offered a sketch of her husband's character as a person of conscience who gave of his knowledge and time to those less fortunate, who attempted to improve society, who promoted a fairer system of distribution out of love of humankind, yet who had not been particularly prominent during the Hungarian Soviet regime.[79]

At the end of 1920, the League of Nations, established in Geneva, Switzerland, became more active internationally in Poland and the Middle East.[80] In early December 1920, the lawyer and writer Paul Mühsam sent Einstein a copy of his recent book, *From the Book of Human Destiny* (*Aus dem Schicksalsbuch der Menschheit*). In the cover letter he mentioned that Einstein was acquainted with his cousin Hans Mühsam, the physician at the Jewish Hospital in Berlin who in the final days of the War had described to Einstein conditions on the Belgian front. Having worked with the Red Cross and become a pacifist during the War, Paul Mühsam expressed great pleasure in their shared identity as pacifists: "A few days ago, when I was in Berlin . . . with my cousin Hans Mühsam, he told me . . . that you are a convinced pacifist." He described his enclosed book, "in which I as a similarly fervent pacifist have examined the insanity of war in poetic form." He then declared, "It would fill me with pride and joy to know that thc words with which I lament the fate of self-destructive humankind in my book also reverberate in your heart."[81]

Meanwhile, Einstein continued his communications with fellow pacifists and internationalists from various countries. In late December, he reiterated his identity as an internationalist, but also his loyalty to his German colleagues, in his reply to Ernest Pickworth Farrow, a Cambridge University (Trinity College) botanist who had asked whether he might be interested in a position at Trinity College:

> The difficulties and attacks that became my burden here are not as bad as they may appear abroad, and, given the great kindness that I have always experienced from German professional colleagues and administrators, it would be an act of base ingratitude if I were to turn my back on

> Germany in this time of poverty and degradation. That I am Swiss and of thoroughly internationalist convictions alters nothing in this personal relationship.[82]

In early 1921, the Treaty of Versailles continued to be a heated issue in Germany, with the Allies establishing reparations figures and processes and the Germans protesting them.[83] Bavaria's refusal to disarm their Home Guards continued to be debated.[84] Einstein was now well known to be a pacifist, enough so that Nuremberg gynecologist Paul Zacharias wrote to him in early January 1921 expressing his worry that a recent newspaper article by New Fatherland League cofounder Otto Lehmann-Rußbüldt might lead to attacks against him.[85] In that article, Lehmann-Rußbüldt had urged France and Germany to revive their economic relations and Germany to stop its resistance to disarmament; he contended that disarmament would bring about the end of the French occupation of the Rhineland and move France toward disarmament.[86] Zacharias's fears proved to be justified, at least with respect to threats to Einstein: on January 9, 1921, the journalist Rudolf Lebius published an article calling for the murder of members of the New Fatherland League, including Einstein, as traitors to Germany during the War.[87] That same month, Einstein heard directly from Lehmann-Rußbüldt in a circular letter to fifty-three recipients, which asked whether they would agree to have their names listed on the left-hand column of the stationery of the New Fatherland League.[88]

Einstein was recognized as a pacifist abroad as well. In January, George B. Jeffery wrote to him from England to thank him for a book on physics that Einstein had sent him a month earlier. Jeffery, like Einstein, greatly hoped for a resumption of friendly international collaboration among scientists. He explained that he had been imprisoned as a pacifist and conscientious objector to military service during the War, and he expressed feeling encouraged by finding that people in Germany also promoted good will and peace.[89] He also mentioned that he had corresponded with the Berlin mathematical statistician Emil J. Gumbel. Gumbel had joined the New Fatherland League in 1915 (the same year as Einstein), had been a member of its political advisory council since 1919, and was a member of the pacifist USPD during the entire period of its existence, from 1917 to 1922.[90]

Two months later, Einstein replied to Jeffery, extolling England and declaring his confidence in its influence. He also discussed the intensification of reactionary sentiment in Germany in response to the Allies' harsh peace terms:

> It is a highly gratifying fact that England's scholars in a not small proportion hold high the pacifist ideal. Through the supreme power that England has in the world today, a gradually healing effect on the intelligence of other countries cannot fail to arise. Our local scholars are unfortunately in the greatest proportion biased by traditional prejudices, which, through the heavy pressure that the country is currently experiencing from abroad, are only being further reinforced.[91]

In late January, in a letter to Max Born, Einstein showed a savvy political understanding of the purpose of the punitive reparations figure in the Versailles treaty, as well as foresight into its eventual fate:

> The large reparations figures and the threats are merely moral sustenance for the beloved public in France, in order to make the situation appear to them in a rosier light. The more impossible the conditions are, the more certain it is that they will not be realized.[92]

Also in January, the Austrian sociologist, political economist, and logical positivist Otto Neurath asked Einstein to collaborate in a series of scholarly "people's books" (*Volksbücher*) to convey knowledge in all fields to broader circles of the population, especially the working class.[93] In a draft of his reply in early March, Einstein responded favorably and offered his help. He suggested four additional collaborators, among them the two prominent and controversial pacifists Georg Friedrich Nicolai (for his professional expertise in biology and physiology) and Emil J. Gumbel (for his expertise in mathematical statistics).[94]

Early in March 1921, as Einstein prepared to depart for his first trip to the United States, he wrote to the prominent French pacifist Romain

Rolland, with whom he had been in contact since early in the War, thanking him for a copy of his new book that Rolland had sent him in February and mentioning that he was planning to read it during his trip to America. In one passage of the book (also mentioned by Rolland in his autograph of this copy), Rolland praised Einstein for his effect on others.[95]

From March to May 1921, heated debates took place and ultimatums and counterproposals flew back and forth between London and Germany until, finally, a reparations agreement was reached between Germany and the Allies.[96] Between March and June, a vote in Upper Silesia over joining Poland versus Germany was violently contested; eventually, the Allies settled matters by partitioning Upper Silesia.[97] War crimes trials in Germany in May resulted in light sentences to none at all.[98]

In May 1921, the sociologist, philosopher and fellow New Fatherland League member Rudolph Goldscheid wrote to Einstein to thank him for an autographed photograph he had sent. Goldscheid expressed his deep respect and admiration.[99] In the summer of 1921, Einstein's relationship with the New Fatherland League was strengthened through advocacy he had undertaken on its behalf while in the United States. This was evidenced in a letter from Hellmut von Gerlach, fellow Lille commission member and a former undersecretary in the German Ministry of the Interior, who was also editor-in-chief of the newspaper *Die Welt am Montag* and cofounder of the German Democratic Party (DDP) and of the German Peace Society (Deutsche Friedensgesellschaft). After Einstein returned from the United States and England on June 17, Gerlach notified members of the New Fatherland League that, while in New York, Einstein had met with philosophy professor and Columbia University president Nicholas M. Butler, a cofounder, trustee, and executive committee member of the Carnegie Endowment for International Peace. Einstein had pleaded with Butler to extend support to German pacifists, in particular the New Fatherland League, and apparently elicited a favorable response.[100]

In July and August 1921, the United States and Germany concluded their separate peace.[101] In late August, Matthias Erzberger was murdered by Kapp Putsch participants. Chancellor Joseph Wirth responded by declaring a state of emergency, which would last almost four months.[102] In September 1921, Käthe Kollwitz, the renowned graphic artist and sculptor who advocated peace and social justice through her art, asked Einstein for help with a communist-sponsored appeal on behalf of the victims of the famine in Russia. In this famine, which had been caused by the Russian civil war along with bad weather, thirty million people suffered and four to five million died.[103] Kollwitz's request followed four other appeals to Einstein for the same cause, from writer Arthur Holitscher, journalist and KPD member Willi Münzenberg (the primary impetus behind this appeal in Germany), book publisher and USPD member Otto Gäbel, and poet Max Barthel.[104] Selective as usual with his endorsements, Einstein wrote to Kollwitz respectfully declining to participate, citing the political character of the appeal, while also expressing his great admiration for her as an artist and a human being.[105]

Einstein continued to devote energies to the hardships resulting from the War in its aftermath. In September 1921, he was contacted by a relief committee composed of Americans of German ancestry to write a testimonial regarding poverty and child mortality in Germany, in order to help them in their fundraising.[106] For this committee, Einstein wrote a moving plea on behalf of hungry children and hardworking Germans in a time of economic crisis:

> The suffering of children in the cities of Germany is still getting ever more terrible, so that a considerable proportion of the future generation is growing up under conditions that are causing them permanent harm. If this were a matter of a lasting situation, foreign aid would be exhausted by such mass suffering. But a great proportion of the poverty results from the unstable economic conditions, which for their part arise from the continual falling of the currency. I do not doubt that this industrious people will overcome any consequences of the War and rebuild stable economic conditions, and therefore I can hardly think of a better and at the same time more worthwhile work of compassion in this hard transitional period than to save these children from premature infirmity and to aid working people, who could be held least responsible for the course of world history in these terrible years.[107]

In the autumn or early winter of 1921, the French League for Human Rights (Ligue des Droits de l'Homme) invited the New Fatherland League to a conference in Paris, held in December 1921 and January 1922. In a draft reply the French League's invitation, Einstein regretted that he could not attend the conference himself due to professional obligations; he did, though, applaud the conference, believing it would improve French-German relations, a central concern of European pacifists between 1919 and 1929:

> It would have been my desire to be present at the first meeting of democratic and pacifist-oriented members of both nations on French soil for the purpose of expressing that the cultural collaboration of both countries is in the interests of all of humankind and that the primary duty of the knowledgeable on both sides is to eliminate the psychological obstacles to collaboration.[108]

The German representatives who attended the conference were Hellmuth von Gerlach, Georg Friedrich Nicolai, Otto Lehmann-Rußbüldt, and the writer, diplomat, and patron of the arts Harry Count Kessler. At this meeting, the New Fatherland League (Bund Neues Vaterland) changed its name to the German League for Human Rights (Deutsche Liga für Menschenrechte). Einstein was among the signatories of a joint declaration by the two leagues, which was published in March 1922.[109]

Thus, Einstein started 1920 with his hopes for Germany and the world greatly reduced from where they had stood a year earlier, at the War's end. Even so, in the ensuing two years Einstein continued to advocate on behalf of political prisoners, in support of the beleaguered pacifist professor Nicolai, in support of opportunities for East European Jewish university students (see chapter 6), and in favor of employment opportunities for pacifist and internationalist German scientists. He did, though, remain selective in his public stances, declining to support the Monist League and other campaigns mildly or blatantly incompatible with his convictions. At the start of 1920, Einstein was sharply disappointed in the reception of the revised (second) edition of *Lille* by the German public, disgusted with the rightist postwar reaction in Germany, and, notwithstanding his public stature, even subjected to disruption of his lectures (see chapter 6).

In March 1920, Einstein noted the failure of the Kapp Putsch with relief. He sharpened his critique of the military's use of violence domestically, the Weimar government's weakness of will, and Germany's disrupted living conditions, poverty, and high mortality rates. He continued to write reflections on the situation of Jews and antisemitism (see chapter 6), wrote a statement to encourage international exchanges of scholarly literature, and signed a declaration of university professors in support of the republican constitution. Even as political violence and murders against supposed "traitors" intensified in the summer of 1920, he continued to identify himself as a pacifist and an internationalist, and he wrote two public statements praising American and English Quakers for their efforts to feed German children in the first years after the War. In the early fall of 1920, politically motivated attacks on Einstein's scientific contributions at the Philharmonie in Berlin and at Bad Nauheim elicited an outpouring of expressions of support for Einstein from a wide spectrum of Germans. This may have further inspired his public statement shortly afterward extolling collaboration in science and art as the best avenue to international reconciliation. Late in 1920, as he had done previously when Friedrich Adler was sentenced to death, Einstein interceded in support of József Kelen, then on trial for his involvement in the short-lived Hungarian Soviet Republic, thereby expressing the usual pacifist concern for political prisoners, for social justice, and for proponents of a populist republic.

Knowledge of Einstein's identity as a pacifist was sufficiently widespread in January 1921 that the rightist journalist Rudolf Lebius listed his name among those he was inciting his readers to murder as traitors. Undaunted, Einstein continued to correspond enthusiastically with English, Austrian, and French pacifists, and upon his return from his first trip to the United States and England, his involvement with the New Fatherland League continued unabated. Thus, in the first three years after World War I ended, despite many personal and societal hindrances and sources of intimidation, Einstein showed in his private and public statements that the pacifism the First World War had awakened in him had become deeply rooted as a carefully considered world view and firm conviction. Moreover, he repeatedly demonstrated the courage of his convictions in outward expressions of this pacifist commitment and its logical corollaries in promoting social justice and meeting human needs.

6 Social Justice Advocacy for Jews, and Confident Jewish Identity

1914–1921

Pacifists typically advocate for social justice, as discussed earlier, seeing a psychological and economic violence inherent in social injustice. Einstein's passion for social justice, compassion for the underdog, and identification with his fellow Jews—with special feeling for East European Jews—was, like his pacifism, awakened by World War I. Woven into his concern for social justice was an impressive confidence in his identity as simultaneously a Jew and a pacifist.[1] In October 1916, in response to false accusations that Jews were not serving in proportionate numbers, the German army conducted a census of its soldiers by religion. That census, which reinforced stereotypes about Jews as weaklings, cowards, shirkers of military service, and even traitors, sparked protests by the German Jewish community and inspired them to found the National Association of Jewish Frontline Soldiers (Reichsbund Jüdischer Frontsoldaten) after the War.[2] Despite pressure to defy such stereotypes, Einstein did not hesitate to be known as a Jew who was also a pacifist.

Two and a half months before the War began, Einstein expressed solidarity with fellow Jews facing persecution. Petr Petrovich Lazarev, vice president of the Russian Imperial Academy of Sciences in St. Petersburg and a physics professor at the Shanyavsky City University of Moscow, invited him to visit, probably to participate in some joint research. Einstein declined due to feelings of solidarity with Russian Jews suffering from pogroms and other antisemitic incidents:

> I thank you for the friendly invitation, which, however, I cannot accept. I am reluctant to travel unnecessarily to a country in which my kinfolk have been so brutally persecuted.[3]

Four years later, toward the end of the War, Einstein's Jewish identity and advocacy for fellow Jews had risen to a more central position among his public stances and commitments. In the spring of 1918, he was maintaining his connections with the broader Jewish community and his reputation for supporting Jewish higher education. In May of that year, Arthur Hantke, head of the Zionist Association of Germany, and Otto Warburg, president of the international Zionist organization and a University of Berlin botany professor, invited Einstein to attend a meeting to help expand Jewish secular education in Poland. They told him that Rabbi Dr. Markus Braude of the Great Synagogue in Lodz was coming to Berlin to raise interest in the Institutes of Jewish Instruction that he had founded—specifically, on behalf of a future Jewish Teacher Training College in Lodz. They emphasized an extreme shortage of Jewish teachers in Poland and the great cultural poverty of Polish Jews resulting from this. They hoped to utilize Einstein's attendance to attract interest in this college among more extended groups of German Jews who might want to elevate the cultural level of Polish Jews.[4]

In December 1918, Einstein was already grappling with the tension between Zionism, or Jewish nationalism, and his internationalism. On December 9, the Zionist Association of Germany invited him to a meeting to be held on December 19, during which the Jewish concept of nationality would be discussed in relation to the international idea.[5] Einstein agreed to attend.[6]

By late March 1919, while he was expressing to Ehrenfest his disillusionment with the Allies and with German political developments, Einstein took comfort in Zionist aspirations:

> [M]any courses here [in Berlin] had to be canceled as a result of the university's closing following the unrest. . . . Politics disappoint me very much these days. Those states whose victory I had felt during the War would be incomparably the lesser evil are now proving to be only slightly the lesser evil. Added to that are the excessively dishonest local domestic

> politics. Reaction with all disgraceful deeds in repulsive revolutionary disguise. One doesn't know where one can look in order to have joy in human activity. I get the most pleasure from the realization of the Jewish state in Palestine. It seems to me that our tribal kinfolk really are more sympathetic (at least less brutal) than these dreadful Europeans.[7]

On September 18, in a speech at Osnabrück, Gustav Stresemann, whose wife was Jewish, denied any link between Jews and radicalism, anarchism, Bolshevism, or cosmopolitanism. He also argued that most German Jews opposed extremist movements and that a liberal party could not support antisemitism.[8] On September 11, 1919, mathematician, physicist, and fellow Jewish colleague Paul Epstein—who had been briefly imprisoned as an "enemy alien" in Munich during the War, and whose assets in Russia had been seized by the Bolsheviks—wrote to Einstein about his negative experience of the current chauvinism. Epstein expressed an eager willingness to consider joining the planned Hebrew University in Jerusalem, an institution that Einstein would strongly support for the rest of his life. Epstein explained that primarily his wartime imprisonment, though also combined with the current nationalist chauvinism, had awakened him to the negative sides of his former cosmopolitanism, teaching him to value being among his own people, where he would not be regarded as an alien or an intruder.[9]

In early October 1919, recognizing his need to reconcile his pacifist internationalism with Jewish nationalism (or Zionism), Einstein replied to Paul Epstein:

> [T]he Zionist cause is very close to my heart. . . . I have strong confidence in the positive development of the Jewish colony and rejoice that there should be a little patch of earth on which our kindred brethren [*Stammesbrüder*] should not be considered foreigners. . . . One can be internationally oriented without being indifferent toward one's tribal comrades [*Stammesgenossen*].[10]

Zionist organizations and leaders were happily aware of Einstein's support. Also in October, at Hermann Struck's suggestion, the Relief and

Works Agency for Palestine (Hilfswerk für Palästina), also known as the Palestine Foundation Fund (Palästina-Aufbaufonds), contacted him to sign their appeal "For the Construction of Jewish Palestine" ("Für den Aufbau des jüdischen Palästina") as part of their fund drive.[11]

At the end of December 1919, in a stand for social justice that reflected his connection with the Jewish people, Einstein published a statement in a major Berlin daily newspaper defending East European Jews in Germany against directly and indirectly antisemitic attacks and allegations against them. He took this stance at a time when German Jews tended to distance themselves from East European Jews.[12] His statement offered an orderly, detailed, and insightful analysis of facts about East European Jews in Germany as well as savvy, nuanced political thinking. In it, he logically refuted various allegations, advocated on behalf of the most powerless among them, and reminded Germans of their reputation abroad:

> Among the German public voices are increasing that demand legal measures against the East European Jews. It is claimed that 70,000 Russians, i.e., East European Jews, live in Berlin alone; these East European Jews are alleged to be black marketers, illicit traders, Bolsheviks, or elements averse to work. All of these arguments call for the most severe measures, i.e., cramming them together into concentration camps or expulsion of all immigrants.
>
> A rule that annihilates so many existences must not be derived from slogan-like claims, even less so when a factual examination shows that this is a case of demagogic agitation, which neither justly reflects the factual situation nor offers an appropriate means for countering real existing wrongs. Agitation in particular against East European Jews from the outset raises the suspicion that quiet judgment is being clouded by strong antisemitic instincts and simultaneously a method is also being chosen which, through influencing the mood of the people, distracts attention from the real problems and the true causes of the general crisis.
>
> An inquiry from official quarters, which, as far as is known, has never yet been attempted, would doubtless prove the invalidity of the accusations that have been raised. It may very well be correct that 70,000 Russians live in Berlin; according to the expert judgment of specialists,

Jews comprise only a tiny fraction of them, while the overwhelming majority are of German descent. Since the peace agreement, according to authoritative estimates, no more than 15,000 Jews have immigrated from the east. Almost without exception they were forced to flee by the horrible conditions in Poland and seek to find a refuge here until they are given the opportunity to emigrate elsewhere. Hopefully many of them will find a true homeland as free sons of the Jewish people in the newly arising Jewish Palestine.

There may very well be Bolshevik agents in Germany, but they doubtless have foreign passports, ample funds at their disposal, and cannot be seized by any organizational measures. The big black marketers among the East European Jews have certainly long since taken precautions to evade arrest by the authorities. Those affected would be exclusively those poor and unfortunate ones who in recent months in the midst of inhumane deprivations found their way to Germany and seek work here. Only these elements, who are certainly harmless to the German economy, would fill the concentration camps and degenerate there physically and spiritually. Then people will complain about the self-cultivated "parasitic beings" who have lost the ability to take their place in a normal economic life. The mistaken policy of suddenly laying off thousands of East European Jewish workers who were forcibly brought to Germany during the War, thereby depriving them of food and systematically blocking them from job opportunities, must in fact force people, who don't want themselves and their families to starve, into illicit trade. German economic life is also certainly best served if the public supports the efforts of those circles (like for example the already often mentioned "Jewish Labor Office") who pursue the goal of channeling East European Jewish immigrants into productive work. Any "Order of Expulsion," which is now so stormily demanded, would only have the effect of the bad and harmful elements staying in the country, while driving those willing to work into bitter misery and desperation.

The conscience of the public is so dulled toward appeals for humanity that it no longer directly feels the terrible injustice that is being sought. . . . [I]t is disconcerting that even leading politicians do not consider how seriously Germany's political and economic position is impaired through their proposed treatment of East European Jews.

> Have people already forgotten how the deportation of Belgian workers undermined the credibility of Germany? And today Germany's situation is incomparably more critical. Despite all efforts, it is extremely difficult to reestablish the disrupted international relationships; a few intellectuals from all nations are making initial attempts; the hope for new economic links (e.g., the material help of America) is still very weak today. The expulsion of the East European Jews, which would result in unspeakable misery, would appear to the whole world as new evidence of "German barbarism," and would offer a new excuse, in the name of humanity, to hinder the reconstruction of Germany.
>
> The recovery of Germany truly cannot be achieved through the use of force against a small, defenseless fraction of the population.[13]

Thus, Einstein's final public statement of 1919 defended East European Jews in Germany.

Compared to University of Berlin medical school professor Georg Friedrich Nicolai, whose lectures had been disrupted by rightist German nationalist students in January 1920, Einstein was more insulated, both by his greater fame and by his lesser inclination to take controversial public stands. Nonetheless, on February 12, 1920, a lecture of Einstein's was disrupted by some registered students protesting his practice of open admission at his lectures to unregistered students, some of whom were East European Jews. There were differing views on the centrality of antisemitism in the incident. Both a rightist and a leftist newspaper portrayed antisemitic motives as playing a key role.[14] However, University of Berlin rector Eduard Meyer maintained that antisemitism was not central.[15] A liberal newspaper also contended that the protests were not antisemitic.[16] A petition to Einstein signed by almost three hundred students, expressing regret over the incident, claimed that press accounts had exaggerated and distorted the events.[17] Einstein, whose position at the Kaiser Wilhelm Institute did not require him to give lectures, explained in an article in a Berlin newspaper his desire to keep his lectures open by some creative means. Even while trying to calm allegations of antisemitism, nonetheless he indicated his awareness of possible antisemitism. He then stated his pragmatic plan for what he intended to do:

> My popular lectures being held on the theory of relativity were attended not only by students but also by many other people who were not actually authorized to attend. Because of this, the student council declared that it would no longer allow this. I drew their attention to the fact that the large hall has space for all who want to listen and that this should hardly cause any inaccessibility. The student council expressed that it was not satisfied with this, and turned to the *rector* [of the University of Berlin] with this question. The rector wrote me a *letter* in which he pointed out that, according to existing regulations, these people do not have the right to enter the hall. This is *formally correct*. I, however, took the perspective that it is morally reprehensible to me to make it impossible for these people to continue listening without good reason, and therefore yesterday, instead of lecturing, I held a discussion with my audience which, however, did not lead to a definite result. I therefore felt obliged to cancel further lectures and declare to the students that they could request a refund of their tuition fees. However, I do not intend to terminate my lectures in general; I plan, rather, to resume them in another form. In which lecture hall is, however, still undecided. Should such an incident as yesterday's occur again, then I will cease my lecturing altogether. What happened yesterday cannot be called a scandal, but all the same some remarks that were made demonstrated a definitely hostile attitude toward me. *Antisemitic remarks* as such did not take place, but their *undertone* could be interpreted that way.[18]

One week later, at Einstein's lecture on February 19, a student read aloud a statement of support, signed by 291 students, expressing that they recognized "the enormous significance of your work for our entire scientific life and for academic instruction in Berlin," characterizing him as an "irreplaceable" teacher, and declaring "unanimously that we staunchly condemn the exaggeration of the incident of February 12, 1920, and very much regret the event."[19] Einstein responded that he was not offended by the matter.[20]

Also on February 19, Einstein expressed his ongoing sense of connection with East European Jews and concern over their hardships (which ultimately developed into his active support for the Hebrew University in Jerusalem), when he and gynecologist Leopold Landau raised with

Prussian minister of education Konrad Haenisch the problem created by quotas limiting numbers of Jews at most eastern European universities. They proposed, with Einstein's quintessential pragmatism, to make university-accredited courses in Germany available to refugees of the War:

> There are currently approximately 200 young foreigners from Russia, Poland, Bulgaria, Rumania, and Lithuania, mostly East European Jews, in Berlin for the purpose of studying, for whom it is impossible to be matriculated at the local university. Because this involves mostly people who for various reasons also cannot study in their own countries, we the undersigned have decided to meet the need and arrange special courses for them. These should be started without delay, so that the young people, whose studies have already been exceptionally impeded by war and poverty, don't lose even more time. Admissions should extend to include temporarily resident foreigners with academic preparation corresponding to the requirements in their own countries. A list of university lecturers participating so far and the courses planned by them is enclosed.
>
> Absolutely no burden should accrue for the state or the community. Also there is no question of these courses counting toward academic credit domestically.
>
> We petition the ministry for only one thing. In order that the lectures also be of external use to our students, the grades given to them must indicate the quality of instruction in a way that provides sufficient basis for the later recognition of these courses toward academic credit abroad. Therefore we petition the ministry to be allowed to describe our arrangement as "state accredited courses by Berlin University instructors."[21]

In early April 1920, in another expression of pride in his Jewish identity and commitment to social justice, Einstein produced an essay exploring the nature of xenophobia, in particular antisemitism. He declared that it was the moral duty of individuals to control such feelings and the legal and ethical responsibility of states to protect minorities. He added that unity among different sectors of the Jewish community was of crucial importance:

> The aversion to Jews in my opinion is simply based on the fact that Jews and non-Jews are different. It is the same feeling of aversion that is always found when nationalities have to deal with one another. This aversion is a consequence of the existence of Jews, not their particular qualities. The reasons given for the aversion are flimsy and ever-changing. Where the feeling is sufficiently intense, people are never at a loss for reasons; and the feeling of aversion toward people of a foreign race with whom one more or less has to share life is of necessity nudged into existence.
>
> Here lies the psychological root of all antisemitism, but absolutely no justification for the agitation of antisemites. A feeling of aversion may be natural, but to follow it without restraint signifies a low level of moral development. A morally upright person will guide his actions by rational insight, not by dull instinct.
>
> How do things stand, however, with society, with the state? Can it tolerate national minorities without fighting against them? These days there is no state that does not officially include toleration and the protection of national minorities as one of its duties. Hopefully it takes this duty seriously! This includes stopping its prior custom of making governmental employment of Jews in many cases dependent on an unprincipled and humiliating act (baptism); it would be all the more advisable to drop this tradition, as it brings with it a really unfortunate selection. . . .
>
> The task of the Jews cannot be to obtain "immunity" from the antisemites through accusation against some part of their own people. In this conception lies a severe misjudgment of the law and of the significance of antisemitism, whose presumption of sitting in judgment over the Jewish people we will never accept.[22]

Shortly thereafter, decades before it had become common to advocate diversity as adding value to a nation, Einstein wrote another essay on this theme. In it, he focused more directly on the duty of the German state to protect internal freedom for its minorities—in particular by not requiring baptism for government employment—as befitting a dignified, civilized nation. He also addressed the potential contribution of Jews to this end, if they would carry themselves with confidence and self-respect:

> The state and especially the German state must come to understand clearly that through its treatment of Jews it will demonstrate whether it can do justice to its true future tasks, whether it can tolerate minorities in its midst without fighting against them. Is it not its highest duty to ensure that people of any origin and of any type can live in complete freedom within its borders? Today there is no state that has not made protection and the right of self-determination of minorities part of its political program. The task of the Jews is to ensure that the state, for the benefit of its own development, turns this forward-looking program into reality. The general public needs to know that it harms itself most severely when it requires from the Jews the surrender of their human individuality and demands characterless conformity to their surroundings, for example, through baptism. Self-confident Jews of strong character will however contribute, through their presence, to an ever stronger feeling of true humanity. They have the task of fostering the triumph of the idea that only respect for human individuality guarantees a dignified life to the nations, and that it is a sign of the deepest lack of civilization to make one group of people into a scapegoat because they are of different blood.[23]

In late July 1920, Einstein received a letter from the German News Agency for Foreign University and Student Affairs, whose goals were to revive international connections between student organizations and to propagate German culture, requesting Einstein's collaboration.[24] In his enthusiastic response, Einstein interweaved his Jewish identity and advocacy with his internationalism by suggesting that they expand their goals to include encouraging more foreign students to study in Germany. This he envisioned as also an avenue to mitigating antisemitism:

> Despite an overwhelming amount of work, I cannot refrain from applauding your effort most warmly. On the occasion of a trip to Scandinavia, I saw how important the relationships are that German academia acquired through the hospitality of its universities before the War. This is the source of the fact that students of otherwise hostile Norway feel friendly to Germany and, after [the War] as before, mainly use German books. In my opinion, one of your main purposes should be to convince young students that it is of immeasurable advantage to Germany if

> many foreign students receive their academic training in Germany. I do know very well that shortages of space and money in German universities forbid a lot, because we naturally must take care of German youth first. But some expulsions of foreigners take place out of political narrow-mindedness, which regrettably developed as a result of unfortunate attitudes among local youth. In particular, foreigners should be liberally enrolled in truly scholarly studies, that is, those without direct practical application, where no overcrowding can be spoken of. As a Jew, I often have occasion, for example, to hear the complaints of East European Jewish students, who very often are turned away on trivial pretexts out of sheer antisemitism. Through this, people are embittered who could form a bridge to the east, which is so important for the future economic development of Germany. But far from wanting to place economic perspectives at the center, I see the main value of your effort in its moral and purifying effect. May your cause attain abundant success.[25]

In early August, the lawyer Heinrich Stern, who nine years later in 1929 would join Einstein in the newly founded Jewish Peace League (Jüdischer Friedensbund), wrote to consult with Einstein about the wording of a statement he intended to publish in the newspaper. This consultation exemplified the early roots and long duration of Einstein's relationships with fellow Jewish pacifists.[26]

Einstein's world renown made his job security stronger than that of his more outspoken, more abrasive, but less famous fellow pacifist, University of Berlin medical school professor Georg Friedrich Nicolai. (In mid-1920, Nicolai was partway through a process of professional rejection that by 1922 would lead him to emigrate to Argentina.[27]) Even so, Einstein's groundbreaking scientific advances sparked attacks by weaker scientists—attacks that were sometimes laced with antisemitic undertones. On August 24, 1920, at the first meeting of the Association of German Natural Scientists for the Preservation of Pure Science (Arbeitsgemeinschaft deutscher Naturforscher zur Erhaltung der reinen Wissenschaft) in Berlin's Philharmonie building, Einstein's contributions to the field of physics, particularly his theories of special and general relativity, were assailed by the German nationalist Paul Weyland and spectroscopist Ernst Gehrcke of the German Physical and Technical Institute

(Physikalisch-Technische Reichsanstalt); both used language with antisemitic implications. It was reported, also, that antisemitic leaflets and swastika lapel pins were displayed and distributed at the entrance to this event.[28] When a second meeting in opposition to the theories of relativity was held on September 2, again at the Philharmonie, University of Prague philosopher Oskar Kraus canceled his attendance, explaining that he did not want to identify himself with the organizers' political orientation. At that event, the engineer Ludwig Glaser spoke, making less virulent criticisms of relativity. A month after the first gathering at the Philharmonie, between September 19 and 26, 1920, a meeting of the Society of German Natural Scientists and Doctors (Gesellschaft Deutscher Naturforscher und Ärzte) and the German Physical Society (Deutsche Physikalische Gemeinschaft) was held at Bad Nauheim. There, with some of the same critics participating, Einstein engaged in a fifteen-minute objective debate with the University of Heidelberg physicist (and later proponent of Nazi ideology) Philipp Lenard.[29]

On August 27, three days after the first Philharmonie meeting, at which he had been present, Einstein had responded in an article in the *Berliner Tageblatt* by listing ten prominent physicists who embraced his work and indeed had built upon it (including Lorentz, Planck, and Eddington). In the same article, he noted Weyland's lack of professional credentials and dearth of meaningful statements, and he refuted Gehrcke's scientific assertions in some detail. In this context, he again reiterated his identity as an internationalist:

> Under the pretentious name "Association of German Scientists," a motley society has come together whose initial purpose of existence is to denigrate the theory of relativity and me as its creator in the eyes of non-physicists. Mr. Weyland and Mr. Gehrcke recently held a first lecture along these lines at the Philharmonie, at which I myself was present. I am very well aware of the fact that both speakers are unworthy of an answer from my pen, because I have good reason to believe that other motives than the striving for truth form the basis of this enterprise. (If I were a German nationalist with or without a swastika instead of a Jew of liberal international convictions [*freiheitlicher,*

internationaler Gesinnung], then . . .). [ellipsis in original] I only answer because well-meaning people have repeatedly urged that I make my opinion known.

First, I want to note that today, to my knowledge, there is hardly a research scientist who has contributed something substantial to theoretical physics who would not agree that the entire theory of relativity is logically constructed and is consistent with reliably established experimental facts available to date. The most important theoretical physicists—I name H. A. Lorentz, M. Planck, Sommerfeld, Laue, Born, Larmor, Eddington, Debye, Langevin, Levi-Civita—support the theory, and most have made valuable contributions to it. As an outspoken opponent of the theory of relativity among physicists of international prominence I could only name Lenard. I admire Lenard as a master of experimental physics; in theoretical physics, however, he has not yet achieved anything significant, and his objections to the general theory of relativity are of such superficiality that until now I did not consider it necessary to answer them in detail. I intend to make up for this.

I am accused of conducting a tasteless advertising campaign for the theory of relativity. Yet I can say that all my life I have been a friend of carefully considered, rational words and of concise presentation. Pretentious phrases and words give me goose bumps, whether they deal with the theory of relativity or with anything else. . . .

Now to the lectures. Mr. Weyland, who does not seem to be an expert at all (physician? engineer? politician? I couldn't find out), has presented nothing whatsoever of pertinence. He spewed forth coarse invectives and base accusations. The second speaker, Mr. Gehrcke, in part presented direct falsehoods, and in part attempted, through a one-sided selection of material and distortion of it, to create a false impression among uninformed laypeople. The following examples show this:

[In the intervening paragraphs, Einstein described and refuted Gehrcke's arguments in detail.]

. . . The personal attacks in this matter which Mr. Gehrcke and Mr. Lenard directed against me have been generally regarded by real experts as unfair; I have until now considered it beneath my dignity to waste even one word on it. . . .

Finally, I would like to add that, on my suggestion, a discussion on the theory of relativity will be held at the conference of scientists in

> Nauheim. Anyone who dares to face a scientific forum can present his objections there.
>
> It will make a strange impression abroad, especially among my Dutch and British colleagues, H. A. Lorentz and Eddington, both of whom have worked intensively in the field of relativity and repeatedly lectured on the subject, when they see how the theory and its creator are disparaged in such a manner in Germany.[30]

On August 28, the day after this article appeared, a rumor was published that Einstein planned to leave Germany in response to the hate campaign against him—an assertion that he refuted shortly thereafter.[31] However, the suggestion of this possibility, alongside all of the publicity about the hostility against him, led to an outpouring of expressions of support from a broad spectrum of both ordinary people and prominent political, intellectual, and cultural figures, including fellow pacifists, starting at the end of August and continuing well into the fall of 1920. Fellow New Fatherland League member Hellmut von Gerlach wrote an article supporting Einstein that was published in the newspaper *Die Welt am Montag*.[32] Two days later, a group of nine writers, actors, and theater people involved in Austria's Salzburger Festspiele, a theatrical endeavor aimed at promoting internationalism in the arts, sent Einstein a telegram of support. They expressed themselves as "outraged at the pan-German rabble-rousing against your excellent character" and assured him in "international spirit" of the "sympathy of all liberal people, who are proud to know you are among their ranks and to count you among the leaders of world science." The group included Oscar Bie, editor-in-chief of the *Neue Deutsche Rundschau*, director Max Reinhardt, and pacifist novelist and playwright Stefan Zweig. This telegram was published in at least three German and Austrian newspapers.[33]

On the same date, an ordinary Jewish citizen of Berlin, Helmut Bloch, wrote to Einstein expressing his support and exemplifying one Jewish perspective on antisemites in Germany in 1920. He conveyed his "heartfelt empathy at the great injustice done to you," declaring that "even a shake of the head would be too much and unworthy of this category of people" and noting that their newest symbol was "the swastika, which means, from the perspective of its wearers, rabble-rousing, hatred, falsehood, and profanity." He found it incomprehensible "that the swastika, which causes incitement, is not forbidden by the authorities," believing that "this sort of people . . . could be simply eradicated through the full force of the law." His inspiring words of advice for Einstein: "Let's ignore these pests. . . . esteemed professor, always hold your head high . . . stand above the primitive, and trust in God and in good honorable people, as thereby one finds vibrant joy in existence and in one's creative work, for oneself and for humanity!"[34]

The Munich physicist Arnold Sommerfeld, in a letter of support in early September 1920, expressed attachment to Einstein as a scientific colleague. He deeply hoped Einstein would stay in Germany. Regarding the Philharmonie meetings, Sommerfeld expressed "real feelings of rage" at allegations that he characterized as "truly a mockery of all justice and reason." Yet, he continued, "But you must not leave Germany! Your entire work is rooted in German (and Dutch) science; nowhere will you find as much understanding as in Germany. To abandon Germany now, when it is so unspeakably abused from all sides, would not seem like you." In a display of German chauvinism, Sommerfeld then suggested that, with Einstein's views, he "certainly would have been locked up in France, England, or America, during the War." With some insensitivity in his casual allusion to pogroms, he expressed the hope that Einstein had rediscovered his "philosophical sense of humor, and compassion for Germany, whose sufferings express themselves, as everywhere, in pogroms."[35]

Prussian minister of education Konrad Haenisch also wrote to Einstein expressing his pain and embarrassment at the public attacks on him, his satisfaction that prominent scholars were standing by him, and his hope that rumors that these attacks might drive Einstein from Berlin were baseless.[36] Einstein reassured him that he was deeply rooted in Germany. He did, however, hint at circumstances that could motivate him to depart:

> . . . during these days I have come to see that Berlin is the place in which I am most deeply rooted through personal and scholarly relationships. I would accept a position abroad only in the event that external conditions compel me to do so.[37]

In this context of politicized attacks on Einstein's professional achievements, two prominent pacifists expressed their enthusiastic support for him in the second half of September. One was Minna Cauer, the octogenarian progressive-oriented women's rights proponent and fellow New Fatherland League member, who with feminist socialist Helene Stöcker had coauthored an appeal to the German government in April 1919. Cauer wrote expressing her dismay and disgust at the attacks: "Now the whole world is speaking of you, of course the envious and the querulous also weighing in." Regarding "those primitive elements . . . men in Germany who do not thankfully appreciate the work and knowledge of researchers and the highly talented . . . I feel it as a disgrace that impacts the entire nation." She rallied behind Einstein: "You, highly esteemed Professor, will not be touched by it. We, however, suffer painfully."[38] The other pacifist was Stefan Zweig, the Austrian Jewish playwright, who followed up the theatrical artists' telegram (published in newspapers three weeks earlier) with a more personal and empathetic comment: "I can imagine what you now must be suffering from the intrusiveness and hostile spitefulness." He wished for Einstein, "may the peace in which you can further develop your work soon be restored to you."[39]

On November 10, 1920, Einstein replied to Zweig expressing warm appreciation for the declaration of support by the Salzburg artists and for Zweig's own letter.[40] On November 19, he wrote a cordial, respectful reply to Cauer thanking her for her letter two months earlier. Einstein showed that his bemused acceptance and humility remained intact in the context of his new fame and the recent attacks on his work:

> I know very well that as a person and as an intellectual I lag far behind the image that you have of me. But what does it matter? There must be people who serve others as the carriers of their illusions, and strangely enough I have become one of these. As long as one clearly recognizes the situation oneself, it does no harm.
>
> The attacks on several occasions do not oppress me, I even understand them. They are a natural reaction to the extensive adulation, which in a way helps to restore the natural balance.[41]

In mid-December, the Viennese pacifist, prolific writer on sociology and philosophy, and fellow New Fatherland League member Rudolf Goldscheid wrote thanking Einstein for a fiftieth birthday gift. He expressed his enthusiasm for the theory of relativity and its potential broader applications in fields beyond physics. He lamented the attacks of August and September on Einstein and admired his courage in embracing internationalism, notwithstanding his vulnerability as a Jew, moreover of high visibility due to his incontestable prominence in his profession.[42]

During this time of politicized challenges to his scientific integrity, Einstein's reputation as a supporter of Jewish causes remained alive, and he made it clear what kind of approach he saw as effective. In September 1920, the Association for Combating Antisemitism (Verein zur Abwehr des Antisemitismus) offered him a position on its executive board. Einstein asked his step-daughter and secretary Ilse Einstein to reply declining the offer; he favored Jews embracing a positive Jewish identity but did not believe that direct campaigns by Jews against antisemitism would be effective:

> Prof. Einstein instructed me to inform you that in his opinion we Jews cannot contribute to combating antisemitism through a direct campaign. Since your view on this point differs from that of Prof. Einstein, I respectfully request in his name that you kindly refrain from your plan—of electing Mr. Einstein onto the board of your association.[43]

Ten days after he rejected the approach of Jews combating antisemitism directly, Einstein expanded on his views in an article in a Swiss Jewish newspaper, *Israelitisches Wochenblatt*. Here, he expressed his support for the socially underprivileged East European Jews and showed a leaning toward cultural Zionism and the cultivation of pride in a positive Jewish identity:

> Today I received your invitation to a meeting . . . dedicated to the fight against antisemitism in academic circles. I would gladly attend if I believed in a possible success in such an enterprise. But first the antisemitism and subservient attitudes among us Jews ourselves must be fought through enlightenment. More dignity and independence in our

> own ranks! Only when we dare to look upon ourselves as a nation, only when we respect ourselves, can we gain the respect of others, or otherwise stated, will they develop this respect themselves. Antisemitism will exist as a psychological phenomenon as long as Jews come in contact with non-Jews—what harm is there in that? Perhaps it is thanks to antisemitism that we preserve ourselves as a race; at least I believe this.
>
> Whenever I happen to read "German citizens of the Jewish faith," I cannot suppress a pained smile. What is hidden in this cute expression; what, then, is Jewish faith? Is there a kind of non-faith that causes one to stop being a Jew? No. In this expression two confessions of refined souls are hidden, namely (1) I don't want to have anything to do with my poor Jewish (East European Jewish) brothers; and (2) I don't want to be regarded as a child of my people, but rather as only a member of a religious community. Is this sincere? Can the Aryan have respect for such pussyfooters? I am not a German citizen . . . but I am a Jew, and I am glad to belong to the Jewish people, although in no way do I consider them to be somehow a chosen people. Let's quietly leave the Aryan his antisemitism; and let's keep ourselves for the love of our own people.[44]

During and after World War I, as Einstein's pacifist world view was solidifying, his Zionist convictions were also clarified and strengthened. In June 1921, Einstein claimed a Zionist identity in the article "How I Became a Zionist" ("Wie ich Zionist wurde"), published in the prominent Jewish periodical *Jüdische Rundschau.* His Zionism tended to be a cultural Zionism and was largely motivated by the oppression of East European Jews. Einstein opened this article by comparing German antisemitism with its Russian and British variants. He contended that, while German Jews were better assimilated than Russian Jews and enjoyed greater legal equality, strong social antisemitism continued in Germany, especially in the educated class, which had created a "science" of antisemitism. By contrast, he maintained, Russian intellectuals opposed the antisemitic movement. In his view, the economic power of German Jews was overstated, although he did concede that Jewish influence in the press, in literature, and in science was significant and highly visible. He noted that non-Jewish Germans who saw Jews as a nationality distinct from Germans feared that this influence might threaten the German national character.

He suggested that although the Jews in England were a similarly tiny percentage of the population, they exerted less influence on English culture and society than in Germany, and as a result they sparked less antisemitism. This, he proposed, made it possible for Jews to attain the highest civil service positions in England, such as lord chief justice and viceroy of India—something that was unimaginable in Germany. Einstein went on to contrast German antisemites who were socialist, who would not express their antisemitism because it was not consistent with their party program, with antisemitic members of German conservative political parties, who exploited antisemitic sentiments in the population. In England, by contrast, he argued that the combination of less Jewish cultural influence and old liberal traditions hindered the rise of antisemitism. Clearly still disgusted by the politicization of his theory of relativity the previous year, Einstein maintained that, in Germany, a newspaper's political orientation determined its judgment of his theory of relativity, whereas English scientists judged it with objectivity. Einstein regarded Jews as having a "racial particularity" (*rassenhaften Besonderheit*) and advised them to display "dignified restraint" (*vornehme Zurückhaltung*) and show less eagerness to mix socially with non-Jews. Yet he also saw antisemitism as having some welcome consequences, arguing that with religious forms disappearing, only the antagonism of antisemitism was preventing rapid assimilation and hence the disappearance of German Jewish identity.

He concluded this lengthy article with a discussion of how the plight of East European Jews in Germany had fostered his Jewish identity, pointing to German calls for expulsion or internment (the Nazi policies two decades later did not develop in a vacuum). He also touched on the question of Zionism as a form of nationalism and how he reconciled himself with this, seeing Jewish national feeling as valuable to building Jewish self-confidence and vitality. (A few years later, in 1925, he would make a more concerted attempt to reconcile his pacifism with the Jewish nationalism inherent in Zionism, as discussed in the epilogue.) He also explained why he saw the establishment of a Jewish community in Palestine and the founding of a Hebrew University as great positives. He began his concluding thoughts in this article by continuing his point about antisemitism fostering Jewish identity:

I have observed this in myself. Until seven years ago I lived in Switzerland, and as long as I was there I was not conscious of my Jewishness and there was nothing in my life that would have stirred my Jewish feeling and stimulated it. This changed as soon as I moved my residence to Berlin. There I saw the distress of many young Jews. I saw how their antisemitic surroundings had made it impossible for them to pursue regular studies and to attain a secure existence. This is especially the case with East European Jews, who are constantly subjected to harassment. I do not believe that their numbers in Germany are large. Only in Berlin are there perhaps a greater number. Nevertheless, their presence has become an issue that increasingly occupies the German public. In meetings, conferences, and newspapers, people press for their swift expulsion or internment. The housing shortage and economic depression are used as arguments to justify these harsh demands. Facts are deliberately exaggerated in order to prejudice public opinion against East European Jewish immigrants. East European Jews are made the scapegoats for the malaise in present-day German economic life, which in truth is a painful after-effect of the War. Taking stances against these unfortunate refugees, who have escaped from the hell that East Europe is today, has become an effective political weapon that is successfully used by demagogues. . . . [Here Einstein described the public stance he took on behalf of East European Jews in his December 30, 1919, *Berliner Tageblatt* article "Die Zuwanderung aus dem Osten" ("Immigration from the East") and the action he took in holding university courses for them.]

These and similar experiences have awakened a Jewish national feeling in me. I am not a nationalist Jew in the sense that I call for the preservation of the Jewish or any other nationality as an end in itself. Rather, I see Jewish nationality as a fact . . . I consider the elevation of Jewish self-confidence to be essential, also in the interest of a natural coexistence with non-Jews. This was the main motive for my joining the Zionist movement. For me, Zionism is not merely a colonizing movement directed toward Palestine. The Jewish nation is a living fact in Palestine as well as in the Diaspora, and the Jewish national feeling must be kept alive everywhere that Jews live. The members of a tribe or a people must under today's living conditions have a lively consciousness of kinship in order not to become insecure and lose their sense

> of dignity. The unbroken vitality of the masses of American Jewry first made me clearly recognize how sickly German Jewry is.
>
> We live in an age of exaggerated nationalism and must as a small nation take account of this circumstance. But my Zionism does not preclude cosmopolitan views. I start out with the reality of Jewish nationality, and believe that every Jew has responsibilities toward his fellow Jews. The meaning of Zionism, by the way, is widely varied. It opens prospects for a dignified human existence to many Jews today who languish in the Ukrainian hell or degenerate economically in Poland. Through leading Jews back to Palestine and their return to a healthy, normal economic life, Zionism represents a productive activity that enriches all of human society. However, the main point is that Zionism strengthens the dignity and self-confidence that are necessary for Jewish existence in the Diaspora, and, through the Jewish center in Palestine, creates a strong bond that gives Jews a sense of support. I have always found repulsive the undignified addiction to conformity of many of my peers.
>
> Through the founding of a free Jewish community in Palestine, the Jewish people will come back into a position to bring their creative capabilities to unhindered and full fruition. Through the establishment of the Hebrew University and similar institutions, the Jewish people will not merely bring about their own national renaissance, but also have the opportunity to contribute to the spiritual-intellectual [*geistigen*] life of the world on a freer basis than ever before.[45]

As an advocate of cultural Zionism, in particular the establishment of the Hebrew University in Jerusalem, Einstein allowed himself to be used by proponents of political Zionism (whose aim was to establish a Jewish state in Palestine) when he traveled with Chaim Weizmann on a fundraising tour through the United States in April and May 1921. On June 27, 1921, he gave a speech in Berlin in which he laid out his views on establishing a Jewish Palestine. A handwritten draft written prior to the speech and a printed version published in a German Jewish newspaper after the speech, while overlapping in some respects, emphasized different points. In the handwritten draft, he emphasized his view of the Hebrew University as the most important goal of Zionism:

For two thousand years the common heritage of the Jewish people has consisted of only its past. Our people, who were dispersed throughout the world, shared nothing other than the carefully guarded tradition. Yes, a few individual Jews created great cultural works, but the Jewish people as a whole seemed no longer to have the strength for great collective achievements.

This has now changed. History has assigned us a great and noble task in the form of active participation in the construction of Palestine. Outstanding kinfolk are already working with all of their strength on the realization of this goal. We are offered the opportunity to establish cultural centers that the entire Jewish nation can regard as their own work. We foster the hope to create in Palestine a homeland of our own national culture [*Heimstätte eigener nationaler Kultur*], which should contribute to the awakening of the Near East to a new economic and intellectual life.

The goal that the leaders of Zionism have in mind is not a political one but rather social and cultural. The community in Palestine should draw closer to the social ideal of our ancestors, as it is laid out in the Bible, and simultaneously become a place of modern intellectual life. An intellectual center for Jews of the entire world. In accordance with this concept, the establishment of a Jewish university in Jerusalem constitutes one of the most important goals of the Zionist Organization. In recent months I was in America in order to help create the material basis for this university. The success of this effort was superb. Thanks to the tireless activity of Jewish physicians of America and their outstanding willingness to make sacrifices, we succeeded in collecting sufficient means for the creation of a medical faculty, and the preparatory work toward its realization is starting immediately. Based on the successes thus far, I harbor no doubts that the material basis for the remaining faculties will be created in a short time. . . . I close with a warm appeal to the Jews of Germany, despite the current difficult economic situation, to contribute to the best of their abilities to the building of a Jewish homeland in Palestine.[46]

In the published version of this article, Einstein explored theories about how nationalities function, what was happening to the Jewish nation, what had changed with emancipation, and what should be done now. In making his case for Zionism, he expressed the typical Zionist antipathy toward Jewish assimilation to majority cultures in the Diaspora, as well as a somewhat romanticized view of Jewish communities of the past:

> The rebuilding of Palestine is for us Jews no mere charity or colonial concern, but rather a problem of central importance for the Jewish people. Palestine is not primarily a refuge for East European Jews, but rather the embodiment of the reawakening national feeling of community of all Jews [*die Verkörperung des wiedererwachenden nationalen Gemeinschaftsgefühls aller Juden*]. Is it now timely and necessary *to revive and strengthen this community feeling*? I believe I must answer this question with an unconditional Yes, based not only on my spontaneous feelings but also on rational grounds.
>
> Let us cast a brief glance at the development of German Jews during the last hundred years. A century ago, our ancestors, with few exceptions, still lived in the ghetto. They were poor and politically disenfranchised, separated from non-Jews by a wall of religious traditions, daily lifestyle, and legal restrictions, and in their intellectual development limited to their own literature, only slightly influenced by the tremendous revival that European intellectual life experienced during the Renaissance. However, these little noticed, modestly living people had one advantage over us: *every one of them belonged to a community with every fiber of his being*, around which his whole life revolved, in which he felt himself an equal, full-fledged member, which demanded nothing of him that conflicted with his natural way of thinking. Our ancestors in those days were physically and intellectually rather atrophied, but in social respects enjoying an enviable spiritual equilibrium. Then came the emancipation. It offered undreamed-of possibilities for development. Individuals rapidly established contact with the higher economic and social levels of society. They absorbed eagerly the magnificent achievements that the arts and sciences of the West have created. They threw themselves with ardent passion into this development, creating lasting contributions of their own. In the process they took on the forms of life of the non-Jewish world, increasingly turned away from their religious and social tradition, and adopted non-Jewish customs, manners, and ways of thinking. It seemed as if they were being completely absorbed into the numerically far greater, politically and culturally better-organized host nations,

> so that after a few generations no visible trace of them would remain. A *complete dissolution of the Jewish nation* in Central and Western Europe appeared inevitable. However, it turned out differently. There seem to be instincts in racially distinct nations that counteract their assimilation. The adaptation of Jews to the European peoples among whom they lived, in language, customs, and even in part in religious forms *could not extinguish the feeling of alienation* that existed between Jews and their European host nations. In short, this spontaneous feeling of alienation is in the final analysis the source of this oppression. This also, therefore, *cannot be eradicated from the world by well-meaning tracts.* Nationalities do not want to mix, but rather to go their own way. A peaceful situation can only be brought about if they *show mutual tolerance and respect.*
>
> To this end it is most important that we Jews *again become conscious of our existence as a nationality* and that we *regain the self-respect* that we urgently need for this existence. We must again learn to dedicate ourselves to our ancestors and to our history, and we must *as a people again accept cultural duties* that serve to strengthen our feeling of community. It is not enough that we as individuals participate in the cultural development of humankind, we must also tackle those tasks that *only national unity can promote.* This is the only way that Jewry can again become socially healthy.
>
> I ask you to view the Zionist movement from this perspective. *History has today assigned us the task* of actively participating in the economic and cultural reconstruction of Palestine. Enthusiastic and highly talented men have laid the groundwork. Many excellent kinfolk are ready to devote themselves completely. May every one of you fully appreciate the importance of this work and contribute to its success to the best of your abilities![47]

Einstein concluded this article with a paragraph about his trip to America, the Jewish community there hailing from many countries, especially eastern Europe. He extolled that community as retaining a healthy national feeling, and he envisioned a future Jewish homeland in Palestine as offering this same sense of community.

For an article in a Swiss Jewish publication in August 1921, Einstein was interviewed regarding his thoughts and feelings about the efforts underway to found a Hebrew University in Palestine. His statement was a passionate yet reasoned plea on behalf of the Hebrew University as benefitting the entire Jewish people in a variety of ways:

> Two reasons exist for the necessity of establishing a Hebrew University in Palestine. First, a Jewish homeland in Palestine is not at all conceivable without a university. For the rehabilitation of the country, *bacteriological* and other branches of research are necessary. For the training as well as the intellectual and material support of physicians, especially the training of doctors *becoming familiar with the country,* and further in reference to all things concerning cultivation of the soil, industries for the chemical analysis of the soil and plant cultivation are necessary; all of this pertains to the *natural sciences* faculty. In order to give these institutions an independent character, one also needs, of course, pure *physics* and *chemistry,* and with that we already have an almost complete *philosophical* faculty, as far as the natural sciences are concerned. No less important, of course, are institutes for the humanities, especially dedicated to *history, Jewish culture,* and the *Near East* in general, particularly the *Hebrew language,* in order that the teachings of the country and overall the people settled there have an *intellectual center.* A large-scale Jewish colonization without such an intellectual center is hardly imaginable, and it is self-evident that these special goals have as a prerequisite the cultivation of philosophy, archaeology, and the full range of philosophies in general, if they are pursued in a scientific manner. All I have said thus far briefly summarizes the needs pertaining to the construction of the country and the modern development of a scientific terminology for the Hebrew language.
>
> The Hebrew University in Palestine shall have a *national* character insofar as the *language of instruction* will generally be *Hebrew*; difficulties of language during the first years will be overcome by the fact that the university initially will be mainly a research institute without teaching activity, especially in the fields of natural sciences.
>
> The second major task of the Hebrew University in Palestine must be to offer an opportunity for study to the Jewish youth of eastern Europe, because many talented young Jewish people today are generally locked out of university study. I am convinced that a very large number of East European Jews who are ready for university have tried in vain to

gain admission into central European universities, and in my opinion it is an *obligation of honor* of the *whole* to lend help in this. We even hope to bring the Hebrew University in Palestine to such a level that Jewish students from the Diaspora will later seek admission to the university, and without just being compelled by need.

Finally, we think it highly significant that Jewish people as a *whole*, and through their *own institutions* and by their *own strength*, take part in international scientific life, since the abundant and successful participation of individual Jews in scientific life and the scientific achievements of so many of their sons in the Diaspora have demonstrated their love and aptitude for the cultivation of knowledge.

I believe that, through the creation of a thriving Hebrew University of international reputation, we can at the same time also contribute to counteracting the shameful inclination of many successful intellectuals among the Jews to deny and fearfully hide their Jewish ancestral affiliation. In my opinion this phenomenon is by no means always due to a lack of character, but rather is because individuals succumb to the influence of their non-Jewish milieu, especially when antisemitic feelings are predominant.

Not only in the current founding period of the Hebrew University in Palestine, but also all my life, I have considered it a sacred duty to contribute, to the best of my ability, to the success of the Hebrew University in Palestine, and I know that many other Jewish research scientists also feel the same as I do in this matter.[48]

Einstein's advocacy on behalf of his fellow Jews and his statements reflecting comfort and pride in his Jewish identity increased in frequency and substance in the immediate postwar years, especially during 1921, the year he joined Chaim Weizmann on a Zionist speaking tour of the United States. His comfort with his identity as a member of an oppressed cultural group was clearly demonstrated in the confidence with which he applied the typical pacifist advocacy for social justice to the unique situation of his own people. As the epilogue to this book shows, Einstein's savvy, nuanced, and confident advocacy for peace and social justice continued throughout the rest of his life.

Epilogue

Pacifist Stances and Professed Identity through 1955

The preceding chapters have shown the awakening of Einstein's incipient pacifism in response to the events of the First World War and interactions with fellow pacifists and other people in his world. Einstein's identity as a pacifist and actions propelled by that identity continued throughout his life.[1] Pacifist sentiments and goals that he acquired during World War I—ranging from disgust at the absurd horror of war, to distrust for political leaders in their warmaking, to a passion for international collaboration as essential to the advancement of science, to a belief in international organization (with enforcement powers) as the means to prevent war—became part of a world view that lasted for the remaining three and a half decades of his life.

At first glance, Einstein's lifelong identity as a pacifist might seem questionable because of a variation of his stances across time: He advocated pacifism consistently from 1914 to 1933, then from 1933 condoned and from 1939 encouraged military opposition to the Nazis, and finally from 1945 until his death in 1955 reverted again to inspiring, high-profile pacifist stances. Absolute pacifism in the context of the Holocaust is an extremely difficult position for Jews to embrace. Einstein was after all only human, and, despite not being religious, he held a strong Jewish identity throughout his life.[2] So, his Nazi-era hiatus is a very understandable variation in his lifelong stances on war and peace. Another challenge to Einstein's identity as a pacifist in the popular mind is based on Einstein's contributions in the field of physics at the beginning of the twentieth century, which are seen as having made nuclear weapons possible, as well as

Einstein's 1939 letter to US president Franklin D. Roosevelt urging him to develop such weapons before the Nazis would.[3] Because of these contributions, some skeptics question whether he really was a pacifist, or do not acknowledge his pacifist identity in terms of the totality of his lifelong commitment to peace advocacy.[4] Yet significant scholarly opinion on Einstein's life and politics supports Einstein's self-characterization as a pacifist.[5]

In support of Einstein's identity as a pacifist are the facts that, across the last half of his life and the last two-thirds of his adult life (starting in 1914 at age thirty-five), Einstein consistently advocated international organization with an executive enforcement power as the means to maintain international peace, he publicly took many other pacifist stands (often more prominently or persistently than other pacifists), he repeatedly (except from 1933 to 1945) advocated refusal of military service, and he continuously described himself explicitly as a pacifist, even during 1933–45. Einstein defined pacifism in his own terms, not according to the standards of others, and this self-defined pacifism included the flexibility to designate the Nazis as a special case that had to be opposed through the use of military violence, in his view. In offering himself as an example of a pacifist with flexibility, yet who unceasingly regarded war as evil and supranational organization as an essential step to its eradication, Einstein provided a more nuanced and compelling definition of pacifism, and one more palatable to Jews. Einstein's pacifism began early in his life; it was a deep-seated conviction that he expressed repeatedly across the years; it involved a thorough exploration of the implications of proposals (such as an international court of arbitration needing executive powers of enforcement); it flowed from his own responses to events around him; and it contained some original and impressively forward-thinking elements—particularly vis-à-vis Arab-Jewish conflicts in Palestine. Einstein's pacifism, as is the case for most pacifists, also involved a strong social justice component, which included his own sense of personal connection with the Jewish people and support for their need to develop a positive Jewish identity.

So, Einstein's activism, which began in 1914 and increased with his growing conviction, insight, experience, and public stature, continued across the following forty years. In late 1921, Einstein wrote a statement for a German book on the peace movement, in which he argued that developments in science required both international collaboration and the economic stability of peacetime, and that, in the universal character of its subject matter, science encouraged pacifist goals.[6] In 1922, Einstein and other intellectuals from various countries, including fellow physicist Marie Curie, accepted an invitation to join the League of Nations' Committee on Intellectual Cooperation. The committee was dedicated to restoring international collaboration among intellectuals, a concrete enactment of Einstein's faith in international organization as the solution to conflict between nations.[7]

In the late 1920s and early 1930s, Einstein frequently took public stances on behalf of conscientious objectors' resistance to military service. In 1928, when the League of Nations codified rules of warfare and banned use of poison gas, Einstein responded, in a statement for the Women's International League for Peace and Freedom (WILPF), that it was pointless to impose rules and restrictions on war, that war was not a game that could be conducted according to rules, and that the only effective way to oppose war was through "refusal of military service" (*Kriegsdienstverweigerung*) by the masses.[8] Over the years, Einstein repeated this advocacy of the refusal of military service several times. That same year, he wrote:

> No person has the right to call himself a Christian or a Jew so long as he is prepared, at the command of an authority, to engage in systematic murder or to allow himself to be misused in any way whatever in service of such an enterprise or the preparation for it.[9]

In January 1928, Einstein was elected to the board of directors of the German League for Human Rights, the former New Fatherland League.[10] In August 1928, Einstein signed a manifesto opposing military conscription.[11] In November 1928, Einstein sent a message to the No More War Movement in London, the British section of the War Resisters International, supporting refusal of military service.[12] The following month, Einstein wrote:

> Peaceful settlement of international conflicts will be possible only when all armies and all forms of compulsory military service are abolished. As a beginning, it would be most helpful if prominent citizens were to advocate the principle of refusing military service.[13]

In April 1929, he wrote another letter for the publication of the No More War Movement, arguing that the people must take the initiative to prevent themselves from being led into slaughter, and that to depend on their governments to protect them was foolishness.[14] In December 1930, speaking publicly in New York while in the United States for the first time in almost ten years, Einstein deplored the current system in which people could be compelled to commit murder in the name of their country. He argued that the one effective response was refusal of military service and that if 2 percent of all those called up refused to serve, governments would be powerless because they would not dare to jail such a large number of people. This "2 Percent" speech became a slogan among pacifists for several years.[15] In March 1931, speaking in Chicago, he argued that the struggle against armaments and the spirit of militarism could not be safely left in the hands of governments; instead, the only way to be effective was through refusing military service. He also made similar comments to a delegation of the War Resisters' League in New York.[16] In August 1931, Einstein sent a message to Belgians commemorating their dead from World War I, stating that the people must declare themselves in favor of total disarmament.[17]

Alongside refusal of military service, the other main thrust of Einstein's pacifism was the maintenance of peace by supranational governmental structures, which was an outgrowth of some of the thought that he developed during World War I. In September 1931, he wrote to a Dutch journal that disarmament could be successful only when every country was assured that the mandates of an international court of arbitration would actually be carried out.[18] In a mid-1932 exchange of public letters with Sigmund Freud on the topic of war and peace (published the following year), Einstein's comments echoed an essay that he had written in 1915 for the Berlin Goethe League (discussed in chapter 1). In the 1915 essay, Einstein contended that a supranational state-like organization was necessary to prevent wars between nations.[19] In the 1932 exchange with Freud, Einstein argued that the first step to eliminate "the menace of war" was "the setting up, by international consent, of a legislative and judicial body to settle every conflict arising between nations."[20]

Like many pacifists, Einstein linked his advocacy for peace with a concern for social justice, and he included the oppression of the group of which he was a member by opposing antisemitism and advocating Zionism. His support for Zionism or Jewish nationalism might seem inconsistent with his identity as a pacifist because nationalism has almost always been conceptually and concretely linked with militarism. Einstein repeatedly struggled to reconcile his compassion and identification with the motivations of the Zionist movement with the inconsistencies between Jewish nationalism and pacifism. He claimed that, among nationalisms, Jewish nationalism was uniquely ethical and internationalist in its nature and mission. In June 1921, as discussed in chapter 6, the main Zionist newspaper in Germany, the *Jüdische Rundschau*, published a statement by Einstein in which he identified himself as a Zionist and declared that he had become one because of the deeply embedded political antisemitism in Germany. While he did not believe in the preservation of any nationality as an end in itself, Jewish nationality required group loyalty, without losing a cosmopolitan awareness, for the sake of Jewish dignity and a healthy existence both in Palestine and in the Diaspora.[21] In July 1921, speaking to the German Zionist organization, Einstein argued that when Jews supported Zionism they simultaneously promoted world peace because the Zionist renewal of the Jewish world also restored Judaism's ancient peace mission.[22] In 1925, Einstein wrote that nationalism was a danger to peace and a source of injustice, and some Jews such as himself would prefer to belong to humankind as a whole, but Jews had always been treated by others as members of a distinct national group. However, Jews could turn this toward good through Zionism, which freed Jews' souls from an "inexcusable feeling of exaggerated modesty" (*unentschuldbaren Gefühl übertriebener Bescheidenheit*) that could only oppress them and make them unproductive. Zionism would give them back their moral strength and justified pride and, through solidarity with one another, would enable them to fulfill the tasks lying before them. Therefore, he saw no contradiction between Zionism, despite its appearance as a nationalist movement, and the pursuit of humanitarian ideals.[23] In 1929, Einstein joined the small, Jewish-identified peace organization then starting in Berlin, the Jewish Peace League (Jüdischer Friedensbund), allowing his famous name to be used on their leaflets and brochures.[24] In July 1931, Einstein wrote to Berlin chief rabbi Leo Baeck, who was similarly involved with the Jewish Peace League, "Nothing lies closer to my heart than the attempt to mobilize the clergy for the pacifist goal."[25]

Einstein advocated directly for understanding and peaceful relations between Jews and Arabs in Palestine, especially during the year following the violence between Arabs and Jews in Palestine in August 1929, in which hundreds were killed and wounded on both sides.[26] In Einstein's message to a Jewish Agency rally in Berlin, he argued that Jews had to find a way of living together with the Arab people, cooperating, and avoiding friction. Lack of contact inevitably led to fear and mistrust, but the coexistence of the two peoples in Palestine was possible; they did not have irreconcilable differences. He cautioned Jews against blinding chauvinism, reminding them that their national task was at its core an international one, that the strength of their movement depended on its ethical value.[27] In February 1930, in a letter to a Jewish group in Hamburg called the Pro-Palestine Committee, Einstein argued that the Jewish people's most difficult but most important task was to bring their national aspirations into harmony with the rights and needs of the population of the land, to find a path of understanding with their Arab neighbors that would replace enmity and rivalry with a creative cooperation in the interests of both peoples.[28]

Moving beyond statements to the Jewish community, after death sentences were pronounced in late 1929 against anti-Jewish Arab rioters in Palestine, and consistent with typical pacifist opposition to the death penalty, Einstein urged the British High Commissioner in Palestine to commute the sentences.[29] (He had done similarly in the case of Friedrich Adler in 1917, discussed in chapter 2; and in the case of József Kelen in 1920, discussed in chapter 5.) Moreover, Einstein communicated directly with the Arab world in a letter in January 1930 to the editors of the English-language edition of the Arab Executive's newspaper, *Falastin*. In this letter, he pleaded that Jews wished to have friendly relations with the Arab population of Palestine, that future humanity must be built upon a community of nations, and that aggressive nationalism must be overcome. He envisioned a future for Palestine as possible only through peaceful cooperation between the two resident peoples, who should offer sympathetic cooperation and support for one another's national and cultural strivings. He expressed regret at the events of the previous August, not only because they showed human nature at its worst, but also because it had estranged the two peoples and hindered their access to one another.[30] Always ready to suggest concrete, pragmatic steps, in March 1930 Einstein wrote to an Arab newspaper editor in Jaffa, Palestine, proposing the establishment of a committee of eight responsible individuals, four Jews and four Arabs, who would meet weekly to discuss issues and reach decisions with the best interests of the entire population of the land in mind.[31] In the fall of 1930, Einstein expressed his support for the Jewish peace group in Palestine, Brith Schalom, recommending greater contact with Arabs and the study of their language, to tackle the challenge of making contact with the Arab masses and awakening their trust. He argued that the Palestine project would only be successful when the Arab question was resolved with genuine understanding between the two peoples.[32] In these ways, Einstein reached outside of his own community and moved beyond the abstract to the concrete in applying his pacifist convictions to the Arab-Jewish conflict in Palestine. He believed that only the humanitarian idealism that inspired pacifism could bring Zionism to a genuine expression of its true nature.

Thus, Einstein had established an impressive articulated and pragmatic pacifist background across many years prior to the start of the Nazi era in 1933. In March 1933, two months after the Nazis took power in Germany, in a volume of his pacifist speeches and writings published in New York, Einstein condemned the Nazi regime but called on the world to practice "moral intervention" against Nazism.[33] Even in early May 1933, he stated that he saw greater heroism in taking the consequences of refusal of military service than in soldiers participating in a war.[34] However, by July 1933, after the first six months of the Nazi regime, Einstein wrote from Belgium to a Dutch pacifist that in response to recent events he had "changed" his ideas (*umgelernt*). In the context of Germany's rearmament efforts, he would not encourage French or Belgians to refuse military service. "In the current situation" (*beim heutigen Stande*), he would still advocate a "supranational organizational power" (*übernationale Organisation der Gewalt*), but saw the advancement of radical pacifist efforts as disadvantageous at that time.[35] In mid-July, Einstein informed King Albert of Belgium, with whom he was friends, that he accepted the current Belgian army as a means of defense, not aggression, yet still recommended that conscientious objectors not be judged as criminals but instead be offered heavy or dangerous labor, such as in mines, hospitals, and mental institutions, as alternative service.[36] Later in July, Einstein declared that France and Belgium were in such grave danger that, if he were a Belgian, he would not refuse military service but would serve in order to save European civilization. However, he expressly stated, he was not surrendering the (pacifist) principle for which he had stood previously, and he hoped for a time in the future when refusal of military service would again be an effective means of serving human progress.[37]

Many pacifists were initially shocked by this statement. In August 1933, H. Runham Brown, secretary of the War Resisters International, referred to Einstein's statement as "a great blow to our cause."[38] In September 1933, the International Antimilitaristic Commission declared "Einstein's action has done unutterable harm to the fight against militarism."[39] Romain Rolland also privately condemned Einstein as vacillating.[40] But Einstein maintained that he had not betrayed his convictions. Due to the threat of Nazi Germany, with its ideology of force and repression, and due to the nonexistence of an international court of arbitration and an international police force, he argued, refusal of military service was not currently justified, and the lesser evil of the hated military must temporarily be used in the present circumstances.[41] In the spring of 1934, a pacifist rabbi in Rochester, New York, asked Einstein whether he still believed that war was never justifiable or whether he had changed his views to believe that military struggle against Hitler's Germany was justifiable.[42] Einstein responded, "I am the same ardent pacifist today as previously" (*Ich bin derselbe glühende Pazifist heute wie früher*), but he declared that refusal of military service could not be recommended until democratic lands were no longer threatened by dictatorships with military tendencies, and that a military weakening of Germany's neighbors would currently be catastrophic to European civilization. The goal of pacifists remained the same, he explained, but their methods needed to be adapted to the circumstances.[43]

A major feature of this adaptation, for Einstein, was that he retained the goal of supranational organization, acquired with the development of his pacifism during World War I, whereas he relinquished the strategy of refusal of military service for the duration of the Nazi regime and never again vigorously re-embraced it. In response to a similar question in late 1934, Einstein provided the following rationale: A real peace could only be achieved by the establishment of a supranational court of arbitration that would have a permanent military or police force to enforce its decisions, and he favored any measure that would bring humankind closer to this goal. Until recently, refusal of military service was such a measure, which, in democratic countries, could influence governments and educate public opinion, but at present it could no longer be recommended, because in some countries it simply led to martyrdom and death, while in the countries that still respected the political rights of their citizens it weakened the ability of the healthy portion of the civilized world to resist aggression. In sum: "Different times, different methods, although the final goal remains the same." So, he concluded, a convinced pacifist must now work for closer cooperation among the peaceful nations.[44]

During the Nazi regime and through the rest of his life, Einstein increasingly focused on world government—an international organization encompassing all states with sufficiently strong executive and enforcement power—as the structure and strategy for ending war. After his October 1933 move to the United States, Einstein tried to convince Americans to give up their traditional isolationism, arguing that international solidarity was the best defense against fascism.[45] Einstein also did extensive advocacy on behalf of Jewish refugees from Germany, directly sponsored and gave financial help to individual Jews, and took public stands in opposition to the Nazis' antisemitic policies.[46] In August 1939, Einstein wrote his famous letter to President Franklin D. Roosevelt, expressing his concerns as a scientist about Nazi Germany potentially developing an atomic bomb before the United States and urging that resources be allocated toward the

United States reaching that goal first.[47] However, Einstein did not work on the Manhattan Project because of the US government's failure to understand the nature of a moderate European social democrat and its resulting worries about his loyalty to the United States due to what it deemed his "radical background."[48]

Three years after World War II ended, Einstein explained to Roy Kepler of the New York office of the War Resisters League, who had been a conscientious objector during the war, that

> during the Hitler-Mussolini time I recognized that refusal of military service was possible only in the non-totalitarian countries and would contribute to the destruction of freedom and liberty in the whole world. For this reason I changed my standpoint temporarily. . . . [In comparison to] individual resistance to military service . . . I am convinced that a vigorous attempt to bring about security through world government is today by far the more promising procedure.[49]

In 1952, he wrote to a pacifist who questioned apparent inconsistencies in his pacifism:

> I am a pacifist but not at any price. I can identify my views nearly completely with Gandhi. I would . . . resist with violence an attempt to kill me or to take away from me the basic means of subsistence. I was, therefore, of the conviction that it was justified and necessary to fight Hitler. For this was such an extreme attempt to destroy other people. I am of the conviction that a realization of the goal of pacifism is only possible through supranational organization. To stand unconditionally for this cause is, in my opinion, the criterion of true pacifism.[50]

In the ten years after World War II, also the last decade of his life, Einstein responded directly and vigorously to the new atomic era with strong stances against war and nuclear weapons. In September 1945, one month after the United States dropped atomic bombs on Hiroshima and Nagasaki, Japan, in his first public comment after World War II ended, Einstein declared:

> The only salvation for civilization . . . lies in the creation of world government, with security of nations founded upon law. . . . As long as sovereign states continue to have separate armaments and armaments secrets, new world wars will be inevitable.[51]

In December 1945, at a Nobel memorial event in New York, Einstein declared that scientists who were linked with the creation of the atomic bomb had constructed it so that the Nazis would not build it first and bring unlimited destruction and enslavement to the peoples of the world. But now the holders of the bomb had a responsibility to the world. There was no guarantee of peace, and the new situation called for bold action and a radical change of mentality, or else civilization would be doomed.[52] In 1946, Einstein and some physicists from the Manhattan Project who felt concerned about the dangers posed by the bomb formed the Emergency Committee of Atomic Scientists, and across the next few years Einstein became a fundraiser for the group, signed appeals and proclamations, spoke on the radio, and participated in broadcast discussions.[53]

Einstein was horrified by the use of the atomic bomb on the Japanese cities Hiroshima and Nagasaki on August 6 and 9, 1945. In March 1947, Einstein stated publicly that, if he had known that the Germans would not develop the atomic bomb, he would not have supported its construction. He also emphasized that development of the bomb was not greatly speeded by his intervention with President Roosevelt.[54] In September 1952 Einstein maintained in a statement to Katusu Hara, editor of the Japanese journal *Kaizo*, and in follow-up correspondence with Seiei Shinohara, the Japanese pacifist who had served as translator of this statement, that his sole participation in the development of the bomb was the letter to Roosevelt, and that he opposed its use against Japan but could not have prevented it.[55]

In this first decade of the Cold War and nuclear arms race between the United States and the Soviet Union, although he objected to the Soviet Union's violations of civil liberties, Einstein saw the anti-Soviet attitudes of the United States as a greater danger and advocated reconciliation with Soviet Russia as openly and directly as he had addressed Jewish-Arab relations in Palestine. In January 1948, he wrote to the English pacifist Member of Parliament Henry Usborne that

> abolishing war . . . can only be done by establishing world government. . . . we should energetically take our stand for world government, even at the risk of Soviet Russia being driven—temporarily—to self-isolation. However, everything should be done in a way to make it easy and attractive for the Russians to change their isolationist attitude.[56]

In March 1948, in response to criticism by Soviet scientists, Einstein responded that a world government was the only machinery that could prevent war, and even a poor world government was better than none because the first goal must be to avoid total destruction through war.[57] That same month, Einstein stated before the US Senate Committee on Armed Services his conviction that, without the immediate threat of hostile invasion in the near future, universal military training would be harmful to the welfare and security of the US and the rest of the world for the following reasons: It would accelerate the arms race and increase tension between the United States and the Soviet Union; it would undermine the democratic spirit; it would strengthen the influence of the military; it would impose an economic burden by shifting resources into unproductive channels; and it would generate a militaristic spirit.[58] In February 1950, in response to President Truman's announcement of the decision to develop the hydrogen bomb, Einstein addressed the nation on television, arguing that with this development "annihilation of all life on earth" would be made possible.[59] In March 1955, in the final months of his life, Einstein wrote to his long-term friend and fellow physicist Max von Laue, "If I had known that these fears [of Hitler developing a weapon of mass destruction] were groundless, I would not have taken part in opening that Pandora's box."[60] In February 1955, the English pacifist Bertrand Russell invited Einstein to sign an appeal by scientists stating that nuclear war would bring no winners or losers, only total catastrophe. Einstein signed this appeal on April 11, one week before he died on April 18, 1955.[61]

Thus, the pacifism that Einstein developed in response to World War I, particularly its concept of a world organization for the preservation of peace, continued to guide his private convictions and public stances for the rest of his life—throughout the extreme vicissitudes of the Nazi regime, World War II, and the use and development of atomic and nuclear weapons. Among people who regard war as evil yet view the diabolical threat posed by the Nazis as having required a military response, there is a clear distinction between those who are resigned to seeing war as unfortunate but inevitable and those who engage in serious, persistent efforts to eradicate war and establish world peace. Einstein did not permanently abandon his pacifism in response to Nazi aggression and the Holocaust, and he never relinquished his willingness to stand publicly in favor of a supranational organization with executive powers of enforcement. As he wrote to a Seiei Shinohara in 1953, "I am a determined but not absolute pacifist."[62] This flexibility may be related to the concrete pragmatism that he repeatedly demonstrated—in the post–World War I commission to investigate German war crimes, in his proposals to foster Arab-Jewish cooperation in Palestine, and in the Emergency Committee of Atomic Scientists. The rationale behind his advocacy of military violence in the one case of opposing the Nazis was carefully defined and circumscribed, and he backed away from it immediately after the defeat of Nazi Germany, when it ceased to be relevant. He then returned to his long-held principles, augmented by a horror of nuclear weapons. Not only did he state that he would not have encouraged the development of nuclear weapons if not for the Nazi threat, but he joined formally with other atomic scientists as they expressed publicly their ethical concerns about the extreme dangers that had arisen from the products of their work. Echoing the way in which he was a creative thinker in the scientific realm, Einstein was also rational, flexible, independent in spirit, and forward-thinking in his pacifism. Through Einstein's signing of the October 1914 "Appeal to the Europeans," through his efforts to meet Romain Rolland, through his participation in the Lille commission, and through many other public and private actions and statements during and following World War I, the same innovative, courageous, and pragmatic qualities that informed his pacifism through the end of his life were already evident as his pacifist consciousness was first awakening in response to World War I.

Appendixes

Notes

Bibliography

Index

APPENDIX A

German-Language Text of the Manifesto "To the Civilized World" ("An die Kulturwelt!")

Wir als Vertreter deutscher Wissenschaft und Kunst erheben vor der gesamten Kulturwelt Protest gegen die Lügen und Verleumdungen, mit denen unsere Feinde Deutschlands reine sache in dem ihm aufgeszwungenen schweren Daseinskampfe zu beschmutzen trachten. Der eherne Mund der Ereignisse hat die Ausstreuung erdichteter deutscher Niederlagen widerlegt. Um so eifriger arbeitet man jetzt mit Entstellungen und Verdächtigungen. Gegen sie erheben wir laut unsere Stimme. Sie soll die Verkünderin der Wahrheit sein.

Es ist nicht wahr, daß Deutschland diesen Krieg verschuldet hat. Weder das Volk hat ihn gewollt noch die Regierung noch der Kaiser. Von deutscher Seite ist das Äußerste geschehen, ihn abzuwenden. Dafür liegen der Welt die urkundlichen Beweise vor. Oft genug hat Wilhelm II. in den 26 Jahren seiner Regierung sich als Schirmherr des Weltfriedens erwiesen; oft genug haben selbt unsere Gegner dies anerkannt. Ja, dieser nämliche Kaiser, den sie jetzt einen Attila zu nennen wagen, ist jahrzehntelang wegen seiner unerschütterlichen Friedensliebe von ihnen verspottet worden. Erst als eine schon lange an den Grenzen lauernde Übermacht von drei Seiten über unser Volk herfiel, hat es sich erhoben wie ein Mann.

"An die Kulturwelt!" declaration signed by ninety-three German intellectual and cultural figures, October 4, 1914. In Georg Friedrich Nicolai, *Die Biologie des Krieges: Betrachtungen eines deutschen Naturforschers* (Zürich: Verlag Orell Füssli, 1917), 7–10.

Es ist nicht wahr, daß wir freventlich die Neutralität Belgiens verletzt haben. Nachweislich waren Frankreich und England zu ihrer Verletzung entschlossen. Nachweislich war Belgien damit einverstanden. Selbstvernichtung wäre es gewesen, ihnen nicht zuvorzukommen.

Es ist nicht wahr, daß eines einzigen belgischen Bürgers Leben und Eigentum von unseren Soldaten angetastet worden ist, ohne daß die bitterste Notwehr es gebot. Denn wieder und immer wieder, allen Mahnungen zum Trotz, hat die Bevölkerung sie aus dem Hinterhalt beschossen, Verwundete verstümmelt, Ärzte bei der Ausübung ihres Samariterwerkes ermordet. Man kann nicht niederträchtiger fälschen, als wenn man die Verbrechen dieser Meuchelmörder verschweigt, um die gerechte Strafe, die sie erlitten haben, den Deutschen zum Verbrechen zu machen.

Es ist nicht wahr, daß unsere Truppen brutal gegen Löwen gewütet haben. An einer rasenden Einwohnerschaft, die sie im Quartier heimtückisch überfiel, haben sie durch Beschießung eines Teils der Stadt schweren Herzens Vergeltung üben müssen. Der größte Teil von Löwen ist erhalten geblieben. Das berühmte Rathaus steht gänzlich unversehrt. Mit Selbstaufopferung haben unsere Soldaten es vor den Flammen bewahrt.—Sollten in diesem furchtbaren Kriege Kunstwerke zerstört worden sein oder noch zerstört werden, so würde jeder Deutsche es beklagen. Aber so wenig wir uns in der Liebe zur Kunst von irgend jemand übertreffen lassen, so entschieden lehnen wir es ab, die Erhaltung eines Kunstwerks mit einer deutschen Niederlage zu erkaufen.

Es ist nicht wahr, daß unsere Kriegführung die Gesetze des Völkerrechts mißachtet. Sie kennt keine zuchtlose Grausamkeit. Im Osten aber tränkt das Blut der von russischen Horden hingeschlachteten Frauen und Kinder die Erde, und im Westen zerreißen Dumdumgeschosse unseren Kriegern die Brust. Sich als Verteidiger europäischer Zivilisation zu gebärden, haben die am wenigsten das Recht die sich mit Russen und Serben verbünden und der Welt das schmachvolle Schauspiel bieten, Mongolen und Neger auf die weiße Rasse zu hetzen.

Es ist nicht wahr, daß der Kampf gegen unseren sogenannten Militarismus kein Kampf gegen unsere Kultur ist, wie unsere Feinde heuchlerisch vorgeben. Ohne den deutschen Militarismus wäre die deutsche Kultur längst vom Erdboden getilgt. Zu ihrem Schutze ist er aus ihr hervorgegangen in einem Lande, das jahrhundertelang von Raubzügen heimgesucht wurde wie kein zweites. Deutsches Herr und deutsches Volk sind eins. Dieses Bewußtsein verbrüdert heute 70 Millionen Deutsche ohne Unterschied der Bildung, des Standes und der Partei.

Wir können die vergifteten Waffen der Lüge unseren Feinden nicht entwinden. Wir können nur in alle Welt hinausrufen, daß sie falsches Zeugnis ablegen wider uns. Euch, die Ihr uns kennt, die Ihr bisher gemeinsam mit uns den höchsten Besitz der Menschheit gehütet habt, Euch rufen wir zu:

Glaubt uns! Glaubt, daß wir diesen Kampf zu Ende kämpfen werden als ein Kulturvolk, dem das Vermächtnis eines Goethe, eines Beethoven, eines Kant ebenso heilig ist wie sein Herd und seine Scholle.

Dafür stehen wir Euch ein mit unserem Namen und mit unserer Ehre!

Adolf von Baeyer, Exz.,
Professor der Chemie,
München.

Prof. Peter Behrens,
Berlin.

Emil von Behring, Exz.,
Professor der Medizin,
Marburg.

Wilhelm von Bode, Exz.,
Generaldirektor der Kgl.
Museen, Berlin.

Alois Brandl
Professor, Vorsitsender der
Shakespeare-Gesellschaft,
Berlin.

Lujo Brentano,
Professor der
Nationalökonomie,
München.

Prof. Justus Brinkmann,
Museumsdirektor,
Hamburg.

Johannes Conrad,
Professor der
Nationalökonomie, Halle.

Franz von Defregger,
München.

Richard Dehmel,
Hamburg.

Adolf Deißmann,
Professor der protest.
Theologie, Berlin.

Prof. Wilhelm Dörpfeld,
Berlin.

Friedrich von Duhn,
Professor der Archäologie,
Heidelberg.

Prof. Paul Ehrlich, Exz.,
Frankfurt a. Main.

Albert Ehrhard,
Porfessor der kath.
Theologie, Straßburg.

Karl Engler, Exz.,
Professor der Chemie,
Karlsruhe.

Gerhard Esser,
Professor der kath.
Theologie, Bonn.

Rudolf Eucken,
Professor der Philosophie,
Jena.

Herbert Eulenberg,
Kaiserswerth.

Heinrich Finke,
Professor der Geschichte,
Freiburg.

Emil Fischer, Exz.,
Professor der Chemie,
Berlin.

Wilhelm Foerster,
Professor der
Astronomie, Berlin.

Ludwig Fulda,
Berlin.

Eduard von Gebhardt,
Düsseldorf.

J. J. de Groot,
Professor der Ethnographie,
Berlin.

Fritz Haber,
Professor der Chemie,
Berlin.

Ernst Haeckel, Exz.,
Professor der Zoologie,
Jena.

Max Halbe,
München.

Prof. Adolf von Harnack,
Generaldirektor der Kgl.
Bibliothek, Berlin.

Gerhart Hauptmann,
Agnetendorf.

Karl Hauptmann,
Schreiberhau.

Gustav Hellmann,
Professor der Meteorologie,
Berlin.

Wilhelm Herrmann,
Professor der protest.
Theologie, Marburg.

Andreas Heusler,
Professor der nordischen
Philologie, Berlin.

Adolf von Hildebrand,
München.

Ludwig Hoffmann,
Stadtbaumeister, Berlin.

Engelbert Humperdinck,
Berlin.

Leopold Graf
Kalckreuth,
Präsident der Deutschen
Künstler-bundes, Eddelsen.

Arthur Kampf,
Berlin.

Fritz Aug. v. Kaulbach,
München.

Theodor Kipp,
Professor der Jurisprudenz,
Berlin.

Felix Klein,
Professor der Mathematik,
Göttingen.

Max Klinger,
Leipzig.

Alois Knoepfler,
Professor der
Kirchengeschichte, München.

Anton Koch,
Professor der kath.
Theologie, Tübingen.

Paul Laband, Exz.,
Professor der Jurisprudenz,
Straßburg.

Karl Lamprecht,
Professor der Geschichte,
Leipzig.

Philipp Lenard,
Professor der Physik,
Heidelberg.

Maximilian Lenz,
Professor der Geschichte,
Hamburg.

Max Liebermann,
Berlin.

Franz von Liszt,
Professor der Jurisprudenz,
Berlin.

Ludwig Manzel,
Präsident der Akademie der
Künste, Berlin.

Josef Mausbach,
Professor der kath.
Theologie, Münster.

Georg von Mayr,
Professor der
Staatswissenschaft, München.

Sebastian Merkle,
Professor der kath.
Theologie, Würzburg.

Eduard Meyer,
Professor der
Geschichte, Berlin.

Heinrich Morf,
Professor der romanischen
Philologie, Berlin.

Friedrich Naumann,
Berlin.

Albert Neisser,
Professor der Medizin,
Berlin.

Walter Nernst,
Professor der Physik,
Berlin.

Wilhelm Ostwald,
Professor der Chemie, Leipzig.

Bruno Paul,
Direktor der
Kunstgewerbeschule,
Berlin.

Max Planck,
Professor der Physik,
Berlin.

Albert Plehn,
Professor der Medizin, Berlin.

Georg Reicke,
Berlin.

Prof. Max Reinhardt,
Direktor des Deutschen
Theaters, Berlin.

Alois Riehl,
Professor der Philosophie,
Berlin.

Karl Robert,
Professor der Archäologie,
Halle.

Wilhelm Röntgen, Exz.,
Professor der Physik,
München.

Max Rubner,
Professor der Medizin, Berlin.

Fritz Schaper,
Berlin.

Adolf von Schlatter,
Professor der protest.
Theologie, Tübingen.

August Schmidlin,
Professor der
Kirchengeschichte,
Münster.

Gustav von Schmoller,
Exz.,
Professor der
Nationalökonomie, Berlin.

Reinhold Seeberg,
Professor der protest.
Theologie, Berlin.

Martin Spahn,
Professor der Geschichte,
Straßburg.

Franz von Stuck,
München.

Hermann Sudermann,
Berlin.

Hans Thoma,
Karlsruhe.

Wilhelm Trübner,
Karlsruhe.

Karl Vollmöller,
Stuttgart.

Richard Voß,
Berchtesgaden.

Karl Voßler,
Professor der romanischen
Philologie, München.

Siegfried Wagner,
Bayreuth.

Wilhelm Waldeyer,
Professor der Anatomie,
Berlin.

August von Wassermann,
Professor der Medizin, Berlin.

Felix von Weingartner.

Theodor Wiegand,
Museumsdirektor, Berlin.

Wilhelm Wien,
Professor der Physik,
Würzburg.

Ulrich von Wilamowitz-Moellendorff, Exz.,
Professor der Philologie,
Berlin.

Richard Wilstätter,
Professor der Chemie,
Berlin.

Wilhelm Windelband,
Professor der Philosophie,
Heidelberg.

Wilhelm Wundt, Exz.,
Professor der Philosophie,
Leipzig.

APPENDIX B

German-Language Text of the "Appeal to the Europeans" ("Aufruf an die Europäer")

Während Technik und Verkehr uns offensichtlilch zur faktischen Anerkennung internationaler Beziehungen und damit zu einer allgemeinen Weltkultur drängen, hat noch nie ein Krieg die kulturelle Gemeinschaftlichkeit des Zusammenarbeitens so intensiv unterbrochen, wie der gegenwärtige. Vielleicht kommt dies uns allerdings auch nur deshalb so auffällig zum Bewußtsein, weil so zahlreiche gemeinschaftliche Bande vorhanden waren, deren Unterbrechung wir schmerzlich verspüren.

Darf uns also dieser Zustand auch nicht wundernehmen, so wären doch diejenigen, denen jene gemeinsame Weltkultur auch nur im geringsten am Herzen liegt, doppelt verpflichtet, für die Aufrechterhaltung dieser Prinzipien zu kämpfen. Diejenigen aber, bei denen man solche Gesinnung vermuten sollte—also vornehmlich Wissenschaftler und Künstler—haben bis jetzt fast ausschließlich Dinge gesagt, die vermuten lassen, als ob mit der Unterbrechung der tatsächlichen Beziehungen auch selbst der Wunsch zu deren Fortsetzung geschwunden

"Aufruf an die Europäer," declaration signed by Berlin physician and medical school professor Georg Friedrich Nicolai; Albert Einstein; Berlin astronomer, former Berlin Observatory director, and German Ethical Society member Wilhelm Förster; and University of Marburg philosophy graduate Otto Buek, mid-October 1914. In Georg Friedrich Nicolai, *Die Biologie des Krieges: Betrachtungen eines deutschen Naturforschers* (Zürich: Verlag Orell Füssli, 1917), 7–10. Also in Georg Friedrich Nicolai, *Aufruf an die Europäer: gesammelte Aufsätze zum Wiederaufbau Europas*, ed. Hans Wehberg (Leipzig, Vienna, and Zurich: Verlag der Wiener Graphischen Werkstätte, 1921), 21–25.

sei, sie haben aus einer erklärlichen Kampfstimmung heraus gesprochen—zum mindesten nicht zum Frieden geredet.

Solche Stimmung ist durch keine nationale Leidenschaft zu entschuldigen, sie ist unwürdig dessen, was bisher alle Welt unter dem Namen der Kultur verstanden hat, und sollte sie Allgemeingut der Gebildeten werden, so wäre das ein Unglück.

Aber nicht nur ein Unglück für die Kultur, sondern—davon sind wir fest überzeugt—ein Unglück dafür, wofür letzten Endes all diese Barberei entfesselt ist; nämlich für den nationalen Bestand der einzelnen Staaten.

Die Welt ist durch die Technik kleiner geworden, die Staaten der großen Halbinsel Europa erscheinen heute einander so nahe gerückt, wie in alter Zeit die Städte jeder einzelnen kleineren Mittelmeerhalbinsel, und Europa—ja man könnte fast sagen die Welt—stellt bereits durch die mannigfaltigsten Beziehungen eine in den Bedürfnissen und Erlebnissen jedes einzelnen begründete Einheit dar.

Da wäre es doch wohl Pflicht der gebildeten und wohlwollenden Europäer, wenigstens den Versuch zu machen, um zu verhindern, daß Europa infolge seiner mangelhaften Gesamtorganisation dasselbe tragische Geschick erleide, wie einst Griechenland. Soll auch Europa sich durch Bruderkrieg allmählich erschöpfen und zugrunde gehen?

Denn der heute tobende Kampf wird kaum einen Sieger, sondern wahrscheinlich nur Besiegte zurücklassen. Darum scheint es nicht nur gut, sondern bitter nötig, daß *gebildete Männer aller Staaten* ihren Einfluß dahin aufbieten, daß—wie auch der heute noch ungewisse Ausgang des Krieges sein mag—die *Bedingungen des Friedens nicht die Quelle künftiger Kriege werden, daß vielmehr die Sache, daß durch diesen Krieg alle europäischen Verhältnisse in einen gleichsam labilen und plastischen Zustand geraten sind, dazu benutzt werde, um aus Europa eine organische Einheit zu schaffen. —Die technischen und intellektuellen Bedingungen dafür sind gegeben.*

In welcher Weise diese Ordnung Europas möglich ist, soll hier nicht erörtert werden. Wir wollen nur grundsätzlich betonen, daß wir fest davon überzeugt sind, daß die Zeit da ist, in der *Europa als Einheit auftreten muß, um seinen Boden, seine Bewohner und seine Kultur zu schützen.*

Wir glauben, daß dieser Wille latent in vielen vorhanden ist, und wir wollen durch gemeinsames Aussprechen dieses Willens bewirken, daß er eine Macht werde.

Zu diesem Zwecke erscheint es vorerst notwendig, daß sich alle diejenigen zusammentun, die ein Herz haben für die europäische Kultur, die also das sind, was Goethe einmal vorahnend „*gute Europäer*" genannt hat, denn man darf die Hoffnung nicht aufgeben, daß ihr gesammeltes Wort—auch unter dem Klange der Waffen—nicht ganz ungehört verhalle, vor allem, wenn unter diesen „guten Europäern von morgen" alle jene zu finden sind, die heute schon bei ihren gebildeten Standesgenossen Ansehen und Autorität genießen.

Aber es ist notwendig, daß die Europäer erst einmal zusammenkommen, und wenn—was wir hoffen—sich genügend *Europäer in Europa* finden, d. h. Menschen, denen Europa nicht nur ein geographischer Begriff, sondern eine wichtige Herzenssache ist, so wollen wir versuchen, einen solchen Europäerband zusammenzurufen.—Der soll dann sprechen und entscheiden.

Wir selber wollen hierzu nur anregen und auffordern, und so bitten wir Sie, falls Sie uns Gesinnungsgenosse und gleich uns entschlossen sind, *dem europäischen Willen einen möglichst weitreichenden Widerhall zu verschaffen, Ihre Unterschrift zu senden.*

APPENDIX C

"Fourteen Points": United States President Woodrow Wilson, January 8, 1918

I. Open covenants of peace, openly arrived at, after which there shall be no private international understanding of any kind but diplomacy shall proceed always frankly and in the public view.

II. Absolute freedom of navigation upon the seas, outside territorial waters, alike in peace and in war, except as the seas may be closed in whole or in part by international action for the enforcement of international covenants.

III. The removal, so far as possible, of all economic barriers and the establishment of an equality of trade conditions among all the nations consenting to the peace and associating themselves for its maintenance.

IV. Adequate guarantees given and taken that national armaments will be reduced to the lowest point consistent with domestic safety.

V. A free, open-minded, and absolutely impartial adjustment of all colonial claims, based upon a strict observance of the principle that in determining all such questions of sovereignty, the interests of the populations concerned must have equal weight with the equitable claims of the government whose title is to be determined.

VI. The evacuation of all Russian territory and such a settlement of all questions affecting Russia as will secure the best and freest cooperation of the other nations of the world in obtaining for her an unhampered and unembarrassed opportunity for the independent determination of her own political development and national policy and assure her of a sincere welcome into the society of free

Woodrow Wilson, "Fourteen Points," President Wilson's Message to Congress, January 8, 1918; Records of the United States Senate; Record Group 46; National Archives.

nations under institutions of her own choosing; and, more than a welcome, assistance also of every kind that she may need and may herself desire. The treatment accorded Russia by her sister nations in the months to come will be the acid test of their good will, of the comprehension of her needs as distinguished from their own interests, and of their intelligent and unselfish sympathy.

VII. Belgium, the whole world will agree, must be evacuated and restored without any attempt to limit the sovereignty which she enjoys in common with all other free nations. No other single act will serve as this will serve to restore confidence among the nations in the laws which they have themselves set and determined for the government of their relations with one another. Without this healing act the whole structure and validity of international law is forever impaired.

VIII. All French territory should be freed and the invaded portions restored, and the wrong done to France by Prussia in 1871 in the matter of Alsace-Lorraine, which has unsettled the peace of the world for nearly fifty years, should be righted, in order that peace may once more be made secure in the interest of all.

IX. A readjustment of the frontiers of Italy should be effected along clearly recognizable lines of nationality.

X. The peoples of Austria-Hungary, whose place among the nations we wish to see safeguarded and assured, should be accorded the freest opportunity of autonomous development.

XI. Rumania, Serbia, and Montenegro should be evacuated; occupied territories restored; Serbia accorded free and secure access to the sea; and the relations of the several Balkan states to one another determined by friendly counsel along historically established lines of allegiance and nationality; and international guarantees of the political and economic independence and territorial integrity of the several Balkan states should be entered into.

XII. The Turkish portions of the present Ottoman Empire should be assured a secure sovereignty, but the other nationalities which are now under Turkish rule should be assured an undoubted security of life and an absolutely unmolested opportunity of autonomous development, and the Dardanelles should be permanently opened as a free passage to the ships and commerce of all nations under international guarantees.

XIII. An independent Polish state should be erected which should include the territories inhabited by indisputably Polish populations, which should be assured a free and secure access to the sea, and whose political and economic independence and territorial integrity should be guaranteed by international covenant.

XIV. A general association of nations must be formed under specific covenants for the purpose of affording mutual guarantees of political independence and territorial integrity to great and small states alike.

Notes

On Annotation

Archive numbers of documents in the Albert Einstein Archives (AEA) are indicated in brackets; for example, AEA [38 088]. Brackets around dates indicate educated guesses of dates that are uncertain.

References to documents printed in *The Collected Papers of Albert Einstein*, volumes 1–12 (Princeton: Princeton Univ. Press, 1987–2009) are indicated by *CPAE* with volume number and document number, e.g., *CPAE* 7: 57; documents from the time period of an earlier volume printed in a later volume indicated with a slash between the earlier volume and document number, e.g., *CPAE* 10: 9/79a. (A few references to material within *CPAE* editorial explanatory notes will include page numbers, in those cases preceded by p. or pp.)

Introduction

1. Virginia Iris Holmes, "'The Inviolability of Human Life': Pacifism and the Jews in Weimar Germany" (PhD diss., Binghamton University, 2001), chapter 6.

2. Jack Wertheimer, *Unwelcome Strangers: East European Jews in Imperial Germany* (Oxford: Oxford Univ. Press, 1987).

3. Holmes, "'Inviolability of Human Life,'" chapter 1.

4. Virginia Iris Holmes, "Was Einstein Really a Pacifist? Einstein's Forward-Thinking, Pragmatic, Persistent Pacifism," *Peace and Change* 33, no. 2 (April 2008): 274–307.

5. The trope "the war to end war," or "the war to end all wars," to refer to World War I originated in a series of London newspaper articles by H. G. Wells in August 1914 that were collected and published in October 1914 in Wells's book *The War That Will End War* (London: Frank and Cecil Palmer, 1914). It was also sometimes associated with US president Woodrow Wilson, although he used the phrase only once. As time passed, the phrase increasingly came to be used ironically, even fairly early during the First World War, and ultimately also by H. G. Wells himself.

1. Horror at War, Pacifist Affiliation, and Alienation from Colleagues: 1914–1915

1. Philipp Frank, *Einstein: His Life and Times* (New York: A. A. Knopf, 1947), 21; and Anton Reiser (pseudonym of Rudolf Kayser), *Albert Einstein: A Biographical Portrait* (New York: A. & C. Boni, 1930), 26.

2. Maja Winteler-Einstein, "Albert Einstein—Beitrag für sein Lebensbild" (excerpt), *The Collected Papers of Albert Einstein* (*CPAE*) 1, pp. lxiii–lxiv.

3. "Release from Württemberg Citizenship," Jan. 28, 1896, Schweizerisches Bundesarchiv, Bern, Switzerland, E 21/23560, also in *CPAE* 1: 16; Swiss Citizenship, *CPAE* 1, pp. 239–41; Albert Einstein, letter to the Swiss Federal Council, Oct. 19, 1899, Schweizerisches Bundesarchiv, Bern, Switzerland, E 21/23560, also in *CPAE* 1: 60; "Minutes of the Municipal Naturalization Commission of Zurich," Dec. 14, 1900, Akten des Stadtrats Zürich, Bürgerliche Sektion, 1901, Abteilung V. Aa, no. 20.1, Stadtarchiv, Zurich, Switzerland; also in *CPAE* 1: 84; Einstein, letter to Maja Winteler-Einstein, Aug. 31, 1935, AEA [29 417]; Einstein, letter to Michele Besso, Feb. 16, 1936, AEA [73 721]; and Albrecht Fölsing, *Albert Einstein: A Biography*, trans. Ewald Osers (New York: Viking Penguin, 1997), 40–41.

4. On the events of July and August 1914 precipitating the First World War, see Martin Gilbert, *The First World War: A Complete History* (New York: Henry Holt, 1994), 16–34; John Keegan, *The First World War* (New York: Vintage Books, Random House, Inc., 1998), 48–70; Roger Chickering, *Imperial Germany and the Great War* (Cambridge: Cambridge Univ. Press, 1998), 10–17; Marc Ferro, *The Great War 1914–1918* (London: Routledge, 1973; orig. Paris: Éditions Gallimard, 1969), 39–46; and James L. Stokesbury, *A Short History of World War I* (New York: Harper Collins, 1981), 22–34.

5. On the "war-enthusiasm," see Chickering, *Imperial Germany and the Great War*, 13–17; Keegan, 71–73; Gilbert, 29 and 34; Ferro, 25; and Stokesbury, 34. For various public statements and observations in Germany at the start of the War, see Bernd Ulrich and Benjamin Ziemann, eds., *Frontalltag im Ersten Weltkrieg: Wahn und Wirklichkeit* (Frankfurt am Main: Fischer Taschenbuch Verlag, 1994), 29–33.

6. For most of Einstein's correspondence and writings during the First World War while he was living in Berlin, see *CPAE*, vols. 7 and 8. For books focused on that portion of his life, see Hubert Goenner, *Einstein in Berlin* (Munich: Verlag C. H. Beck, 2005); and Thomas Levenson, *Einstein in Berlin* (New York: Bantam Dell, 2003).

7. The German army's plan on the western front of a lightning invasion of France through Belgium, the "Schlieffen plan," was effective at first, but the French ground it to a halt along the river Marne, where both sides fell into trench warfare that would last throughout the War. Meanwhile, the Russians mobilized more quickly than the Germans expected, and this drained German resources from the western to the eastern front, despite some dramatic successes by the Germans in the east early in the War.

After mid-August 1914, a series of battles took place or were started: From August 17 to September 2, at Tannenberg and in subsequent fighting in the east, the Germans took close to 200,000 Russian prisoners. On August 20, German forces occupied Brussels, Belgium. On August 23, Japan declared war on Germany. On August 26 and 27, the battle of Le Cateau took place, and on August 29 and 30, the battle of St. Quentin. On September 2, German troops occupied Moronvilliers in France. On September 5, the London Agreement was reached, stipulating that no member of the Triple Entente (Britain, France, Russia) could seek a separate peace with the Central Powers (Germany and Austria-Hungary). On the same day, the first battle of the Marne began northeast of Paris, with French troops attacking a German advance on the French capital. Over two million soldiers fought and one hundred thousand were killed or wounded in this battle. On September 6, a French and British attack at Marne ended the German advance on Paris. From September 13 to 28, the first battle of the Aisne took place. On October 9, the siege of Antwerp ended with that city's capture by German troops. On the land battles in the first two months of the War on both western and eastern fronts, especially the first battle of the Marne and the beginning of trench warfare, see Gilbert, 35–123; Keegan, 71–174; Chickering, *Imperial Germany and the Great War,* 23–31; Ferro, 49–58; and Stokesbury, 36–74.

At sea, on August 28, in the battle of Heligoland, British cruisers sank three German cruisers. On November 1, in the battle of Coronel, a British naval squadron was defeated by a German squadron. On October 29, German warships flying the Turkish flag shelled Russian Black Sea ports, bringing the Ottoman Empire into the War on the side of Germany and Austria. On November 5, Britain annexed Cyprus. On November 7, Japanese forces seized Jiaozhou Bay in China, the base of the German East Asia Squadron. On maritime battles early in the War, see Gilbert, 101–6; and Stokesbury, 75–89; for background on the naval battles, see Keegan, 257–62; and Chickering, *Imperial Germany and the Great War,* 52–53.

On December 24, British and German soldiers spontaneously interrupted World War I to celebrate Christmas Eve together on the battlefield of the western front in what is known as the Christmas truce. Their commanding officers were extremely displeased and did not permit any gestures of this sort to be repeated later in the War. On the Christmas truce, see Gilbert, 117–19.

8. Einstein, letter to Paul Ehrenfest, Aug. 19, 1914, AEA [9 357]; also in *CPAE* 8: 34; and in Otto Nathan and Heinz Norden, *Einstein on Peace* (New York: Schocken Books, 1968, c1960), 2. (When I mention that a document also appears in Nathan and Norden, it is likewise in English translation, and in the majority of cases it is a smaller excerpt than the one I have provided.)

9. Einstein, letter to Heinrich Zangger, Aug. 24, 1914, AEA [87 055]; also in *CPAE* 10: 8/34a.

10. While it violated the Treaty of London of 1839, in which Britain, France, Prussia, Austria, and Russia guaranteed Belgian neutrality, the German military's Schlieffen Plan to encircle and conquer France depended on a quick pass through Belgium, and German soldiers, infuriated at resistance from the Belgian army and civilians in August 1914, retaliated savagely. In the first incident in Hervé on August 4, German soldiers razed about 500 houses, leaving only 19, and killed large numbers of civilians. In the first three weeks of the War, German soldiers killed 156 Belgian civilians in Aarschot, 50 in Seilles, 211 in Andenne on August 20; 384 in Tamines on August 22; and 612 men, women, and children, including a three-month-old baby, in Dinant on August 23. In Louvain during August 25–30, German soldiers burned a medieval library of 230,000 books and manuscripts, destroyed 1,100 buildings, killed 209 civilians, and forcibly deported the whole population of 42,000. German soldiers repeatedly raped Belgian women, particularly in Aarschot and Brabant province. The German Military Governor on October 5 publicly proclaimed that German army would engage in collective reprisals, such as for sniping incidents. The German army also committed atrocities against French civilians during August 1914, including incidents in Lunéville, Gerbéviller, Vitrimont, and Sarrebourg. As the War continued, German authorities forced large numbers of Belgians to work in Germany, transported Belgian industrial machinery and equipment to Germany, and exploited Belgian industry so intensely that the majority of it was ultimately destroyed. Keegan, 30–34, 69, 78–83; Gilbert, 29, 32, 36, 41–43, 52–53, 88, 247; Chickering, *Imperial Germany and the Great War,* 82–85; J. Lipkes, *Rehearsals: The German Army in Belgium, August 1914* (Leuven, Belgium: Leuven Univ. Press, 2007); John N. Horne and Alan Kramer, *German Atrocities, 1914: A History of Denial* (New Haven: Yale Univ. Press, 2001); Larry Zuckerman, *The Rape of Belgium: The Untold Story of World War I* (New York: New York Univ. Press, 2004); and Alan Kramer, *Dynamic of Destruction: Culture and Mass Killing in the First World War* (Oxford: Oxford Univ. Press, 2007).

11. Manifesto "To the Civilized World" ("An die Kulturwelt!" also known as the "Manifesto of the 93"), in Georg Friedrich Nicolai, *The Biology of War,* trans. Constance A. Grande and Julian Grande (New York: Century, 1919), xi–xiv. For the German-language original of the manifesto, see appendix A and Georg Friedrich Nicolai, *Die Biologie des Krieges: Betrachtungen eines deutschen Naturforschers* (Zurich: Orell Füssli, 1917). For an immediate postwar discussion, see Hans Wehberg, *Wider den Aufruf der 93! Das Ergebnis einer Rundfrage an die Intellektuellen über die Kriegsschuld* (Charlottenburg: Deutsche Verlagsgesellschaft für Politik und Geschichte, 1920). See also Jürgen von Ungern-Sternberg and Wolfgang von Ungern-Sternberg, *Der Aufruf "An die Kulturwelt!" Das Manifest der 93 und die Anfänge der Kriegspropaganda im Ersten Weltkrieg* (Stuttgart: Franz Steiner Verlag, 1996), 144–47; and Bernhard vom Brocke and Hubert Laitko, eds., *Die Kaiser-Wilhelm-/Max-Planck-Gesellschaft und Ihre Institute. Das Harnack-Prinzip* (Berlin: Walter de Gruyter, 1996).

12. Einstein, letter to Arnold Sommerfeld, Dec. 14, 1946, AEA [21 368].

13. David E. Rowe and Robert Schulmann, eds., *Einstein on Politics: His Private Thoughts and Public Stands on Nationalism, Zionism, War, Peace, and the Bomb* (Princeton, NJ: Princeton Univ. Press, 2007), 25–27.

14. Einstein, letter to Hendrik Lorentz, July 21, 1915, AEA [83 432]; also in *CPAE* 8: 98; and in Rowe and Schulmann, 68–69.

15. Einstein, letter to Romain Rolland, Sept. 15, 1915, AEA [33 006]; also in *CPAE* 8: 118; and in Rowe and Schulmann, 71–72.

16. *Zu Max Plancks sechzigstem Geburtstag. Ansprachen gehalten am 26. April 1918 in der Deutschen Physikalischen Gesellschaft von E. Warburg, M. v. Laue, A. Sommerfeld und A. Einstein* (Karlsruhe: C. F. Müllersche Hofbuchhandlung, 1918), 29–32; Einstein, letter to Max Planck, Apr. 6, 1933, AEA [19 391]; also in Rowe and Schulmann, 273–75; and in Nathan and Norden, 217–18.

17. Einstein, letter to Hendrik Lorentz, Aug. 2, 1915, AEA [16 438]; also in *CPAE* 8: 103; and in Rowe and Schulmann, 69–70.

18. "Appeal to the Europeans" ("Aufruf an die Europäer"), in Nicolai, *Biology of War*, xvii–xix; Nicolai, *Biologie des Krieges*; Georg Friedrich Nicolai, *Aufruf an die Europäer: gesammelte Aufsätze zum Wiederaufbau Europas*, ed. Hans Wehberg (Leipzig, Vienna, and Zurich: Verlag der Wiener Graphischen Werkstätte, 1921), 21–25; also in *CPAE* 6: 8; in Rowe and Schulmann, 64–66; and in Nathan and Norden, 4–6. (As with Nathan and Norden, when I mention that a document also appears in Rowe and Schulmann, it is likewise in English translation, and in most cases it is a smaller excerpt than the one I have provided, although not shorter in the case of the "Appeal to the Europeans" and other titled documents in Rowe and Schulmann.) For the German-language original of the appeal, see appendix B. The other two signatories to the appeal besides Nicolai and Einstein were Wilhelm Förster (the father of Friedrich Wilhelm Förster) and Otto Buek.

19. Einstein, letter to Paul Ehrenfest, [beginning Dec. 1914], AEA [9 139]; also in *CPAE* 8: 39.

20. After the German navy sank two British ships off the coast of Chile on November 1, 1914, killing 1,600 sailors, the British navy sank four German battleships in the South Atlantic on December 8, and a fifth one, the *Dresden*, on Mar. 14, 1915; Keegan, 213–15; and Stokesbury, 80–82. On January 1, 1915, a German submarine sank the British battleship HMS *Formidable* off Lyme Regis, Dorset, England, killing 547 sailors; Gilbert, 124. On January 30 in the English Channel, a German submarine attacked and sank a British merchant ship for the first time. Gilbert, 127; for context, see Chickering, *Imperial Germany and the Great War*, 64.

On January 3, 1915, the German military used poison gas on the eastern front against the Russians. At this point it was a nonlethal, tear-producing type of gas; Keegan, 197. On January 19, at Great Yarmouth and King's Lynn, German zeppelins bombed English cities for the first time, killing more than twenty people; Gilbert, 125. A British attack on the Dardanelles on the Gallipoli peninsula, and attempt to pass through the strait with the intention to take Constantinople and establish communications with Russia, had been planned since January 5. The execution of this plan started with a bombardment on February 19, followed by a failed attempt by ten Allied battleships to pass through the strait on March 18 when one French and two British battleships were sunk and at least 667 men drowned; Gilbert, 124–25, 131–38; Keegan, 238–39; Chickering, *Imperial Germany and the Great War,* 59; Ferro, 65–67; and Stokesbury, 115–20.

21. Einstein, letter to Heinrich Zangger, [after Dec. 27, 1914], AEA [86 528]; also in *CPAE* 10: 8/41a.

22. Einstein, letter to Edgar Meyer, Jan. 2, 1915, AEA [75 603]; also in *CPAE* 8: 44.

23. Einstein, letter to Paolo Straneo, Jan. 7, 1915, AEA [70 604]; also in *CPAE* 8: 45.

24. Einstein, letter to Heinrich Zangger, Jan. 11, 1915, AEA [87 054]; also in *CPAE* 10: 8/45a. Peter Debye, Max Born, and Max von Laue were fellow physicists.

25. Einstein, letter to Hans Albert Einstein, Jan. 25, 1915, AEA [75 851]; also in *CPAE* 8: 48.

26. Einstein, letter to Georg Friedrich Nicolai, Feb. 20, 1915, AEA [36 799]; also in *CPAE* 8: 57. Einstein's suggested signatories included Max Planck in Berlin and Zangger in Zurich.

27. New Fatherland League, Minutes of meeting of March 21, 1915, Nachlaß Wehberg, vol. 14, 62–64, Bundesarchiv, Koblenz, Germany; also in AEA [84 165]. See also *CPAE* 8: 65, note 2; Karl Holl, *Pazifismus in Deutschland* (Frankfurt am Main: Suhrkamp, 1988), 113–20; Otto Lehmann-Rußbüldt, *Der Kampf der Deutschen Liga für Menschenrechte, vormals Bund Neues Vaterland, für den Weltfrieden, 1914–1927* (Berlin: Hensel & Co. Verlag, 1927), 9–24; and Richard A. Cohen, "The German League for Human Rights in the Weimar Republic" (PhD diss., State University of New York at Buffalo, 1989), 7–9.

28. Einstein, letter to Romain Rolland, Mar. 22, 1915, AEA [84 165]; also in *CPAE* 8: 65; and in Nathan and Norden, 13.

29. Romain Rolland, letter to Einstein, Mar. 28, 1915, AEA [33 003]; also in *CPAE* 8: 68; and in Nathan and Norden, 13–14.

30. Albert, Elsa, and Ilse Einstein, postcards to Georg Friedrich Nicolai, Apr. 2, 1915, AEA [82 545], and [after Apr. 2, 1915], AEA [82 544]; also in *CPAE* 8, Apr. 2, 1915 Calendar entry; and Wolf Zuelzer, *The Nicolai Case: A Biography* (Detroit: Wayne State Univ. Press, 1982), 27–28.

31. Ernst Reuter, letter to Walther Schücking, Apr. 2, 1915, Nachlaß Hans Wehberg, vol. 15: 47, Bundesarchiv, Koblenz, Germany. The BNV members hoped that Einstein, with his French connections, might encourage French intellectuals to attend.

32. Einstein, letter to Heinrich Zangger, [ca. Apr. 10, 1915], AEA [39 662]; also in *CPAE* 8: 73; and in Rowe and Schulmann, 67.

33. Einstein, letter to Tullio Levi-Civita, Apr. 14, 1915, AEA [16 244]; also in *CPAE* 8: 75.

34. Einstein, letter to Tullio Levi-Civita, May 5, 1915, AEA [16 250]; also in *CPAE* 8: 80.

35. The Ottoman Empire, blaming the local Armenian population for military losses to the Russians, over the next seven months beginning on April 8, 1915, massacred 600,000 Armenians and deported another 500,000, of whom 400,000 died from the brutalities of the forced march over the mountains to Cilicia (Anatolia), Syria, and Mesopotamia. Over one million Armenians were killed in the Turkish genocide of the Armenians. Gilbert, 142–43 and 166–67; Keegan, 223; Ferro, 100–101; and Stokesbury, 102–4.

On April 22, in the second battle of Ypres, German troops used a deadly chlorine gas for the first time on the western front at Ypres, in Belgium. Gilbert, 143–45; Keegan, 197–98; Chickering, *Imperial Germany and the Great War,* 57; Ferro, 63 and 88; Paul Fussell, *The Great War and Modern Memory* (Oxford and New York: Oxford Univ. Press, 2000, c1975), 10; Stokesbury, 95–96.

On April 25, Allied troops, with heavy assignment to the Australian-New Zealand Corps, landed and began their assault on Turkey's Gallipoli Peninsula, hoping for a quick victory that would turn the tide of the War on other fronts. But the Turks, led by a German commander, put up stiff resistance, and the ensuing battles deteriorated into trench warfare. The campaign, which cost the Allies over 500,000 killed, wounded, and missing, led to no clear victory by the time the Allies withdrew at the end of 1915. Gilbert, 146–53; Keegan, 234–49; Chickering, *Imperial Germany and the Great War,* 59; Ferro, 65–67; and Stokesbury, 121–23.

On the seas, from early 1915 to early 1916, when Germany declared unrestricted submarine warfare in response to the Allied blockade of Germany, German submarines sank 37 unarmed merchant ships. They also disabled over 50 English merchant ships per month during 1915. The sinking of the English passenger liner *Lusitania* on May 7 killed 1,201 people, including 128 Americans, and provoked public outrage in England, Canada, South Africa, and the United States; it also almost led to a diplomatic rupture between the United States and Germany. Gilbert, 156–58, 188, and 236–37; Keegan, 265; Chickering, *Imperial Germany and the Great War,* 64 and 90; Ferro, 105–13; Stokesbury, 111–13, 216–18.

Beginning on May 9, in the second battle of Artois, French and British troops launched a coordinated offensive in France. The British tried to capture Aubers Ridge; one week later, the French tried to take Vimy Ridge. By the end of May, after tens of thousands of casualties on both sides, the War on the western front had ground to a stalemate. Gilbert, 159–65; Keegan, 199; Chickering, *Imperial Germany and the Great War,* 57; Ferro, 63–64; Stokesbury, 96. Italy, after diplomatic negotiations with both belligerent sides since April, joined the Allies, declaring war on Austria-Hungary on May 23. Gilbert, 142, 151, and 165–66; Keegan, 200; Chickering, *Imperial Germany and the Great War,* 59; Ferro, 68–72; and Stokesbury, 105–9.

36. Einstein, letter to Heinrich Zangger, May 17, 1915, AEA [39 664]; also in *CPAE* 8: 84.

37. Einstein, letter to Walter Dällenbach, May 31, [1915], AEA [9 070]; also in *CPAE* 8: 87.

38. New Fatherland League (Bund Neues Vaterland), List of Members, June 1, 1915, Nachlaß Hans Wehberg, vol. 14, pp. 109–10, Bundesarchiv, Koblenz, Germany.

39. Einstein, letter to Heinrich Zangger, July 7, [1915], AEA [39 668]; also in *CPAE* 8: 94.

40. Reinhold Seeberg (with 1,347 signatories), "Petition of the Intellectuals" (*Intellektuelleneingabe*), July 8, 1915. Also discussed in Holl, 120; Gordon A. Craig, *Germany, 1866–1945* (New York: Oxford Univ. Press, 1978), 361; Klaus Schwabe, *Deutsche Professoren und die politischen Grundfragen des Ersten Weltkrieges*, PhD diss., Philosophische Fakultät, Albert Ludwigs Universität, Freiburg im Breisgau, 1958, 160 ff.; and Klaus Schwabe, "Zur politischen Haltung der deutschen Professoren im Ersten Weltkrieg," *Historische Zeitschrift* 193 (1961).

41. Hans Delbrück and Bernhard Dernburg (with 140 signatories), petition to German chancellor Theobold von Bethmann Hollweg, July 27, 1915. Also in Lehmann-Rußbüldt, *Der Kampf der Deutschen Liga*, 64–66. See also Hans Delbrück, "Die Differenzen über die Kriegsziele hüben und drüben," in *Preußische Jahrbücher* 162 (1915): 167–72; Annelise Thimme, *Hans Delbrück als Kritiker der wilhelminischen Epoche* (Düsseldorf: Droste, 1955); and Gilbert, 177–78.

42. Einstein, letter to Hendrik A. Lorentz, July 21, 1915, AEA [88 432]; also in *CPAE* 8: 98; and in Rowe and Schulmann, 68–69.

43. Einstein, letter to Heinrich Zangger, [between July 24 and Aug. 7, 1915], AEA [39 670]; also in *CPAE* 8: 101. (Einstein referred to Felix Kaufler, a former chemistry instructor at the Eidgenössische Technische Hochschule [ETH] in Zurich who had moved to Vienna in 1912.)

44. Einstein, letter to Hendrik A. Lorentz, Aug. 2, 1915, AEA [16 438]; also in *CPAE* 8: 103; in Rowe and Schulmann, 69–70; and in Nathan and Norden, 11.

45. On August 6, 1915, in battles with the Turks at Gallipoli, the Allies timed a diversionary attack to coincide with a major Allied landing of reinforcements at Suvla Bay. This led to battles in August that caused extremely high casualties—a significant proportion due to disease and poor medical care—but achieved no great military gains. Gilbert, 180–85, 188–91; Keegan, 248. On September 25, as part of an Allied offensive on the western front, British troops used poison gas for the first time. Gilbert, 197–98. On October 12, 1915, the British nurse Edith Cavell was executed by a Germany firing squad for helping Allied soldiers escape from Belgium—an event that was very effectively exploited by Allied propaganda. Gilbert, 202–3; Ferro, 124; Stokesbury, 110.

46. Einstein, letter to Paul Ehrenfest, [Aug. 23, 1915], AEA [9 361]; also in *CPAE* 8: 112; and in Nathan and Norden, 12.

47. Einstein, letter to Elsa Einstein, [Aug. 30, 1915], AEA [73 043]; also in *CPAE* 8: 114.

48. Einstein, letter to Elsa Einstein, [Sept. 3, 1915], AEA [73 050]; also in *CPAE* 8: 115.

49. Einstein, letter to Heinrich Zangger, Sept. 19, 1915, AEA [39 674]; also in *CPAE* 8: 120.

50. Einstein, letter to Elsa Einstein, [Sept. 13, 1915], AEA [73 048]; also in *CPAE* 8: 117.

51. Einstein, letter to Romain Rolland, Sept. 15, [1915], AEA [33 006]; also in *CPAE* 8: 118; and in Rowe and Schulmann, 71–72.

52. Einstein, letter to Heinrich Zangger, Sept. 21, 1915, AEA [39 676]; also in *CPAE* 8: 121. Théophile Delcassé was French minister of foreign affairs and an advocate of colonial expansion. The Algeciras Treaty of 1906 affirmed the independence of Morocco but not of parts of North Africa held by the Ottoman Empire. Jean Jaurès (1859–1914), a French socialist and pacifist, was assassinated on July 31, 1914. An article condemning annexation had appeared in the organ of the Social Democratic Party of Switzerland, "Gegen den Länderraub," *Volksrecht* 18, no. 195 (Aug. 23, 1915), [1–2].

53. Einstein, letter to Hendrik A. Lorentz, Sept. 23, 1915, AEA [16 440]; also in *CPAE* 8: 122. See also *Die Friedens-Warte* 17, no. 8 (Oct. 1915), 298.

54. Einstein, letter to Heinrich Zangger, [Oct. 4, 1915], AEA [86 537]; also in *CPAE* 10: 8/124a.

55. Einstein, letter to Paul Hertz, [before Oct. 8, 1915], AEA [12 205]; also in *CPAE* 8: 125.

56. Einstein, letter to Paul Hertz, [before Oct. 8, 1915], AEA [12 206]; also in *CPAE* 8: 126.

57. Einstein, letter to Walther Schücking, Oct. 22, 1915, AEA [78 993]; also in *CPAE* 8: 131.

58. Einstein, "My Opinion on the War" ("Meine Meinung über den Krieg"), Oct. 23–Nov. 11, 1915, AEA [70 457]; also in *CPAE* 6: 20; and in Rowe and Schulmann, 73–76. For his letters to the Berlin Goethe League (Berliner Goethebund) around this time, see Einstein, letter to Berliner Goethebund, [after Oct. 23, 1915], AEA [43 756]; also in *CPAE* 8: 132; Einstein, letter to Berliner Goethebund, Nov. 11, 1915, AEA [29 003]; also in *CPAE* 8: 138; and Einstein, letter to Berliner Goethebund, Nov. 16, 1915, AEA [43 760]; also in *CPAE* 8: 146.

59. Einstein, letter to Berliner Goethebund, Nov. 11, 1915, AEA [29 003]; also in *CPAE* 8: 138.

60. Einstein, letter to Heinrich Zangger, [Nov. 26, 1915], AEA [81 178]; also in *CPAE* 8: 152.

61. Einstein, letter to Michele Besso, [Nov. 30, 1915], AEA [81 521]; also in *CPAE* 8: 155.

62. Einstein, letter to Hans Albert Einstein, [Nov. 30, 1915], AEA [75 899]; also in *CPAE* 8: 156. Einstein, letter to Mileva Einstein-Marić, [Dec. 1, 1915], AEA [75 868]; also in *CPAE* 8: 159.

63. Einstein, letter to Heinrich Zangger, [before Dec. 8, 1915], AEA [86 453]; also in *CPAE* 10: 8/159a.

2. Disgust with War, Pessimism, and International Organization: 1916–1917

1. On January 8, 1916, Allied forces withdrew from Gallipoli. More than 60,000 Turks and 40,000 Allied troops from England, France, Australia, and New Zealand had died during the campaign; there were 250,000 casualties on each side including the wounded and missing. Gilbert, 225–26; Keegan, 248; Ferro, 66–67; Stokesbury, 126.

On January 31, nine German Zeppelins dropped 389 bombs on England. By May 1916, this bombing campaign destroyed the air base at Dunkirk, bombed London, and killed 550 English civilians. In June, French and British bombers killed 150 German civilians in Karlsruhe. Gilbert, 230, 238, 255, and 192; Ferro, 94 and 123; and Stokesbury, 252–53.

On February 21, 1916, the battle of Verdun began on the western front in France. This German offensive against almost exclusively French defenders would last through December. The Germans launched three poison gas attacks on Verdun in a four-day period in late April. In the first two months of that battle, 89,000 French and 82,000 German soldiers died at Verdun; by December, 542,000 French and 434,000 German soldiers had died there. Gilbert, 230–35, 237, 240, 246, 255–56, 262, 265, 284–85, 222–300, 303; Keegan, 279–86; Ferro, 76–78; Chickering, *Imperial Germany and the Great War*, 67–70; and Stokesbury, 145–48.

2. Dieter Riesenberger, *Geschichte der Friedensbewegung in Deutschland: Von den Anfängen bis 1933* (Göttingen: Vandenhoeck & Ruprecht, 1985), 108; Holl, 125; Ludwig Quidde, *Der deutsche Pazifismus während des Weltkrieges 1914–1918*, aus dem Nachlaß Ludwig Quiddes, ed. Karl Holl and Helmut Donat (Boppard am Rhein: Harald Boldt Verlag, 1979); Reinhold Lüttgemeier-Davin, *Pazifismus zwishcen Kooperation und Konfrontation: Das Deutche Friedenskartell in der Weimarer Republik* (Köln: Pahl-Rugenstein Verlag, 1982), 25; Lehmann-Rußbüldt, *Der Kampf*, 78–79. On the link between the New Fatherland League's support for the July 1915 Delbrück-Dernburg petition (see chapter 1 in this book) and the outlawing of the New Fatherland League, see Gilbert, 177–78.

3. Einstein, letter to Elsa Einstein, [Apr. 15, 1916], AEA [73 049]; also in *CPAE* 8: 213. On Nicolai's experiences in the Germany army, see Zuelzer, 161–229.

4. Einstein, letter to Paul Ehrenfest, [Apr. 26, 1916], AEA [9 376]; also in *CPAE* 8: 218; see also Doc. 218's note 5.

5. In the North Sea from May 31 to June 1, 1916, in the battle of Jutland, the largest surface naval battle of the War, fourteen British and ten German warships were

destroyed, and 6,097 British and 2,551 German sailors were killed, many by drowning. Despite these devastating losses to both the British and German navies, the maritime strategic balance remained unaltered and the Allied blockade of Germany continued. Gilbert, 251–52; Keegan, 270–74; Chickering, *Imperial Germany and the Great War*, 91; Stokesbury, 168–79. On June 5, on its way to a mission in Russia with British Secretary of State for War Lord Horatio Kitchener on board, the HMS *Hampshire* hit a mine and sank off the Orkney Islands of Scotland. Kitchener was drowned. Gilbert, 253; Keegan, 311.

On June 4 on the eastern front, the Russians began the Brusilov offensive with a massive artillery barrage, achieving a breakthrough into the Austro-Hungarian battle lines and taking more than 400,000 prisoners by September. (The offensive is named after its Russian commander, Alexei Brusilov.) Gilbert, 252–54, 272–73, and 276–77; Keegan, 303–6; Ferro, 78–79; Chickering, *Imperial Germany and the Great War*, 68–71; Stokesbury, 162–64.

6. The battle of the Somme was a joint British-French offensive. Regarding that battle and World War I trench warfare generally, Keegan has written, "The simple truth of 1914–18 trench warfare is that the massing of large numbers of soldiers unprotected by anything but cloth uniforms, however they were trained, however equipped, against large masses of other soldiers, protected by earthworks and barbed wire and provided with rapid fire weapons, was bound to result in very heavy casualties among the attackers" (293). Another historian has emphasized that most British soldiers, burdened by 66 pounds of equipment, made for a bulky, slow-moving target. Peter H. Liddle, *The 1916 Battle of the Somme, A Reappraisal* (London: Leo Cooper, 1992), 39; see also Gilbert, 258–59. From July 1 to November 18, 1916, over one million soldiers died in the battle of the Somme, including 60,000 British casualties on the first day. By November, 420,000 British, 195,000 French, and 650,000 German soldiers had died there. Gilbert, 258–72, 275–76, 280; and 299; Keegan, 290–99; Ferro, 80–81; Chickering, *Imperial Germany and the Great War*, 70–71; and Stokesbury, 156.

7. Within Germany, the failure of the 1916 potato crop, combined with the Allied blockade of Germany in effect since 1915, caused extreme economic hardship on the home front. On the blockade, see C. Paul Vincent, *The Politics of Hunger: The Allied Blockade of Germany, 1915–1919* (Athens: Ohio Univ. Press, 1985); for a local case study, see Anne Roerkohl, *Hungerblockade und Heimatfront: Die kommunale Lebensmittelversorgung in Westfalen während des ersten Weltkrieges* (Stuttgart: Steiner, 1991). Despite censorship, communications between the war fronts and the home front were impossible to suppress. Information on the home front included notification of individual families of two million soldiers' deaths and four million soldiers' injuries, many of these involving extreme disfigurement, as well as interactions by soldiers on leave and letters from soldiers (about seven million letters per day; thirty billion items in both directions during the War). In 1916, the optimism and enthusiasm of the first two years shifted to cynicism

and disillusionment. Chickering, *Imperial Germany and the Great War*, 99–103. On the German home front, see also Ulrich and Ziemann.

8. Einstein, letter to Heinrich Zangger, July 11, [1916], AEA [86 498]; also in *CPAE* 10: 8/232a. The difficulties in his relations with his sons were probably fueled by the fact that in February 1916 Einstein had suggested divorce to Mileva Einstein-Marić, after a separation of a year and a half. Their divorce would not actually be concluded until after the War. One factor in the delay was that Einstein-Marić fell ill during the summer of 1916 and was confined to bed for over a year.

9. Einstein, letter to Heinrich Zangger, July 19, 1916, AEA [86 456]; also in *CPAE* 10: 8/237a.

10. Einstein, letter to Heinrich Zangger, Aug. 3, [1916], AEA [86 499]; also in *CPAE* 10: 8/247a.

11. Einstein, letter to Heinrich Zangger, Sept. 26, 1916, AEA [86 540]; also in *CPAE* 10: 8/261a.

12. On September 2, 1916, ten German airplanes dropped bombs on London. William Leefe-Robinson shot one of the planes down—the first time in history that this had been done. Gilbert, 289. On the development of air warfare more broadly, see Stokesbury, 252–53.

On August 27, encouraged by Russian successes in the Brusilov campaign, the Romanians joined the War on the Allies' side. Keegan, 306–7; Ferro, 79; Chickering, *Imperial Germany and the Great War*, 71; Stokesbury, 164–66. In September and October, German and Bulgarian forces drove Romanian troops back into Romania and through the province of Dobruja, taking most of the territory of Romania and eventually its capital Bucharest on December 5. This victory gave Germany access to crucial oil fields and wheat supplies, which probably prolonged the War. Gilbert, 288–89; Keegan, 307–8; Ferro, 79–80; Chickering, *Imperial Germany and the Great War*, 75, 85; Stokesbury, 165–66.

On November 5, the Kingdom of Poland was proclaimed by a joint act of the emperors of Germany and Austria-Hungary. This attempt to gain Polish support backfired when Germany refused to allow Poland its own independent government. Gilbert, 295–96; Ferro, 102; Chickering, *Imperial Germany and the Great War*, 86.

On November 18, the first battle of the Somme ended when British commander Douglas Haig in France called for his forces to stop fighting it. Gilbert, 297–99; Keegan, 298–99; Ferro, 80–81; Chickering, *Imperial Germany and the Great War*, 70; Stokesbury, 156. On December 12, in the Dolomites in the eastern Alps of northern Italy, an avalanche buried 18,000 Austrian and Italian soldiers. Heinz von Lichem, *Gebirgskrieg, 1915–1918* (Bozen/Bolzano, Italy: Athesia, 1980–1982). In the Sinai desert on December 21, Australian and New Zealand troops pushed back the Turks, capturing the town of El Arish twenty miles from the border of Palestine. On December 23 in the battle of Magdhaba, they captured the Turkish garrison. Gilbert, 304; and Stokesbury, 191.

13. Einstein, letter to Elsa Einstein, [Sept. 30, 1916], AEA [143 035]; also in *CPAE* 10: 8/261c; and Einstein, letter to Elsa Einstein [Oct. 7, 1916], AEA [143 037]; also in *CPAE* 10: 8/262b.

14. Einstein, letter to Wander and Gertruida de Haas, [Oct. 3, 1916], AEA [70 413]; also in *CPAE* 8: 262; and Einstein, letter to Elsa Einstein, [Oct. 5, 1916], AEA [143 036]; also in *CPAE* 10: 8/262a.

15. Einstein, letter to Werner Weisbach, Oct. 14, 1916, AEA [76 155]; also in *CPAE* 8: 264. The Latin, from the book of Ezekiel, translates, "I have spoken and thereby saved my soul," whether those who are warned heed the warning or not. On the Vereinigung Gleichgesinnter, see Holl, 126.

16. Einstein, letter to Paul and Tatiana Ehrenfest, Oct. 18, 1916, AEA [9 386]; also in *CPAE* 8: 268.

17. Einstein, letter to Paul Ehrenfest, Oct. 24, 1916, AEA [9 388.2]; also in *CPAE* 8: 269. In addition to Max Planck, Heinrich Rubens, Wilhelm von Waldeyer-Hartz, and Hermann Diels were fellow Berlin academics.

18. Einstein, letter to Paul Ehrenfest, Nov. 7, 1916, AEA [9 389]; also in *CPAE* 8: 275. Waldeyer-Hartz, as one of two permanent secretaries of the physical-mathematical section of the Prussian Academy of Sciences, had participated in the Academy's appointment of Einstein as head of the Kaiser Wilhelm Institute in Berlin in 1914. For Waldeyer's letter to Lorentz, see Wilhelm von Waldeyer-Hartz to Hendrik A. Lorentz, Nov. 2, 1916, Archief H. A. Lorentz, Rijksarchief Noord-Holland, Haarlem, Netherlands. On Waldeyer's participation in Einstein's appointment to the Kaiser Wilhelm Institute, see Wilhelm Waldeyer, draft of letter to Einstein, July 20, 1914, Archiv der Berlin-Brandenburgischen Akademie der Wissenschaften, Berlin, Germany, II–III, vol. 36: 101.

19. Einstein, letter to Hendrik A. Lorentz, Nov. 13, 1916, AEA [16 455]; also in *CPAE* 8: 276; and in Nathan and Norden, 18. Walther Nernst was another fellow Berlin academic.

20. Einstein, letter to Paul Ehrenfest, [Nov. 17, 1916], AEA [9 391]; also in *CPAE* 8: 277.

21. Einstein, letter to Paul Ehrenfest, Dec. 4, 1916, AEA [9 394]; also in *CPAE* 8: 282.

22. Einstein, letter to Heinrich Zangger, Oct. 25, 1916, AEA [86 504]; also in *CPAE* 10: 8/269a.

23. Einstein, letter to Heinrich Zangger, Nov. 16, 1916, AEA [86 542]; also in *CPAE* 10: 8/276). On Rolland, see David J. Fisher, *Romain Rolland and the Politics of Intellectual Engagement* (Berkeley: Univ. of California Press, 1988), 44; Harold March, *Romain Rolland* (New York: Twayne, 1971), 89; William T. Starr, *Romain Rolland, One Against All: A Biography* (The Hague: Mouton, 1971), 167.

24. Einstein, letter to Hans Albert Einstein, Nov. 26, 1916, AEA [75 860]; also in *CPAE* 8: 279.

25. President Woodrow Wilson of the United States, still the sole major power maintaining neutrality, who was propagating "peace without victors," made an attempt to mediate peace in December 1916 and proposed terms for peace in Europe in his State of the Union address on January 21, 1917. Gilbert, 305–7; Keegan, 319; Ferro, 114; Chickering, *Imperial Germany and the Great War,* 169; Stokesbury, 220. On January 31, Germany announced that its U-boats would engage in unrestricted submarine warfare. On February 3, the United States broke off diplomatic relations with Germany in response to this announcement. Gilbert, 306–11; Keegan, 351–52; Ferro, 114–15; Chickering, *Imperial Germany and the Great War,* 169; Stokesbury, 220–21. On February 24, 1917, the United States ambassador to Britain, Walter H. Page, was given the Zimmermann telegram, an intercepted and decoded German communication directing the German ambassador to Mexico to offer to give Texas, New Mexico, and Arizona back to Mexico in the event that the United States declared war on Germany, if Mexico would declare war on the United States. On March 1, the US government released the text of the Zimmermann telegram to the public, provoking outrage and facilitating growing support among Americans for a US declaration of war against Germany. Gilbert, 308 and 312; Keegan, 351; Ferro, 114–15; Chickering, *Imperial Germany and the Great War,* 171; Stokesbury, 221.

Meanwhile, in Germany, the Allied blockade, in combination with the failed potato harvest of 1916, led to the hunger of the "turnip winter" of 1916–17. Vincent, *Politics of Hunger*; Roerkohl; and August Skalweit, *Die deutsche Kriegsnährungswirtschaft* (Stuttgart: Deutsche Verlags-Anstalt, 1927).

26. Einstein, letter to Heinrich Zangger, Jan. 8, [1917], AEA [86 490]; also in *CPAE* 10: 8/287a.

27. Einstein, letter to Heinrich Zangger, Feb. 1, 1917, AEA [86 439]; also in *CPAE* 10: 8/291a.

28. Einstein, letter to Georg Friedrich Nicolai, [ca. Jan. 22, 1917], AEA [44 556]; also in *CPAE* 8: 289); Georg Friedrich Nicolai, letter to Einstein, Feb. 26, 1917, AEA [82 522]; also in *CPAE* 8: 302; Einstein, letter to Georg Friedrich Nicolai, Feb. 28, 1917, AEA [44 552]; also in *CPAE* 8: 303; and Einstein, letter to Georg Friedrich Nicolai, [after Feb. 28, 1917], AEA [44 554]; also in *CPAE* 8: 304.

29. Einstein, letter to Michele Besso, Mar. 9, 1917, AEA [7 296]; also in *CPAE* 8: 306; and Einstein, letter to Heinrich Zangger, Mar. 10, 1917, AEA [39 680]; also in *CPAE* 8: 309.

30. Victor Adler, note to Einstein on calling card imprinted "Hr. Victor Adler," [early 1917], AEA [34 413].

31. On the process leading to Einstein's appointment at the German University of Prague, see *CPAE* 5: 204; 211; 224; 227; and notes to these documents.

32. Friedrich Adler and Johann Wolfgang Brügel, eds., *Friedrich Adler vor dem Ausnahmegericht. 18. u. 19. Mai 1917* (Vienna: Europa Verlag, 1967); Peter Galison, Gerald Holton, and Silvan S. Schweber, eds., *The Assassin of Relativity,* DVD video (Univ. of

California, Berkeley, Department of Physics, 2006); Rudolf G. Ardelt, *Friedrich Adler. Probleme einer Persönlicheitsentwicklung um die Jahrhundertwende* (Vienna: Österreichischer Bundesverlag, 1984); Ronald Florence, *Fritz: The Story of a Political Assassin* (New York: Dial Press, 1971); Robert William Kern, *Friedrich Adler and the Austrian Social Democratic Party*, AM thesis, University of Chicago, 1961.

33. Einstein, letter to Kathia Adler, Feb. 20, 1917, AEA [6 025]; also in *CPAE* 8: 301.

34. Einstein, letter to Emperor Franz Josef, [between mid-Feb. and Apr. 29, 1917], AEA [91 157]; also in *CPAE* 10: 8/300a.

35. Einstein, letter to Heinrich Zangger, [after Mar. 10, 1917], AEA [39 682]; also in *CPAE* 8: 310. Michele Besso was one of Einstein's most important friends from his university days; Walter Dällenbach was a mechanical engineer at the ETH working on a dissertation under the supervision of Hermann Weyl.

36. Friedrich Adler, letter to Einstein, Mar. 9, 1917, AEA [6 001]; also in *CPAE* 8: 307.

37. Einstein, letter to Friedrich Adler, Apr. 13, [1917], AEA [6 023]; also in *CPAE* 8: 324.

38. Einstein, letter to Walther Rathenau, Mar. 8, 1917, AEA [80 223]; also in *CPAE* 8: 305. Einstein was referring to Walther Rathenau, *Von kommenden Dingen* (Berlin: Fischer Verlag, 1917), of which Rathenau had given him a copy autographed on March 2, 1917. Those who see Einstein's opposition to representatives of economic interests having access to military weapons as a prescient expression of convictions similar to those of 1950s US president Dwight D. Eisenhower's later critique of a "military-industrial complex" might be interested in another exploration of Einstein's foresight: Peter L. Galison, Gerald Holton, and Silvan S. Schweber, eds., *Einstein for the 21st Century: His Legacy in Science, Art, and Modern Culture* (Princeton: Princeton Univ. Press, 2008).

39. Einstein, letter to Heinrich Zangger, Mar. 10, 1917, AEA [39 680]; also in *CPAE* 8: 309.

40. Willem de Sitter, letter to Einstein, Mar. 15, 1917, AEA [20 543]; also in *CPAE* 8: 312; Willem de Sitter, "On Einstein's Theory of Gravitation, and Its Astronomical Consequences. First Paper," *Royal Astronomical Society. Monthly Notices* 76 (1915–16), 699–728; Willem de Sitter, "On Einstein's Theory of Gravitation, and its Astronomical Consequences. Second Paper," *Royal Astronomical Society. Monthly Notices* 77 (1916–17), 155–84.

41. During the middle two weeks of March 1917, the first phase of the Russian Revolution began with a general strike by workers (including those in the major armaments factory), the mutiny of sailors at Kronstadt, the detention of Tsar Nicholas II of Russia, and the abdication of the Tsar. On March 17, power in Russia passed to the provisional government, which declared its intention to continue fighting the War. The Bolsheviks, however, were determined to end what they regarded as an imperialist war—a sentiment widely shared among workers and soldiers. On April 16, 1917, Russian Bolshevik leader

Vladimir Lenin returned to Petrograd from his exile in Zurich, Switzerland, to begin his campaign to transform the more democratic Russian Revolution of March 1917 into the more socialist one of November. In November 1917, after the second phase of the Russian Revolution, Russia would withdraw from the War, putting an end to the eastern front and enabling Germany to focus exclusively on the western front. Gilbert, 313–17, 319, 326–28, 373–74; Keegan, 332–43; Ferro, 153, 161–71, 177–79, 184–95, 207–10; Chickering, *Imperial Germany and the Great War,* 157–58, 168; Stokesbury, 205–15.

On March 26, in the first battle of Gaza, British cavalry retreated after 17,000 Turks blocked their advance. Gilbert, 317; and Stokesbury, 191.

42. On April 2, 1917, President Woodrow Wilson asked the US Congress for a declaration of war on Germany. Congress voted in favor, and on April 6 the United States declared war on Germany. Gilbert, 318; Keegan, 352; Ferro, 115; Chickering, *Imperial Germany and the Great War,* 171; Stokesbury, 221–22.

Between April 9 and 15, on the western front, British and Canadian troops launched simultaneous offensives at Arras and Vimy Ridge, achieving initial successes before being stopped by German forces, with heavy losses on both sides. The cemeteries at Arras and Vimy Ridge memorialize the names of more than 38,300 British soldiers and 11,500 Canadian soldiers killed there. On April 16, the French launched the Nivelle offensive on the river Aisne, which led to massive and unexpected losses—more than 40,000 casualties in the first few days. Gilbert, 320–25; Keegan, 184, 324–29; Ferro, 83; Chickering, *Imperial Germany and the Great War,* 174; and Stokesbury, 229–35.

43. Hendrik A. Lorentz, letter to Einstein, Mar. 22, 1917, AEA [16 458]; also in *CPAE* 8: 315.

44. Einstein, letter to Hendrik A. Lorentz, Apr. 3, 1917, AEA [16 460]; also in *CPAE* 8: 322; and in Nathan and Norden, 18.

45. Einstein, letter to Heinrich Zangger, Apr. 16, [1917], AEA [87 056]; also in *CPAE* 10: 8/326a.

46. Einstein, letter to Heinrich Zangger, [on or before Apr. 29, 1917], AEA [86 544]; also in *CPAE* 10: 8/330b.

47. Einstein, letter to Michele Besso, [Apr. 29, 1917], AEA [7 299]; also in *CPAE* 8: 331.

48. Einstein, letter to Heinrich Zangger, May 4, 1917, AEA [86 459]; also in *CPAE* 10: 8/332a.

49. "Friedrich Adler als Physiker" (interview of Einstein), *Vossische Zeitung,* no. 259 (May 23, 1917), 2–3.

50. Einstein, letter to Michele Besso, May 13, 1917, AEA [7 305]; also in *CPAE* 8: 339.

51. On May 9, 1917, the Nivelle offensive was abandoned at a time of widespread mutiny among French soldiers in the army, who were demanding improved conditions and refusing to return to the front during May and June 1917. On May 27, over 30,000

French troops refused to go to the trenches in Missy-aux-Bois. On June 1, a French infantry regiment seized Missy-aux-Bois and declared an anti-War military government. The French army arrested the members of this regiment shortly afterwards. French general Robert Nivelle was replaced by General Henri-Philippe Pétain, who ended the crisis through a judicious combination of a limited number of court-martials with severe punishments, improved conditions for soldiers (such as better food, leave time), and a promise to refrain from further French offensives in the immediate future. Gilbert, 325, 333–34; Keegan, 329–31; Ferro, 83, 181–84; Chickering, *Imperial Germany and the Great War,* 174–76; Stokesbury, 234–37.

The German bombing of London intensified starting on May 25, when bombing raids killed dozens of civilians. The raids continued, and on June 13 fourteen German bombers dropped over 100 bombs on London, killing 162 civilians and injuring 432 people. Gilbert, 334–35, 339–40.

On May 18, 1917, the US Congress passed the Selective Service Act, giving the president the power to draft soldiers. On June 5, conscription began in the United States. On June 15, domestic wartime restrictions intensified with the passage of the Espionage Act. Gilbert, 336; Keegan, 352–53, 372–73; Ferro, 115.

52. Einstein, letter to Paul Ehrenfest, June 3, 1917, AEA [9 402]; also in *CPAE* 8: 350; and in Nathan and Norden, 18–19.

53. Einstein, letter to Heinrich Zangger, June 17, 1917, AEA [86 462]; also in *CPAE* 10: 8/352a.

54. Einstein, “On the Questionnaire Concerning the Right of National Self-Determination” (“Zum Fragebogen über das Selbstbestimmungsrecht der Völker”), [July 1917–before Mar. 10, 1918], AEA [28 095]; also in *CPAE* 7: 6/45a.

55. On July 6, 1917, Arab troops accompanied by T. E. Lawrence captured Aqaba from the Turks. Gilbert, 344; Keegan, 414–15; Ferro, 101; Stokesbury, 191–92. The first use of mustard gas occurred on the western front when, from July 12 through the end of the month, German troops fired about one million gas shells, killing over six hundred Allied soldiers and injuring tens of thousands. Gilbert, 346, 351.

56. On July 19, Matthias Erzberger spoke before a Reichstag committee on his resolution demanding peace without annexations. The Reichstag passed his resolution by a vote of 212 to 126. The peace resolution stated, “The Reichstag strives for a peace of understanding and the permanent reconciliation of peoples. Forced acquisitions of territory and political, economic, or financial oppression are inconsistent with such a peace.” Gilbert, 347; Ferro, 196; Chickering, *Imperial Germany and the Great War,* 163–64; Stokesbury, 299–300. In the days preceding the vote, the generals Erich Ludendorff and Paul von Hindenburg blamed Chancellor Theobald von Bethmann-Hollweg for this peace resolution and pressured Kaiser Wilhelm II to dismiss Bethmann-Hollweg, which the kaiser did on July 12. On July 13, Bethmann-Hollweg resigned and was succeeded by Georg Michaelis. Gilbert, 347; Ferro, 153, 196; Chickering, *Imperial Germany and the*

Great War, 162–64; Stokesbury, 299–300. For more detail on the replacement of Bethmann Hollweg, see Wolfgang J. Mommsen, "Die deutsche öffentliche Meinung und der Zusammenbruch des Regierungssystems Bethmann Hollweg im Juli 1917," *Geschichte in Wissenschaft und Unterricht* 19 (1968): 656–71.

57. On July 20, the Corfu Declaration, which helped create the postwar Kingdom of Yugoslavia, was signed by the Yugoslav Committee and the Kingdom of Serbia. Gilbert, 348; Ferro, 100. In Russia, public pressure for an end to the War became increasingly compelling: six thousand sailors at the Kronstadt naval base joined an uprising in Petrograd on July 16, and a German-Austrian offensive against Russia starting on July 19 turned into a rout during which six thousand retreating Russian soldiers were captured and tens of thousands more deserted. In the wake of these events, Prince Georgy Lvov resigned as premier of the Russian provisional government on July 20; he was replaced by Alexander Kerensky. Gilbert, 249; Keegan, 339; Chickering, *Imperial Germany and the Great War*, 176; Stokesbury, 211.

58. In August 1917 in the United States, in the Green Corn Rebellion, several hundred central Oklahoma farmers resisted the draft. The IWW (International Workers of the World) opposed the War, and a pacifist demonstration was held in New York on June 15, but generally patriotic sentiment ran high once war was declared and the government shifted from anti-War to pro-War propaganda. Ferro, 116; Peter Brock and Nigel Young, *Pacifism in the Twentieth Century* (New York: Syracuse Univ. Press, 1999), 29–33; Peter Brock, *Twentieth-Century Pacifism* (New York: Van Nostrand Reinhold, 1970), 27–32.

59. Einstein, postcard to Elsa Einstein, [July 4, 1917], AEA [143 040]; also in *CPAE* 10: 8/359d.

60. Einstein, postcard to Elsa Einstein, [July 25, 1917], AEA [143 049]; also in *CPAE* 10: 8/364b; *Frankfurter Zeitung*, July 14, 1917, 2nd morning edition; and Wolfgang Steglich, *Die Friedenspolitik der Mittelmächte 1917/18*, vol 1. (Wiesbaden: Steiner, 1964), 107–8.

61. Einstein, postcard to Elsa Einstein, [Aug. 9, 1917], AEA [143 055]; also in *CPAE* 10: 8/370b.

62. Einstein, postcard to Elsa Einstein, [Aug. 15, 1917], AEA [143 059]; also in *CPAE* 10: 8/371a.

63. Einstein, postcard to Elsa Einstein, [Aug. 17, 1917], AEA [143 060]; also in *CPAE* 10: 8/371b.

64. Einstein, letter to Paul Seippel, [Aug. 19, 1917], AEA [44 885]; also in *CPAE* 8: 372.

65. Near Ypres, Belgium, where the soil was particularly clay-like, two weeks of continuous rain in the first half of August 1917 and more rain in September and October created slippery conditions, and shell holes filled with water, making forward movement by soldiers carrying 60 to 80 pounds of materiel a dangerous undertaking; many Allied soldiers drowned, in addition to those killed by shelling or mustard gas. The largely British

offensive ended when the village of Passchendaele was taken by Canadian divisions on November 10, 1917, after about 300,000 Allied casualties, including 66,000 killed. There were about 400,000 German casualties in that battle, with 83,000 killed and 26,000 taken prisoner. Gilbert, 353–55, 361–65, 367, 370, 377; Keegan, 355–69; Ferro, 84, 152; Chickering, *Imperial Germany and the Great War,* 176; Stokesbury, 239–43. Regarding Allied soldiers killed or wounded at Passchendaele, Gilbert cites the figure 245,000, while Chickering cites 324,000. Passchendaele made such a strong impression on the home front that when the draft began in Canada on October 13, 1917 (the Military Service Act having been passed on August 29), 21,568 Canadian men reported for service while 310,376 applied for exemptions, most of which were granted. By January 1, 1918, 380,510 had applied for exemptions. Gilbert, 377.

66. Einstein was probably referring to Berlin astronomer Wilhelm Foerster, a fellow cosigner of the 1914 "Appeal to the Europeans," and father of the likewise pacifist Munich historian Friedrich Wilhelm Foerster. If not, then he was referring to the latter.

67. Einstein, letter to Heinrich Zangger, [Aug. 21, 1917], AEA [89 523]; also in *CPAE* 10: 8/372a; and in Rowe and Schulmann, 77–78.

68. Presumably, Rolland was referring to Rolland, *Liluli* (Paris: Ollendorff, 1919); and Rolland, *Pierre et Luce* (Geneva: Sablier, 1920).

69. Nicolai, *Die Biologie des Krieges.*

70. Romain Rolland, letter to Einstein, Aug. 21, 1917, AEA [33 008]; also in *CPAE* 8: 373; and in Nathan and Norden, 19.

71. Einstein, letter to Romain Rolland, Aug. 22, 1917, AEA [84 166]; also in *CPAE* 8: 374; in Rowe and Schulmann, 78–79; and in Nathan and Norden, 20–21. German chancellor Theobold von Bethmann Hollweg had resigned on July 13, 1917, and was replaced as chancellor by Georg Michaelis. On the political maneuverings around this event, see Wolfgang J. Mommsen, "Die deutsche öffentliche Meinung."

72. Romain Rolland, letter to Einstein, Aug. 23, 1917, AEA [33 011]; also in *CPAE* 8: 376. Bertrand Russell, head of the British National Committee of the No-Conscription Fellowship, had recently been fired from a position as lecturer in mathematics and logic at Trinity College, Cambridge University, due to his pacifist activity. On Russell, see Jo Vellacott, *Bertrand Russell and the Pacifists in the First World War* (Bath: Harvester Press, 1980); and Gilbert, 253.

73. Einstein, letter to Heinrich Zangger, [Aug. 26, 1917], AEA [86 551]; also in *CPAE* 10: 8/376a; Einstein, letter to Michele Besso, Sept. 22, 1917, AEA [7 317]; also in *CPAE* 8: 381; and Einstein, letter to Michele Besso, Oct. 6, 1917, AEA [86 562]; also in *CPAE* 10: 8/385a.

74. Einstein, letter to Edouard Guillaume, Sept. 24, 1917, AEA [79 001]; also in *CPAE* 8: 383.

75. "Kundgebungen für den Verständigungsfrieden," *Frankfurter Zeitung* 61, no. 284 (Oct. 14, 1917), 1st morning supplement, 4.

76. Einstein, letter to Werner Weisbach, Oct. 15, 1917, AEA [76 156]; also in *CPAE* 8: 391.

77. On October 15, at Vincennes outside of Paris, the Dutch dancer Mata Hari was executed by a firing squad for spying for Germany. Mata Hari's real name was Margueretha Gertruida Zella. Gilbert, 368. On October 26, Brazil declared war on Germany. Gilbert, 158, note 1; Ferro, 205.

78. On November 7, 1917, the Bolsheviks in St. Petersburg, led by Vladimir Lenin, attacked the Kerensky provisional government and took control. Much of their popular support, including that of nine thousand sailors at the Kronstadt naval base, derived from their advocacy of Russian withdrawal from the War. On December 3, Bolshevik and German representatives started meeting in Brest-Litovsk to work out a peace agreement, although the treaty would not be signed until March 3 the following year, after another German offensive against the almost nonexistent Russian army in February 1918. Gilbert, 373–74; Keegan, 341; Ferro, 207–9; Chickering, *Imperial Germany and the Great War,* 176; Stokesbury, 212–15.

79. In the final months of 1917 and the beginning of 1918, while the Russian army was disintegrating and the Russian civil war was breaking out, various nationalities moved toward independence or founding their own countries. On November 20, Ukraine declared itself a republic; on December 6, Finland declared independence from Russia; on December 27, Latvia declared independence; and in January 1918, Romania took steps toward forming its own country. Gilbert, 386; Keegan, 383; Ferro, 209–12; Stokesbury, 211–15.

80. On November 7, 1917, the third battle of Gaza ended with British forces capturing Gaza from the Ottoman Empire. On November 16, British troops occupied Tel Aviv and Jaffa in Palestine. On December 11, they captured Jerusalem from the Ottoman Empire. Gilbert, 370–72, 376–77, 387–88; Keegan, 415; Ferro, 67; Stokesbury, 254–56.

On November 16, Georges Clemenceau, who was more militarily aggressive than his predecessor, became prime minister of France. On November 20, the battle of Cambrai began with about 250,000 British troops attacking about 250,000 Germans. This was the first battle in which the primary force of the attack was carried out by tanks, with 324 British tanks starting the attack. British forces made early gains, leading to elation in England and the United States, but were soon beaten back by a German counterattack on November 30. On December 4, they were obliged to dig in for the winter. The Germans regained most of the territory lost at first and some additional ground previously held by the Allies. Dead and wounded at Cambrai numbered 44,000 British and Canadians and 53,000 Germans. Gilbert, 378–83; Keegan, 369–71; Ferro, 198–200, 218; Chickering, *Imperial Germany and the Great War,* 179, 198; Stokesbury, 244, 248–50.

81. Hans Thirring, letter to Einstein, Dec. 3, 1917, AEA [84 047.2]; also in *CPAE* 8: 401; Einstein, letter to Gustav Mie, Dec. 14, 1917, AEA [17 211]; also in *CPAE* 8: 407; Brigitte Zimmel and Gabriele Kerber, eds., *Hans Thirring. Ein Leben für Physik und*

Frieden (Vienna: Böhlau, 1992), 18; and Gustav Mie, letter to Wilhelm Wien, Oct. 21 to Nov. 1, 1917, Nachlaß Wilhelm Wien, 056/009, Archiv, Deutschen Museum, Munich, Germany.

82. Einstein, letter to Heinrich Zangger, Dec. 6, 1917, AEA [39 689]; also in *CPAE* 8: 403. Exemplifying the significant contribution of disease to high mortality rates from the War, Lemberg mathematical physicist Marian von Smoluchowski had died on September 5, 1917, at age forty-five from effects of dysentery. See also Holmes, "Was Einstein Really a Pacifist?"

83. Einstein, letter to Heinrich Zangger, Dec. 6, 1917, AEA [39 689]; also in *CPAE* 8: 403.

84. Heinrich Zangger, letter to Einstein, [Dec. 17, 1917], AEA [39 691]; also in *CPAE* 8: 412.

85. Heinrich Zangger, letter to Einstein, [Dec. 17, 1917], AEA [39 693]; also in *CPAE* 8: 411.

86. Einstein, letter to Hendrik A. Lorentz, Dec. 18, 1917, AEA [16 462]; also in *CPAE* 8: 413; and in Nathan and Norden, 21.

87. Michele Besso, letter to Einstein, Dec. 27, 1917, AEA [7 070]; also in *CPAE* 8: 419.

88. Max Planck, letter to Einstein, Dec. 29, 1917, AEA [19 261]; also in *CPAE* 8: 423.

3. Hope and Optimism beyond the Turbulence of Defeat and Revolution: 1918

1. Gilbert, 394; Keegan, 413; Ferro, 208; Chickering, *Imperial Germany and the Great War*, 187–88; Stokesbury, 301–3; N. Gordon Levin Jr., *Woodrow Wilson and World Politics: America's Response to War and Revolution* (Oxford: Oxford Univ. Press, 1968), 61, 64, 73. See Woodrow Wilson, "Fourteen Points," President Wilson's Message to Congress, Jan. 8, 1918, records of the US Senate, RG 46, National Archives. Text of the "Fourteen Points" is also presented in Margaret MacMillan, *Paris 1919: Six Months that Changed the World* (New York: Random House, 2001), 495–96.

2. Einstein, "On the Questionnaire"; and Einstein, letter to Heinrich Zangger, [Aug. 21, 1917], AEA [89 523]; also in *CPAE* 10: 8/372a; and in Rowe and Schulmann, 77–78.

3. As frustration with wartime conditions grew in Germany, over 400,000 workers went on strike in Berlin on January 28. Within forty-eight hours, strikes had spread to six other German cities. German authorities declared martial law in Berlin and Hamburg and drafted many striking workers into the army, often sending strike leaders to the front, which officers noted was harmful to morale there. Gilbert, 395; Keegan, 381; Ferro, 215; and Chickering, *Imperial Germany and the Great War*, 184.

4. Einstein, letter to Michele Besso, Jan. 5, 1918, AEA [7 318]; also in *CPAE* 8: 428.

5. Hans Albert Einstein, letter to Einstein, [after Jan. 25, 1918], AEA [144 009]; also in *CPAE* 10: 8/442a.

6. Fritz Haber, letter to Einstein, [before Jan. 29, 1918], AEA [12 310]; also in *CPAE* 8: 445; Einstein, letter to Fritz Haber, Jan. 29, 1918, AEA [12 312]; also in *CPAE* 8: 446. (Haber was director of the Kaiser Wilhelm Institute of Physical Chemistry and Electrochemistry.)

7. Cornelia and Gunnar Nordström, letter to Einstein, Jan. 31, 1918, AEA [18 430]; also in *CPAE* 8: 452.

8. Gustav Mie, letter to Einstein, Feb. 5, 1918, AEA [17 215]; also in *CPAE* 8: 456.

9. Einstein, letter to Erwin Freundlich, [before Jan. 17, 1918], AEA [11 225.1]; also in *CPAE* 8: 438.

10. Christa Kirsten and Hans-Jürgen Treder, eds., *Albert Einstein in Berlin 1913–1933. Part 1, Darstellung und Dokumente* (Berlin: Akademie-Verlag, 1979), Doc. 118, 198–99.

11. Ernst Troeltsch, letter to Einstein, Feb. 4, 1918, AEA [45 133]; also in *CPAE* 8: 455; and Ernst Troeltsch, letter to Einstein, Feb. 7, 1918, AEA [45 129]; also in *CPAE* 8: 458.

12. Between January 27 and April 1918, a civil war between Red and White armies was fought in Finland, parallel to the one happening in Russia. At the same time, Finnish nationalists demanded the withdrawal of all Russian troops from Finnish territory, to which Bolshevik leaders agreed on February 23; on February 26 a Finnish battalion pledged to drive out the rest of the Bolsheviks. On March 6, Finland formed an alliance with Germany, which had helped them drive the Bolsheviks out of Finland. Lithuania on February 16, and Estonia on February 24, declared independence from Russia; meanwhile, Armenia, Azerbaijan, and Georgia declared independence, creating the Transcaucasian Democratic Federative Republic. Gilbert, 399, 401–2; Keegan, 378–80, 383–86, 420; Ferro, 209–11.

On March 3, after peace negotiations had stalled two weeks earlier, while German forces were advancing rapidly though Russian territory, Germany, Austria, and the Soviet Union signed the Treaty of Brest-Litovsk, ending Russia's involvement in World War I. Notwithstanding the conclusion of this treaty, and partly because German troops were near Petrograd, on March 6 the Soviet Union moved its capital from Petrograd to Moscow. Gilbert, 400–402; Keegan, 62, 381–83, and 386–87; Ferro, 209–11; Chickering, *Imperial Germany and the Great War,* 167–68 and 178; Stokesbury, 213–15.

13. In February 1918, British aircraft bombed German barracks and airfields, including one near Metz. From March 4 to 8, Austrian airplanes bombed four Italian cities, including Venice, and German planes bombed London, Paris, and Naples. On March 12, German troops occupied Odessa; at that point, Germany controlled territory from the North Sea to the Black Sea. With Russia out of the war, the German army focused on the western front. On March 9, they hurled 1,000 tons of gas in preliminary artillery and poison gas attacks as a lead-up to what would be known as the second battle of the Somme. Due to material limitations and strained public opinion, this would be their last major

offensive on the western front. On March 21, the offensive began again with bombardment by artillery and about 2 million gas shells. On March 23, three German cannons shelled Paris from 114 kilometers or 74 miles away, killing 256 civilians. Fighting was fierce, with high casualties on both sides. German troops advanced during the first seven to nine days, crossing the Somme on March 24 and getting within fifty miles of Paris by March 27. But the German advance was largely stopped by then, and successful Allied counterattacks began on March 30, continuing until the Germans ended the offensive on April 12. During this offensive, about a quarter of a million German soldiers were killed and wounded; the French and British armies suffered similar combined losses. Gilbert, 403–12; Keegan, 392–404; Ferro, 215–16; Chickering, *Imperial Germany and the Great War*, 178–81; Stokesbury, 262–67.

14. Einstein, letter to Karl Camillo Schneider, Feb. 24, 1918, AEA [44 985]; also in *CPAE* 8: 471; and in Nathan and Norden, 22; and Karl Camillo Schneider, letter to Einstein, [after Mar. 16, 1918], AEA [44 989]; also in *CPAE* 8: 482.

15. Paul Ehrenfest, letter to Einstein, Mar. 27, 1918, AEA [9 410]; also in *CPAE* 8: 494; and Einstein, letter to David Hilbert, [Apr. 12, 1918], AEA [13 108]; also in *CPAE* 8: 503.

16. Wilhelm Schweydar, letter to Einstein, Apr. 14, 1918, AEA [21 566]; also in *CPAE* 8: 504.

17. Einstein, letter to Heinrich Zangger, Apr. 22, 1918, AEA [39 705]; also in *CPAE* 8: 514.

18. Einstein, letter to Hans Albert Einstein, Apr. 26, 1918, AEA [75 865]; also in *CPAE* 8: 520.

19. Einstein, letter to David Hilbert, before Apr. 27, 1918, AEA [13 115]; also in *CPAE* 8: 521; and in Rowe and Schulmann, 80. This idea of a collection of essays to promote international reconciliation was first initiated in early 1917 by the University of Berlin professor of medicine Georg Friedrich Nicolai (who in early 1918, with much fanfare but little success, was still attempting to resist the German military from within its ranks).

20. David Hilbert, letter to Einstein, May 1, 1918, AEA [13 122]; also in *CPAE* 8: 530. Max Lehmann, University of Göttingen professor of history, was another member of the New Fatherland League (Bund Neues Vaterland) and had signed the Delbrück-Dernburg petition of July 1915 opposing German territorial annexations. On the attacks on Richard von Kühlmann, state secretary in the Ministry of Foreign Affairs, who as German representative had signed the treaty with the Soviets at Brest-Litovsk, see *Deutsche Zeitung* 23, no. 204 (Apr. 23, 1918), 1–2. On the difficulties of Antonius Count von Spee, a Center Party representative in the Prussian legislature, in his efforts regarding an electoral reform bill, see *Deutsche Zeitung* 23, no. 218 (Apr. 30, 1918), 2; and *Deutsche Zeitung* 23, no. 209 (May 1, 1918), 1–2.

21. Ernst Troeltsch, letter to Einstein, May 1, 1918, AEA [45 131]; also in *CPAE* 8: 531.

22. Einstein, letter to Ilse Einstein, [May 12, 1918], AEA [44 559]; also in *CPAE* 8: 536; and Einstein, letter to Georg Friedrich Nicolai, [May 12, 1918], AEA [44 559.1]; also in *CPAE* 8: 537. On Nicolai's court-martial and detention, see Zuelzer, chapter 12.

23. Georg Friedrich Nicolai, letter to Einstein, May 18, 1918, AEA [44 561]; also in *CPAE* 8: 541; and in Nathan and Norden, 7–8.

24. Einstein, letter to David Hilbert, May 24, 1918, AEA [13 124]; also in *CPAE* 8: 548.

25. On the western front on April 9, 1918, the German army launched another fierce offensive along the Lys River, which included firing 2,000 tons of mustard gas and other poison gases. By then the armies on both sides were approaching exhaustion, and the first American troops were arriving and on the verge of entering the War. The Germans captured significant territory in the first twenty days of this offensive. On April 20, they unleashed another 2,000 tons of poison gas. On April 24, the Allies began retaking territory and driving the Germans back, until the German army ended its offensive on April 29. By then, over 20,000 Allied and 30,000 German soldiers had been killed. Throughout May and early June 1918, German planes bombed London, Paris, and other targets, and British planes bombed Cologne and other targets. On May 27, the Germans launched the third battle of the Aisne, again with initial successes, advancing to within forty miles of Paris. However, by June 3, strengthened by American troops, the Allies halted the German advance, most famously at Belleau Wood. On May 16, 1918, the US Congress passed and Woodrow Wilson signed the Sedition Act, aimed at curbing pacifist and union opponents of the War, many of whom were then arrested. Gilbert, 412–15, 418, 423–24, and 425–30; Keegan, 405–8; Ferro, 116 and 216; Chickering, *Imperial Germany and the Great War,* 180–83; Stokesbury, 268–70, 273–77.

26. In the east, throughout April 1918 the German army occupied major cities in and around the Ukraine, including Kharkov, Simferopol, and Sebastopol, eventually controlling all of the Ukraine and later confiscating its 1918 harvest. German troops invaded the Don province in Russia on May 1 and captured Rostov on May 6. On May 7, Romania and the Central Powers signed the Treaty of Bucharest; the Central Powers rewarded their ally Bulgaria with coastal lands it had lost to Romania in 1913. Gilbert, 419, 421; Keegan, 386–88; Ferro, 209–11, 215; Chickering, *Imperial Germany and the Great War,* 167; Stokesbury, 259. On June 8, 1918, the German army launched another offensive on the western front, firing 15,000 tons of poison gases. This time, the Allies were better prepared. The Germans called off the offensive on June 12; the Allies responded with their own offensives. On July 15, the Germans launched yet another offensive, known as the second battle of the Marne, in France near the River Marne. After fierce fighting that has been described as carnage, the tide turned; July 18, French and American troops began advancing and would continue to predominate through July and into August. Gilbert, 431–32, 440–48; Keegan, 408–9; Ferro, 216–17; Chickering, *Imperial Germany and the Great War,* 184–86; Stokesbury, 277–80.

On May 26, the Transcaucasian Democratic Federative Republic was abolished. Georgia declared its independence, and on May 27 the first Republic of Georgia was founded. On May 28, Armenia and Azerbaijan each declared independence. Gilbert, 425; Keegan, 383; Ferro, 211; Stokesbury, 214. On July 16, at Ekaterinburg, the Bolsheviks killed Tsar Nicholas II, the Empress Alexandra, their five children, and their household employees. Gilbert, 439; Ferro, 186. On July 13, the National Czechoslovak Council was established, which Britain recognized on August 9 as the trustee of the future Czech government; on October 14, the Allies would recognize it as the provisional government of the future Czechoslovakia. Gilbert, 450–51, 479; Stokesbury, 298.

27. Einstein, letter to Adolf Kneser, June 7, 1918, AEA [81 562]; also in *CPAE* 8: 560.

28. Divorce Agreement, June 12, 1918, AEA [81 362]; also in *CPAE* 8: 562; and Divorce Decree, Feb. 14, 1919, AEA [81 374]; also in *CPAE* 9: 6.

29. Einstein, letter to Michele Besso, [July 9, 1918], AEA [7 321]; also in *CPAE* 8: 586.

30. Einstein, letter to Michele Besso, [July 29, 1918], AEA [7 322]; also in *CPAE* 8: 591; Rudolf Goldscheid, *Reine Vernunft und Staatsvernunft* (Leipzig: Suschitzky, 1918).

31. Friedrich Adler, letter to Einstein, July 6, 1918, AEA [6 007]; also in *CPAE* 8: 582; Einstein, letter to Friedrich Adler, Aug. 4, 1918, AEA [6 024]; also in *CPAE* 8: 594; and Friedrich Adler, letter to Einstein, Aug. 9, 1918, AEA [6 008]; also in *CPAE* 8: 596. For more on the Adler case, see Florence, *Fritz*.

32. On August 8, in the battle of Amiens, Canadian and Australian troops broke through the German front lines and began a series of victories. German general Erich Ludendorff would later call this the "black day of the German army." Because the German military leadership saw the arrival of American troops as the greatest threat to their continuing War effort, on August 14, Ludendorff recommended to the German Crown Council immediate negotiations for peace. British and French leaders had until then been deeply pessimistic, planning for a War into 1919 or possibly even 1920. But on August 20, in response to Allied advances, French general Ferdinand Foch told Georges Clemenceau that they could achieve victory in 1919, and on August 21 British general Douglas Haig stated that they could attain victory before the end of 1918. Allied territorial gains continued throughout August and September. Fighting was still fierce and losses were heavy, but German morale was plummeting, and increasing numbers of German soldiers were surrendering. In September, with a new expectation of forward movement, the Allies started broader offensives, which included the firing of poison gas shells by American and other Allied troops. Gilbert, 450–52, 454–60, 465–66; Keegan, 410–12; Ferro, 217–18; Chickering, *Imperial Germany and the Great War*, 186–87; Stokesbury, 283–91.

33. The "Spanish Flu" had begun to spread in June 1918, and by August it became a pandemic. Over the following six months more than 25 million people would die of the disease, almost twice as many as died in the War. Because of poor nutrition, the disease hit German soldiers and civilians disproportionately hard. More than 62,000 American

soldiers died of influenza in 1917–18, which was more than the 48,900 who died in combat. Gilbert, 437; Keegan, 408; Chickering, *Imperial Germany and the Great War,* 278.

On August 1, British forces occupied Archangel, on the White Sea, as part of an anti-Bolshevik offensive in eastern Russia and Siberia in the summer of 1918, along with Czech, French, Japanese, and American troops. As part of this campaign, American troops landed in Vladivostok on August 16 and in Archangel on September 11. Gilbert, 438–39, 449, 452–53, 460; Keegan, 385–92; Ferro, 210–13; Stokesbury, 298, 313.

34. Einstein, letter to Walter Dällenbach, Aug. 5, 1918, AEA [9 078]; also in *CPAE* 8: 595.

35. Heinrich Zangger, letter to Einstein, [before Aug. 11, 1918], AEA [39 701]; also in *CPAE* 8: 598.

36. Kurt Hiller, letter to Einstein, Sept. 7, 1918, AEA [43 896]; also in *CPAE* 8: 611. The Westend Resolutions were taken at the League for the Goal's founding meeting during August 10–12, 1917, with twenty-eight signatories. Lewis D. Wurgaft, "The Activists: Kurt Hiller and the Politics of Action on the German Left 1914–1933," *American Philosophical Society: Transactions* 67 (1977): 51–52; and Kurt Hiller, ed., "Anhang: Dokumente," in *Das Ziel: Jahrbücher für geistige Politik,* vol. 3, part 1 (Leipzig: Wolff, 1919), 218–23.

37. Einstein, letter to Kurt Hiller, Sept. 9, 1918, AEA [43 898]; also in *CPAE* 8: 613; and in Nathan and Norden, 22–23.

38. On September 14, Austria requested a confidential meeting with the Allies to exchange views on seeking peace; the United States rejected the request on September 16, France on September 17. On September 19, Belgium rejected a German separate peace offer that was predicated on no restitution claims. On September 28, General Erich Ludendorff strongly recommended to General Paul von Hindenburg, and on September 29 to the civilian leadership, that Germany immediately request an armistice based on Wilson's "Fourteen Points." On October 2, Kaiser Wilhelm II appointed Prince Max von Baden as chancellor. Von Baden accepted only on condition that the kaiser withdraw from direct control over the military. As Allied armies continued to advance, Hindenburg and Ludendorff pushed again for an immediate peace. On October 4, a German and Austrian "Peace Note" was sent to US president Woodrow Wilson, asking for an armistice that was not surrender, to end the War without preconditions harmful to Germany or Austria. On October 8, Wilson rejected the "Peace Note," stating that a first condition of an armistice would be withdrawal of German troops from all occupied territories, including Belgium, France, and Serbia. Gilbert, 460–63, 466, 471–75; Keegan, 412–13; Chickering, *Imperial Germany and the Great War,* 187–88; Ferro, 218–19; Stokesbury, 303–4; Levin, 131–36; Detlev J. K. Peukert, *The Weimar Republic: The Crisis of Classical Modernity,* trans. Richard Deveson (New York: Hill and Wang, 1993, c1987), 26.

39. Gilbert, 477, 479; Keegan, 408; Chickering, *Imperial Germany and the Great War,* 186; Stokesbury, 278.

40. By September 30 in Germany, 34 billion paper German marks were in circulation. Because of the ways in which Germany had financed its War, an extreme inflation was already present in Germany at the end of the War, although the German public was not yet fully aware of it or the reasons for it. Erich Eyck, *A History of the Weimar Republic*, trans. Harlan P. Hanson and Robert G. L. Waite (Cambridge, MA: Harvard Univ. Press, 1967), 130–32; Ferro, 171–72; Chickering, *Imperial Germany and the Great War*, 193.

On October 6, Serbs, Croats, and Slovenians formed the provisional government of Yugoslavia. Over the following days, Serbs drove Austrian and German troops out of Serbia; Croatians and Dalmatians followed suit on October 29. On October 7, a Polish state was declared in Warsaw. On October 14, the Allies recognized the Czechoslovak National Council, and on October 28 in Prague the Washington Declaration proclaimed an independent Czechoslovak Republic. Gilbert, 474, 477, 479, 486, 488; Keegan, 416; Ferro, 219.

41. *Mitteilungen des Bundes Neues Vaterland*, Neue Folge Nr. 1 (Nov. 1918), [3].

42. *Mitteilungen des Bundes Neues Vaterland*, Neue Folge Nr. 1 (Nov. 1918), [4]; also in Nathan and Norden, 23–24.

43. Einstein, letter to Paul Ehrenfest, Sept. 27, 1918, AEA [9 423]; also in *CPAE* 8: 625.

44. Einstein, letter to Hans Albert Einstein, Oct. 17, 1918, AEA [75 896]; also in *CPAE* 8: 634.

45. Hans Mühsam, letter to Einstein, Oct. 24, 1918, AEA [38 321]; also in *CPAE* 8: 639.

46. Max Planck, letter to Einstein, Oct. 26, 1918, AEA [19 280]; also in *CPAE* 8: 640.

47. On October 24, 1918, Ludendorff defied his civilian superiors by sending a telegram to all army commanders ordering them to fight to the finish. The telegram was stopped; however, its contents were leaked to the newspapers and published the next day, creating an uproar in Berlin. On October 26, an angry kaiser and chancellor forced Ludendorff's resignation. On October 27, the Austrian emperor informed the German kaiser that he would be seeking an immediate armistice, and the next day Austria communicated this request to the Allies. On October 30, the German emperor, Kaiser Wilhelm II, left Berlin for the Belgian town of Spa. An armistice with the Ottoman Empire began on October 31. On November 3, the armistice between the Allies and Austria-Hungary took effect. Gilbert, 482–86, 489–91; Keegan, 413–17; Ferro, 219; Stokesbury, 299, 304–5.

On October 28, when the German fleet was scheduled to leave port, sailors at Wilhelmshaven went on strike. On November 3, a sailors' mutiny began in the German fleet at Kiel. In the following days, this would fuel rebellions, which spread from city to city, beginning what is known as the German Revolution. On November 7, Kurt Eisner proclaimed the Bavarian Republic in Munich, declaring Ludwig III and the Wittelsbachs deposed. Gilbert, 485, 490–91, 493–95; Keegan, 416–17; Ferro, 220–21; Eberhard Kolb,

The Weimar Republic, trans. P. S. Falla (London and New York: Routledge, 1988, c1984), 9; Eyck, 57–58; Stokesbury, 305.

On November 9, 1918, Prince Max von Baden resigned as chancellor, handing the position to the Social Democratic Party leader Friedrich Ebert, who became head of first Weimar Republic government. On the same day, Karl Liebknecht and the Spartacists (the more radical leftists) seized the Imperial Palace and proclaimed the establishment of a German Soviet Republic; another SPD leader, Philip Scheidemann, standing on the steps of the Reichstag, declared a Socialist Republic. Also on November 9, Kaiser Wilhelm II abdicated; the following day he traveled to exile in the Netherlands. On November 10, the German Council of People's Representatives (Rat der Volksbeauftragten) was formed, consisting of three SPD representatives, Friedrich Ebert, Philipp Scheidemann, and Otto Landsberg, and three USPD (Independent Social Democratic Party) representatives, Hugo Haase, Wilhelm Dittmann, and Theodor Barth. All were members of the Reichstag except Barth. However, the early Weimar Republic's Social Democratic leadership failed to invite collaboration from the more radical leftists, nor did they make reforms beyond those that would establish a constitutional order. For instance, they failed to replace monarchists dominating the civil service. Instead, they focused largely on maintaining order. This failure had a far-reaching negative impact on the German people's introduction to democratic and republican structures and processes, thus on acceptance and support of the Weimar Republic, and hence on Germany's political development during the next three decades. Gilbert, 498; Keegan, 418–19; Ferro, 221; Chickering, *Imperial Germany and the Great War*, 188–89; Peukert, 6, 28–30; Kolb, 7–15; Eyck, 45, 49; Stokesbury, 306.

On November 11, 1918, Emperor Karl I of Austria abdicated; Austria became a republic on November 12. Also on November 11, in a railroad car outside of Compiège, France, after concluding that they had no other choice, German representatives signed the Armistice agreement, whose hard terms were a foretaste of the harsh conditions the Germans would be compelled to accept in the Versailles treaty a few months later. Gilbert, 498, 500; Keegan, 418; Ferro, 221; Chickering, *Imperial Germany and the Great War*, 197–98, 202; Stokesbury, 299, 306–7; Kolb, 9–10; Eyck, 50.

The Allied blockade of Germany still continued, but the Allies assumed responsibility for feeding Germany in the event of emergency. The blockade continued officially until the Versailles treaty was signed on June 28, 1919; it was lifted on July 11, 1919. On the Allied blockade of Germany, see Vincent, *Politics of Hunger*; and Roerkohl. On the continuation of the blockade past the end of the War, see Suda Lorena Bane and Ralph Haswell Lutz, eds., *The Blockade of Germany after the Armistice 1918–1919: Selected Documents of the Supreme Economic Council, Superior Blockade Council, American Relief Administration, and Other Wartime Organizations* (Stanford: Stanford Univ. Press, 1942). For a Berlin newspaper article objecting to the ongoing blockade, see Theodor Wolff, "Hunger, Angst und Verrohung," *Berliner Tageblatt*, Mar. 10, 1919.

48. Estimates of combat casualties in World War I:

Germany: Over 2 million killed; over 4.1 million wounded (41 percent of those mobilized were killed or wounded); 624,000 civilians died as a result of the War.

Russia: 1.7 million killed; over 4.1 million wounded; over 5 million civilians died as a result of the War.

France: 1.4 million killed; 1.7 million wounded (60 percent of those mobilized were killed or wounded); 500,000 civilians died as a result of the War.

Austria-Hungary: Over 1.2 million killed; 5.3 million wounded (38 percent of those mobilized were killed or wounded); over 2.3 million civilians died as a result of the War.

Great Britain: Over 744,000 killed; over 1.6 million wounded (37 percent of those mobilized were killed or wounded); 292,000 civilians died as a result of the War.

Italy: 615,000 killed.

Romania: 335,000 killed.

Turkey: 325,000 killed.

Bulgaria, Canada, Australia, India, the United States, Serbia, Belgium, New Zealand: Each of these countries had between 10,000 and 100,000 killed. Other countries suffered significant losses. Gilbert, 540–41; Keegan, 423; Ferro, 227; Chickering, *Imperial Germany and the Great War,* 195; Stokesbury, 310.

49. Einstein, handwritten text of a speech supporting the German Revolution delivered by Einstein at a meeting of the New Fatherland League on November 13, 1918, AEA [28 001]; also in *CPAE* 7: 14; in Rowe and Schulmann, 82–83; and in Nathan and Norden, 25–26. See also Max Born, *My Life: Recollections of a Nobel Laureate* (New York: Scribner's Sons, 1978), 184–86; and Kolb, 7–8.

50. Einstein, letter to Mileva Einstein-Marić, [ca. Nov. 9, 1918], AEA [75 876]; also in *CPAE* 8: 647.

51. Heinrich Zangger, letter to Einstein, [ca. Nov. 10, 1918], AEA [7 071]; also in *CPAE* 8: 648.

52. Michele Besso, letter to Einstein (appended to Heinrich Zangger letter), Nov. 10, 1918, AEA [7 071]; also in *CPAE* 8: 649.

53. Einstein, letter to Pauline Einstein, Nov. 11, 1918, AEA [29 352]; also in *CPAE* 8: 651; and in Nathan and Norden, 24. The building housing the apartments of Einstein, Elsa and her daughters, and Elsa's parents was on Haberland Street.

54. Einstein, letter to Paul Winteler and Maja Winteler-Einstein, Nov. 11, 1918, AEA [29 388]; also in *CPAE* 8: 652.

55. Einstein, letter to Leo Arons, Nov. 12, 1918 or later, in Leo Arons, *Universitäten heraus!* (Berlin: Verlag der Sozialistische Monatshefte, 1918), 3; also in *CPAE* 8: 653.

56. *Berliner Tageblatt* 47, no. 583 (Nov. 14, 1918), morning edition, [3].

57. Einstein, handwritten text of a speech supporting the German Revolution delivered by Einstein at a meeting the New Fatherland League on November 13, 1918, AEA [28 001]; also in *CPAE* 7: 14; in Rowe and Schulmann, 82–83; and in Nathan and

Norden, 25–26. (Nathan and Norden incorrectly identified the context of the speech as taking place four days earlier at the Reichstag.)

58. Einstein, letter to Svante Arrhenius, Nov. 14, 1918, AEA [73 209]; also in *CPAE* 8: 654.

59. Georg Count von Arco, Einstein, Georg August Gaul, Wilhelm Herzog, and René Schickele, telegram to Ludwig Quidde, Nov. 15, 1918, AEA [83 589]; also in *CPAE* 8: 655.

60. Ludwig Quidde, letter to New Fatherland League (Bund Neues Vaterland), Nov. 16, 1918, AEA [84 161]; also in *CPAE* 8: 656. Gerlach, a writer for the Berlin newspaper *Welt am Montag* had, like Einstein and Quidde, been a member of the New Fatherland League since 1915.

61. *Berliner Tageblatt* 47, no. 587 (Nov. 16, 1918), morning edition, 1; and *Vossische Zeitung* (Nov. 16, 1918), morning edition, 4.

62. On October 28, the National Council of Czechoslovakia assumed governmental powers and proclaimed independence; on November 14, Czechoslovakia became a republic, incorporating Bohemia, Moravia, Slovakia, and a large German-speaking population in the Sudetenland. On October 25, a Hungarian National Council was formed; on October 30, Hungary established a government; and on November 16, the Hungarian People's Republic was established. The Poles and Ruthenes declared independence on October 15 and 19, respectively. In a declaration by the Croatian parliament on October 29 and a vote by the Podgorica Assembly or Serbian People's National Assembly on November 26, the people of Croatia, Dalmatia, Bosnia, Herzegovina, Slovenia, Montenegro, and Serbia united to form Yugoslavia. Gilbert, 481, 486, 488, 506, 508, 520–21; MacMillan, 245; Kolb, 24, 55–56.

63. On November 15, 1918, the German Council of People's Representatives appointed Hugo Preuss, a professor of constitutional law, as minister of the interior, and Chancellor Friedrich Ebert assigned Preuss the task of preparing a draft constitution. On November 29, the Council of People's Representatives decided that elections for a National Assembly would be held on January 19, 1919; the Congress of Workers' and Soldiers' Councils voted to confirm this on December 19. Peukert, 28–36; Kolb, 15–18; and Eyck, 51, 54–56.

64. On November 15, 1918, German industrial owners and unions created a Central Labor Association (Zentralarbeitsgemeinschaft), a corporatist agreement that satisfied a number of perennial workers' demands, such as for an eight-hour workday, some social welfare commitments, and collective bargaining procedures. These concessions were a means for employers to avoid the total nationalization or socialization of industry. Peukert, 31–32; Kolb, 13–14; Richard Bessell, *Germany after the First World War* (Oxford: Oxford Univ. Press, 1993), 107.

On November 22, the Spartacus League founded the German Communist Party. On November 26, the communist organ *Die Rote Fahne* published the "Spartacus

Manifesto," which argued that socialism was the only avenue to real and lasting peace and that a socialist revolution was imminent in Germany but required the solidarity of workers in all countries. Kolb, 10–11, 15–16; for the "Spartacus Manifesto," see "Richtlinien für die Arbeiter– und Soldatenräte Deutschlands," *Die Rote Fahne* (Nov. 26, 1918); for an English translation, see "Spartacus Manifesto," in Anton Kaes et al., eds., *The Weimar Republic Sourcebook* (Berkeley: Univ. of California Press, 1994), 37–38.

65. Paul Winteler, letter to Einstein, Nov. 22, 1918, AEA [144 790]; also in *CPAE* 10: 8/659a. The nine demands were included in the proclamation of the Swiss national strike of November 11, 1918. Willi Gautschi, ed., *Dokumente zum Landesstreik 1918* (Zurich: Benziger, 1971), 237–40. Robert Grimm, a Swiss Social Democratic government official, and Arthur Hoffmann, a Swiss Liberal Democratic Party politician, had privately attempted to organize a separate peace between Germany and Russia.

66. Hans Albert Einstein, letter to Einstein, [ca. Nov. 25, 1918], AEA [144 013]; also in *CPAE* 10: 8/659b.

67. Einstein, letter to Hermann Weyl, Nov. 29, 1918, AEA [24 051]; also in *CPAE* 8: 661.

68. On December 4, 1918, President Woodrow Wilson sailed from New York, arriving in Europe on December 13 as the first US president to travel to Europe while in office. Although his "Fourteen Points" speech of the previous January had inspired many Europeans and Germans with a vision of a more just postwar order, and had helped foster the propagation of concepts like the self-determination of nations (a term Wilson never clearly defined), protection of minorities, a League of Nations, and a nonpunitive peace, the subsequent real sacrifices of American soldiers in the War and other developments had reduced Wilson's fervor for advocating these ideals. These concepts would to some degree be incorporated into the Treaty of Versailles when it was being written in the first half of 1919, and Wilson relied on the League of Nations as a structure and venue for later modification of the treaty. However, Wilson's influence at the Paris conference was limited by his failures as a diplomat and by contentious debates with and between the more impassioned and more directly affected French and British delegations and their prime ministers, Georges Clemenceau and David Lloyd George. MacMillan, 3–16; Levin, 123–82; Gilbert, 507–18; Peukert, 42–46; Kolb, 23–33; Eyck, 80–128.

69. In late December 1918, there were armed clashes in Berlin between rightist militias (the beginnings of the Freikorps) and leftists. On December 24, Friedrich Ebert further alienated the SPD's potential supporters and collaborators, the USPD, and the Spartacists (the incipient German Communist Party), by ordering troops to Berlin for an attempt to clear the palace of the "people's sailors." On December 28, in response to Ebert's action, the USPD members resigned from the Council of People's Representatives. Peukert, 28–33; Kolb, 15–16; Eyck, 51. On December 31, 1918, the "Founding Manifesto of the German Communist Party (KPD)," published in *Die Rote Fahne*, criticized the mainstream SPD for ineffective half-measures that betrayed the true purpose

of socialism, and urged the further development of workers' and soldiers' councils as the sole means to achieve a real socialist revolution. Rosa Luxemburg, "Konstituierung der Kommunistische Partei," *Die Rote Fahne*, no. 45 (Dec. 31, 1918); for an English translation, see Rosa Luxemburg, "Founding Manifesto of the Communist Party of Germany (KPD)," in *The Weimar Republic Sourcebook*, ed. Anton Kaes et al. (Berkeley: Univ. of California Press, 1994), 40–46; Peukert, 32, 35; and Kolb, 16.

The SPD has been criticized for failing to enact basic democratic reforms from the start of the republic, which might have garnered it stronger loyalty from members of the USPD and incipient KPD. In its defense, the limited possibility for structured political reform was demonstrated in December 1918 by the failed effort by Adolf Hoffmann, Prussian minister for education, culture, and church affairs, to achieve a separation of schools and the church. Peukert, 31.

70. Einstein, letter to Michele Besso, Dec. 4, 1918, AEA [7 328.1]; also in *CPAE* 8: 663.

71. Einstein, letter to Paul Ehrenfest, Dec. 6, 1918, AEA [9 425]; also in *CPAE* 8: 664; and in Nathan and Norden, 29.

72. Arnold Sommerfeld, letter to Einstein, Dec. 3, 1918, AEA [21 331]; also in *CPAE* 8: 662. Sommerfeld referred to the physicist Walther Kossel.

73. Einstein, letter to Arnold Sommerfeld, Dec. 6, [1918], AEA [21 390]; also in *CPAE* 8: 665. Food shortages were severe in Berlin at this time. Bane and Lutz, 652–53.

74. Einstein, letter to Hans Albert and Eduard Einstein, Dec. 10, 1918, AEA [75 897]; also in *CPAE* 8: 667.

75. Einstein, letter to Mileva Einstein-Marić, [mid-Dec. 1918], AEA [75 907]; also in *CPAE* 8: 672.

76. Einstein, "To the Society 'A Guaranteed Subsistence for All'" ("An den Verein 'Allgemeine Nährpflicht'"), Dec. 12, 1918, AEA [32 754]; also in *CPAE* 7: 16; and in Rowe and Schulmann, 409. The society was founded in February 1918 by friends of Josef Popper-Lynkeus, a Viennese philosopher and engineer. See Ingrid Belke, *Die sozialreformischen Ideen von Josef Popper-Lynkeus (1838–1921) im Zusammenhang mit allgemeinen Reformbestrebungen des Wiener Bürgertums um die Jahrhundertwende* (Tübingen: Mohr, 1978), 219. The "subsistence army" (*Nähr-Armee*) was one of Popper-Lynkeus's ideas. See Josef Popper-Lynkeus, *Nach dem Kriege. Ein Auszug aus dem Werke Die allgemeine Nährpflicht als Lösung der sozialen Frage*, 2nd ed. (Dresden: Reißner, [1915]), 26–33.

4. Uprisings, Paris Peace Conference, Treaty of Versailles, and the Publication of *Lille*: 1919

1. On December 24, 1918, Chancellor Friedrich Ebert used army troops in a failed attempt to drive revolutionary sailors from the palace. On December 31, the Spartacists founded the German Communist Party. During January 5–15, 1919, workers' protests

and socialist demonstrations in Berlin, known as the Spartacus uprising, were attacked by radical rightist Freikorps militias. This led to street fighting, violently suppressed by army troops, who killed over one thousand protesting workers. On January 13, the Berlin's workers' councils ended their general strike and the Spartacus uprising. On January 15, 1919, Spartacist communist leaders Karl Liebknecht and Rosa Luxemburg were murdered. Peukert, 32; Kolb, 16; Eyck, 51–53.

2. On January 1, 1919, the sinking of the *Iolaire* took place, which resulted in the drowning of 205 Scottish soldiers returning from the western front toward Stornoway harbor. *Sea Sorrow: The Story of the Iolaire Disaster* (Stornoway: Stornoway Gazette, 1972). On January 16, in granting a second extension of the armistice, the Allies said they could not spare shipping to bring food to Germany and that the German merchant marine would have to be put under their control to ensure the relief of Germany and the rest of Europe. On January 18, the Paris Peace Conference opened at Versailles, France. On January 25, the League of Nations was founded. Peukert, 42–46; Kolb, 24–27; Eyck, 81–88.

3. On January 3, the German constitutional law professor and minister of the interior Hugo Preuss published a draft of Germany's new federal constitution. (See events of August 1919 for the completed German constitution.) On January 12, elections were held in Bavaria. On January 19, elections for German National Assembly took place. Consistent with the Erfurt Declaration of the Social Democratic Party, women as well as men voted, with the voting age set at twenty. Parties of the left and center gained over 70 percent of the vote in this election, and the National Assembly elected Friedrich Ebert as president. Of 421 seats, the SPD won 163 (37.9%); the USPD won 22 (7.6%); the German National People's Party (DNVP) won 42 (10.3%); the German People's Party (DVP) won 21 (4.4%); the Center Party won 88 (19.7%); and the DDP won 75 (18.5%). On January 25, a conference was held to establish the relationship between the German federal government and the states. On January 26, elections for the Prussian Assembly took place. Peukert, 33, 36–42; Kolb, 16–21; Eyck, 56, 61–62.

4. Einstein, letter to Hedwig and Max Born, Jan. 15, 1919, AEA [74 314]; also in *CPAE* 9: 2.

5. Einstein, letter to Hedwig and Max Born, Jan. 19, [1919], AEA [8 138]; also in *CPAE* 9: 3.

6. Einstein, letter to Arnold Sommerfeld, [Feb.] 5, 1919, AEA [21 393]; also in *CPAE* 9: 5. See also Einstein, letter to Mileva Einstein-Marić and Hans Albert Einstein, [Jan. 10, 1919], AEA [75 767]; also in *CPAE* 9: 1.

7. Theodor Vetter, letter to Einstein, Jan. 28, 1919, AEA [81 532]; also in *CPAE* 9: 4.

8. Excerpts from Liga zur Beförderung der Humanität (Menschheitsbund), "Erklärung in Sachen Liebknecht-Luxemburg," [Jan. 23, 1919], Nachlaß Nicolai, ED 184/15, Institut für Zeitgeschichte, Munich, Germany. Dated by letter and draft from Georg Friedrich Nicolai to Hans Delbrück, Jan. 23, 1919, Nachlaß Hans Delbrück, Mappe G. Nicolai, Bl. 1, Handschriftenabteilung, Deutsche Staatsbibliothek, Berlin.

9. In February 1919, the German National League for Defense and Defiance (Deutsch-Völkischer Schutz- und Trutz-Bund) was founded by leaders of the nationalist Pan-German League. On the impact of the Pan-German movement, see Peukert, 67; Eyck, 142–43. On February 6, the newly elected National Constituent Assembly convened in Weimar; on February 11, it established a provisional government and elected Friedrich Ebert of the SPD as president by a vote of 277 to 102; Ebert then selected Philipp Scheidemann as chancellor. On February 24, Hugo Preuss delivered his draft constitution to the National Assembly, which debated it for three days, then referred it to a committee, which examined it in detail from March through July 1919. Peukert, 35–42; Kolb, 18–22; Eyck, 64–68.

10. On February 3, Soviet troops occupied the Ukraine. It was not until October 1920 that Poland and Russia attained an armistice, later leading to the Treaty of Riga in 1921. Gilbert, 523, 526–27; Kolb, 24.

11. Freundes-Rat des Internationalen Jugend-Bundes et al., "Aufruf an die freie Jugend aller Stände und Völker," *Republik* (Feb. 8, 1919); reprinted in *Die Friedens-Warte. Blätter für zwischenstaatliche Organisation* 21, no. 3 (Apr. 1919): 74–75.

12. Harry Graf Kessler, *Tagebücher* (Frankfurt am Main: Insel-Verlag, 1961), 125.

13. Divorce Decree, Feb. 14, 1919, AEA [81 374]; also in *CPAE* 9: 6.

14. Einstein, letter to Heinrich Zangger, Feb. 28, 1919, AEA [39 713]; also in *CPAE* 9: 7.

15. On February 10, 1919, the German National Assembly revoked the Law of Temporary National Authority and passed the Law Concerning Provisional Authority in the Reich, drafted by a committee of the states. This established the basic division of power in the new republic. Peukert, 36; Kolb, 17; Eyck, 77–78.

On February 21, Kurt Eisner, the USPD prime minister of Bavaria, was murdered in the street in Munich while on his way to the Bavarian state legislature (Landtag) to announce his resignation. That same day, two more people were killed in the Landtag. A wave of revolt swept Bavaria and a general strike was declared. Eyck, 57, 78. For a eulogy to Eisner, see Wilhelm Hausenstein, "Erinnerungen an Eisner," in *Der neue Merkur* 3 (1919–20): 56–57; for an English translation, see Wilhelm Hausenstein, "Remembering Eisner," in *The Weimar Republic Sourcebook*, ed. Anton Kaes et al. (Berkeley: Univ. of California Press, 1994), 52–53. See also Richard Bessell, "Violence as Propaganda: The Role of the Storm Troopers in the Rise of National Socialism," in *The Formation of the Nazi Constituency, 1919–1933*, ed. Thomas Childers, 131–46 (London: Croom Helm, 1986).

During this volatile period, Gustav Noske, the defense minister under the Ebert-Scheidemann government, declared a state of emergency in response to the communist-led uprising in early January, with an order allowing armed people to be shot on sight (this was revoked by mid-March). The Bavarian Landtag established a constitutional government in mid-March under the SPD teacher Johannes Hoffmann. On April 6, a Soviet Republic was proclaimed in Munich, which refused to collaborate with the federal government. Hoffmann fled to the Catholic city of Bamberg. Between April 30 and May 3, army troops and independent militias crushed the Bavarian Soviet government, which executed ten people just prior to its defeat. The Freikorps killed over 1,000 people during street fighting and arrested and executed more than 700, including the imprisoned intellectual Gustav Landauer. Hoffman was reinstalled, and he formed a coalition government that lasted for about a year until immediately after the Kapp Putsch, at which time it was replaced by a right-wing government and Bavaria developed into a milieu more conducive to right-wing politics. In addition to the Bavarian Soviet Republic, other left-wing and working-class activity in Germany in the first few months after the German Revolution included a general strike that began on March 31 in the industrial Ruhr region. Peukert, 32–34; Kolb, 16, 21–22, 38; Eyck, 52–53, 78–79.

16. On March 2, 1919, the first Communist International meeting was held in Moscow. On March 21, the Hungarian Soviet Republic was established by Béla Kun; it lasted for nine months, delaying the signing of a peace treaty in that region until 1920. On March 23, in Milan, Italy, Benito Mussolini founded his Fascist political movement. Kolb, 39, 54–55; Gilbert, 520–21, 530.

17. On March 5, at Spa, Belgium, the German government announced that it would make Germany's ships available to the Allies as demanded only if the Allies formally assumed responsibility for feeding Germany until the next harvest in September. On March 8, the Allied Supreme Council, with David Lloyd George and Georges Clemenceau present, discussed the matter of providing food to Germany. On March 13 in Brussels, the Allied delegation promised the Germans a steady food supply as long as Germany delivered the ships. The blockade remained officially in force, albeit looser in many ways. Kolb, 27; Eyck, 88–89; MacMillan, 159–60. On the blockade, see Vincent, *Politics of Hunger.*

18. Arnold Sommerfeld, letter to Einstein, Mar. 25, 1919, AEA [21 332]; also in *CPAE* 9: 12.

19. Einstein, letter to Aurel Stodola, Mar. 31, [1919], AEA [22 257]; also in *CPAE* 9: 16. Einstein was offended by the *Berliner Tageblatt*'s abandonment during the War of its prewar pacifism. See Zuelzer, 243; and Roger Chickering, *Imperial Germany and a World Without War: The Peace Movement and German Society, 1892–1914* (Princeton: Princeton Univ. Press, 1975), 190.

20. On April 14, the Allies invited the German delegation to Versailles to receive the terms of the peace treaty. Although they would be allowed to present responses, they were never permitted actually to negotiate the terms. On April 20, German foreign minister, former ambassador, and DDP member Count Ulrich von Brockdorff-Rantzau replied to the Allies that an official of the foreign office would be sent to Paris to receive the treaty, but Clemenceau insisted that only officials empowered to sign the treaty be sent. On April 21, the German government gave the Allies the names of the empowered delegates: Brockdorff-Rantzau, Minister of Justice Otto Landsberg, Christian Workers' Movement

leader Johann Giesberts, Social Democrat Robert Leinert, DDP Reichstag member Walther Schücking, Carl Melchior, and their legal advisor, Walter Simons. Peukert, 42–46; Kolb, 23–33; Eyck, 90–91; MacMillan, 276.

21. During April 1919, providing fuel for German rightist propaganda, German army troops still stationed in the Baltics as a defense against the Bolsheviks (with Allied permission, later revoked) helped the Latvians temporarily depose their prime minister, Karlis Ulmanis. Eyck, 139–40.

On April 13, in the Amritsar massacre, British imperial troops fired on a meeting of nationalist Indians, killing 379 and wounding over 1,000. This event further fueled the independence movement in India. Dennis Dalton, *Mahatma Gandhi: Nonviolent Power in Action* (New York: Columbia Univ. Press, 1993), 237, note 13; Daniel R. Brower, *The World in the Twentieth Century: From Empires to Nations*, 4th ed. (Upper Saddle River, NJ: Prentice Hall, 1999), 85; and William J. Duiker, *Twentieth-Century World History* (Belmont, CA: Wadsworth Publishing, 1999), 102–3.

22. Helene Stöcker, letter to Einstein, and enclosed appeal, Apr. 9, 1919, AEA [45 073]; also in *CPAE* 9: 20.

23. Georg Count von Arco, letter to Einstein, Apr. 12, 1919, AEA [31 005]; also in *CPAE* 9: 21. For the Monists' appeal, see *Mitteilungen des Bundes Neues Vaterland*, Neue Folge Nr. 4 (Aug. 1919): 19.

24. Einstein, letter to Emil Zürcher, Apr. 15, 1919, AEA [80 817]; also in *CPAE* 9: 23.

25. Georg Count von Arco, Albert Einstein, Walburga Geiger, Hellmut von Gerlach, Maximilian Harden, Max Hodann, Luise Kautsky, Elisabeth Rotten, Erich Schlesinger, and Helene Stöcker, *Lille: Beiträge zur Naturgeschichte des Krieges* (Berlin: Engelmann, 1919). [First/original version.]

26. Walburga Geiger, Max Hodann, Elisabeth Rotten, and Erich Schlesinger, *Lille: Beiträge zur Naturgeschichte des Krieges*, 2nd ed. (Berlin: Engelmann, 1920). [Second/revised version.]

27. Einstein, letter to Hendrik A. Lorentz, Apr. 26, 1919, AEA [16 464]; also in *CPAE* 9: 28; and in Nathan and Norden, 32–33. The four fellow commission members to whom he referred almost certainly included Walburga Geiger, Max Hodann, Elisabeth Rotten, and Erich Schlesinger, signatories of the introductions of both the first and second versions of *Lille*.

28. Georg Friedrich Nicolai, letter to Konrad Haenisch, Apr. 23, 1919, Nachlaß Nicolai, ED 184, Bd. 45, Institut für Zeitgeschichte, Munich, Germany.

29. Georg Friedrich Nicolai, letter to Max Reinhardt, Apr. 30, 1919, Nachlaß Nicolai, ED 184, Bd. 50, Institut für Zeitgeschichte, Munich, Germany.

30. At the end of April 1919, the German delegation arrived at Versailles. On May 7, the Allies handed the German delegation the draft peace treaty, in which Germany was forced to cede territory to France and to Poland, limit the size of its army and navy, accept

responsibility for the War, and pay reparations to Allied governments. The Allies gave the Germans fifteen days, later extended by one week, to prepare in writing any objections, questions, and proposals. Gilbert, 508–18; Peukert, 42–48; Kolb, 23–33; Eyck, 90–98; MacMillan, 460–67. On that same day, Count Brockdorff-Rantzau made his outraged response. Ulrich von Brockdorff-Rantzau, "Ansprache des Reichsaußenministers Grafen Brockdorff-Rantzau bei Überreichung des Friedensvertrags-Entwurfs durch die Alliierten und Assoziierten Mächte," *Dokumente* (Charlottenburg: Deutsche Verlagsgesellschaft für Politik und Geschichte, 1920), 113ff.; for an English translation, see Ulrich von Brockdorff-Rantzau, "Speech of the German Delegation, Versailles, May 7, 1919," in *The Weimar Republic Sourcebook*, ed. Anton Kaes et al. (Berkeley: Univ. of California Press, 1994), 9–12. See also MacMillan, 463–64; Gilbert, 514–15; Eyck, 93–95. For newspaper coverage in Berlin, see "Die unannehmbaren Friedensbedingungen: Nein!" *Berliner Tageblatt*, May 8, 1919, morning edition; "Die Gebietsforderungen der Entente," "Überreichung des Entwurfts," and "Die Rede des Grafen Brockdorff-Rantzau," *Vossische Zeitung*, May 8, 1919, morning edition; and "Abgabe der Friedensbedingungen," *Die Rote Fahne*, May 8, 1919. On May 12, the German National Assembly met at the University of Berlin, where Prime Minister Philipp Scheidemann made a speech pronouncing the treaty unacceptable to the German government. Kolb, 31; MacMillan, 465; Eyck, 97–98.

31. On May 23, the German delegation to Versailles met with members of German cabinet at Spa, Belgium. The delegates felt Germany could not sign the treaty unless the Allies granted several concessions, but the USPD's Hugo Haase felt that Germany had to sign if it did not wish to resume the War. Center Party member Matthias Erzberger (who became finance minister in June) also became convinced that Germany should sign. In late May, the Germans presented the Allies with a memorandum on war guilt by experts Hans Delbrück, Max Weber, Albrecht Mendelssohn-Bartholdy, and General Max Montgelas. On May 27, the German counterproposals were given to the press in Berlin and published there the next day. These included a reduced reparations obligation, admission into the League of Nations, a plebiscite in Silesia, and no ceding of territory to Poland. On May 29, the German delegation presented its "Observation" on the treaty to the Allies along with a cover letter by journalist Bernhard Guttmann summarizing its points. "Scheidemann Goes to Spa Conference; Bernstorff, Dernburg, and Erzberger Also There to Decide Reply," *New York Times*, May 24, 1919; "Reply to Allies in German," dateline May 29, *New York Times*, May 30, 1919; Eyck, 95–98; Kolb, 31; Peukert, 43; MacMillan, 468.

32. When news reached China on May 4 that the Paris Peace Conference had granted Japan the formerly German territorial rights in the Shandong province in northern China as a reward for aiding the Allies in the War, the May Fourth Movement arose in China, in conjunction with the Chinese nationalism and in opposition to foreign colonizers. Brower, 94–96; Duiker, 111.

The Ottoman Empire had shared in Germany's defeat and lost large portions of its territory. On May 19, 1919, as part of the "Young Turks" nationalist movement, the army officer and future Turkish leader Mustafa Kemal Ataturk landed at the port city of Samsun on the Turkish Black Sea coast. Thus began the Turkish war of independence, which would lead to the fall of the Ottoman Empire and the founding of the Republic of Turkey under Ataturk in 1922. Brower, 76–79; Duiker, 104–6.

33. Hendrik A. Lorentz, letter to Einstein, May 4, 1919, AEA [16 469]; also in *CPAE* 9: 34; and in Nathan and Norden, 33. Jean Massart, a professor of biology at the Free University of Brussels, in March 1916 proposed the idea of creating a mixed German-Belgian commission to investigate allegations of German war crimes in Belgium. See Jean Massart, "Les intellectuels allemands et la recherche de la verité," *Revue de Paris* 25 (1918): 643–72, esp. 645–66. In August to September 1914, German troops burned down the city of Louvain and force-marched and killed many of its people. On German war crimes and allegations in the early part of the War, see John N. Horne and Alan Kramer, "War between Soldiers and Civilians, 1914–1915," in *Great War, Total War: Combat and Mobilization on the Western Front, 1914–1918*, ed. Roger Chickering and Stig Förster, 153–68 (Cambridge: Cambridge Univ. Press, 2000).

34. Georg Friedrich Nicolai, letter to David Hilbert, May 6, 1919, Nachlaß Nicolai, ED 184, Bd. 44, Institut für Zeitgeschichte, Munich, Germany.

35. "Wir unterzeichneten Deutschen . . . ," enclosure in Georg Friedrich Nicolai and Hans Wehberg, letter to Wilhelm Mühlon, June 27, 1919, Nachlaß Wilhelm Mühlon, ED 142, Band 26, Blatt 4753, Institut für Zeitgeschichte, Munich, Germany.

36. "Un Appel: Fière Déclaration d'Intellectuels", *L'Humanité* (Paris), June 26, 1919.

37. New Fatherland League (Bund Neues Vaterland), "Appeal Signed by Cultural Associations of All of Germany and by Individuals" ("Aufruf der von Kulturvereinen ganz Deutschlands und von Einzelpersonen unterzeichnet werden soll"), postdated May 7, 1919, ZSg. 1–13/1 (8), Bundesarchiv, Koblenz, Germany; also AEA [82 572].

38. Einstein, letter to Wander de Haas, May 9, 1919, AEA [70 418]; also in *CPAE* 9: 36. Max Planck's ambivalence about Germany and the War was compounded by the death of his wife before the War and of his daughter and son during the War; the son, a soldier, had been killed in action. See John L. Heilbron, *The Dilemmas of an Upright Man: Max Planck as Spokesman for German Science* (Berkeley: Univ. of California Press, 1986), 82–84.

39. "Appeal of the German Intellect for Socialism" ("Aufruf des deutschen Geistes zum Sozialismus"), and "To the Intellectuals of the Proletariat" ("An die Geistigen des Proletariats"), both attached to a circular signed by Paul Natorp, Wilhelm Schäfer, Hermann Herrigel, Alfons Paquet, and Leopold Ziegler, dated April 1919, Nachlaß Paul Natorp, Ms. 831, Universitätsbibliothek Marburg, Germany.

40. Einstein, letter to Paul Natorp, May 11, 1919, AEA [72 204]; also in *CPAE* 9: 37.

41. For newspaper coverage in Berlin, see "Das Verbot der Lustbarkeit," *Berliner Tageblatt*, May 9, 1919, evening edition; "Die Landestrauer," *Vossische Zeitung*, May 9, 1919, evening edition; and "Die Trauerwoche: Die amtliche Verfügung," *Berliner Tageblatt*, May 10, 1919, morning edition.

42. Einstein, letter to Pauline Einstein, [May 14, 1919], AEA [29 355]; also in *CPAE* 9: 39.

43. Tankred Dorst, *Die Münchner Räterepublik: Zeugnisse und Kommentar* (Frankfurt am Main: Suhrkamp, 1967), 155–56. Leviné was later sentenced to death on June 3 and executed on June 5, 1919.

44. Karl A. Aner et al., telegram to Munich Military Tribunal, [May 19, 1919], AEA [86 060]; also in *CPAE* 9: 44).

45. Einstein, postcard to Pauline Einstein, [June 16, 1919], AEA [29 357]; also in *CPAE* 9: 61; Hendrik A. Lorentz, telegram to Einstein, [Sept. 22, 1919], AEA [16 478]; also in *CPAE* 9: 110; and Einstein, postcard to Pauline Einstein, Sept. 27, 1919, AEA [29 364]; also in *CPAE* 9: 113. See also Matthew Stanley, "'An Expedition to Heal the Wounds of War': The 1919 Eclipse and Eddington as Quaker Adventurer," *Isis* 94 (2003): 57–89.

46. Marriage Certificate, June 2, 1919, AEA [29 163]; also in *CPAE* 9: 54; and Civil Registry Record, June 2, 1919, AEA [76 551]; also in *CPAE* 9: 55.

47. Einstein, letter to Heinrich Zangger, June 1, 1919, AEA [39 715]; also in *CPAE* 9: 52.

48. Einstein, letter to Max Born, June 4, 1919, AEA [8 140]; also in *CPAE* 9: 56. On France's postwar attitude toward Germany and influence on the Treaty of Versailles, see John M. Keynes, *The Economic Consequences of the Peace* (New York: Harcourt, 1920), chapters 3 and 4. The comment on Austria referred to three semesters in 1911 to 1912 when Einstein was a physics professor at the German University in Prague, then part of the Austro-Hungarian Empire.

49. Leonhard Grebe, letter to Einstein, June 6, 1919, AEA [78 052]; also in *CPAE* 9: 57.

50. "24-stündiger Generalstreik in Berlin geplant," *Berliner Tageblatt*, June 6, 1919, evening edition; "Der 24-Stunden-Streik: Berlin wieder ohne Verkehr," *Vossische Zeitung*, June 7, 1919, evening edition; and Georg Bernhard, "Berlin Proteststreik," *Vossische Zeitung*, June 7, 1919, evening edition.

51. Einstein, letter to Hans Albert and Eduard Einstein, June 13, 1919, AEA [75 709]; also in *CPAE* 9: 60. The exchange rate had declined from 1.4 German marks to 1 Swiss franc at the end of the War to 2.8 marks to 1 franc half a year later (*Vossische Zeitung*, Nov. 26, 1918, evening edition; and June 13, 1919, evening edition), largely due to reparations demands by the Allies. See Frank D. Graham, *Exchange, Prices, and Production in Hyperinflation: Germany, 1920–1923* (New York: Russell and Russell, 1967), 9.

52. In early June 1919, the Allies also conveyed preliminary peace terms to the Austrian delegation at Saint-Germain-en-Laye outside Paris. There was popular protest in Austria against southern Tirol being given to Italy; there was also sentiment in favor of western Austria and parts of Bavaria joining Switzerland or forming a Central European Catholic confederation. The Allies did not allow the defeated nations to combine or form confederations. However, the fact that the Hapsburg Empire had dissolved, leaving Austria a tiny country, and that the new emperor Karl had sought peace in 1917, as well as the politeness of the Austrian delegation, influenced the Allies to offer more lenient terms to Austria than to Germany. MacMillan, 243–56; Kolb, 30, 32; Eyck, 106, 109; Gilbert, 520; Peukert, 45.

53. On June 16, the Allies delivered a strong response to the Germans, deprecating German diplomatic and military policy but making a few concessions: the fate of Upper Silesia would be determined by a plebiscite; the Saar would be returned to Germany after fifteen years; occupation troops in the Rhineland would be limited in number; and the German civil administration would continue there. However, the war guilt clause was emphasized even more strongly. The Allies stated that this was their final, unchangeable position, and that Germany had to respond in five days, later changed to one week, or the armistice would end and an Allied invasion of Germany would be launched. On June 18, when Brockdorff-Rantzau returned to the German National Assembly at Weimar, he found the German cabinet sharply divided over whether to sign the treaty. This included splits within political parties, divisions that reflected feelings among the German people, who were offended by the "dictated peace" (*Friedens-Diktat*). Philipp Scheidemann was opposed, while Matthias Erzberger favored signing. On June 20, when the Center Party and SPD decided in favor of signing, with the provision that neither acceptance of war guilt (Article 231) nor the obligation to extradite so-called war criminals be binding, Scheidemann and his cabinet resigned. They were succeeded by an SPD-Center Party coalition under Social Democrat Gustav Bauer. On June 22, after intense debate and strong statements opposing the treaty, the German National Assembly voted 237 to 138 to sign it, with a few provisos—primarily, that it not acknowledge Germany's responsibility for the War. When the Allies demanded that Germany sign the treaty unconditionally, the National Assembly on June 23 reauthorized the government to sign. On June 21, Admiral Ludwig von Reuter scuttled the German fleet in Scapa Flow, Orkney, off the Scottish coast, prior to its scheduled delivery to the Allies; during that action, nine Germans were killed. On June 28, at the Hall of Mirrors at Versailles, Foreign Minister Hermann Müller (SPD) and Transport Minister Johannes Bell (Center Party) signed the Treaty of Versailles. MacMillan, 470–74; Kolb, 31–32; Peukert, 43–46; Eyck, 98–105. For the war-guilt clause, Article 231, and the reparations clauses, Articles 232–236, see "The Treaty of Versailles: The Reparations Clauses," in Kaes et al., 8–9; see also *The Treaty of Peace Between the Allied and Associated Powers and Germany* (London: His Majesty's Stationery Office, 1919).

54. Heinrich Zangger, letter to Einstein, [before June 18, 1919], AEA [39 718]; also in *CPAE* 9: 62.

55. Einstein, letter to Heinrich Zangger, June 18, [1919], AEA [86 095]; also in *CPAE* 9: 63.

56. Paul Natorp, letter to Einstein, June 26, 1919, AEA [44 519]; also in *CPAE* 9: 64. The appeal was published three days later, signed by Alfons Paquet, Natorp, Wilhelm Schäfer, and Hermann Herrigel. Paquet was a journalist with the *Frankfurter Zeitung*; Schäfer was a novelist and publisher of the journal *Rheinlande*; and Herrigel was a literary critic with an interest in education for the masses. "An die Geistigen des Proletariats," *Räte-Zeitung: Organ der Hand- und Kopfarbeiter-Räte Deutschlands*, June 29, 1919. Georges Sorel advocated the use of violence by the proletariat in its struggle against the inherently violent economic force of the bourgeoisie. Georges Sorel, *Réflexions sur la Violence* (Paris: Librairie de "Pages libres," 1908); and Sorel, *Reflections on Violence*, 3rd ed. (New York: Huebsch, 1912). Natorp's article described his vision of true democracy. Paul Natorp, "Ein Weg der Rettung," *Frankfurter Zeitung*, Mar. 30, 1919, morning edition; Natorp, "Ein Weg der Rettung," *Kunstwart und Kulturwart. Halbmonatschau für Ausdruckskultur auf allen Lebensgebieten* 32, no. 3 (Apr.–June 1919): 100–107.

57. On the development and dissemination of the appeal, see Georg Friedrich Nicolai, ed., *Romain Rollands Manifest und die deutschen Antworten* (Charlottenburg: Mundus, [1920]); for texts of the appeal in French and German, see 22–25; also in Nathan and Norden, 30–31.

58. On Natorp, see Norbert Jegelka, *Paul Natorp: Philosophie, Pädagogik, Politik* (Würzburg: Könighausen und Neumann, 1992), 141–42.

59. Adolf Schmidt, letter to Einstein, July 1, 1919, AEA [44 958]; also in *CPAE* 9: 68. Einstein was one of twenty-four Germans among the original 102 signatories of the appeal; by the end of 1919, that number had increased to 617 German signatories. Nicolai, *Romain Rollands Manifest*, 26–27, 35–47. Rolland and Henri Barbusse, a writer, communist, and wartime pacifist, were cofounders of the Clarté movement, which aimed to establish a just political order. Einstein was briefly connected to it in this immediate postwar period, although he later distanced himself.

60. Robert Holtzmann, letter to Einstein, July 10, 1919, AEA [43 945]; also in *CPAE* 9: 71. This Breslau group probably was affiliated with the Berlin-based Council of Intellectual Workers (Rat geistiger Arbeiter), in which the pacifist and socialist writer and editor Kurt Hiller played a leading role. For Hiller's unique political perspective and creative use of language, see Kurt Hiller, *Geist werde Herr. Kundgebungen eines Aktivisten vor, in und nach dem Kriege*, 3rd ed. (Berlin: Reiß, 1920); and Holmes, "'Inviolability of Human Life.'"

61. Elsa Einstein, letter to Einstein, [July 12, 1919], AEA [44 519.1]; also in *CPAE* 9: 72. For a report on the railway and transport workers' strikes, see "Berlin ohne Kohlen. Die Folgen des Eisenbahner-Streiks," *Vossische Zeitung*, July 1, 1919, morning edition.

62. Adriaan D. Fokker, letter to Einstein, July 26, 1919, AEA [11 035]; also in *CPAE* 9: 75. On Jan Christiaan Smuts of South Africa advocating reconciliation with Germany at the Paris Peace Conference, see *Vossische Zeitung*, June 30, July 3, and July 19, 1919, morning editions.

63. Einstein, letter to Adriaan D. Fokker, July 30, [1919], AEA [73 264]; also in *CPAE* 9: 78. US president Woodrow Wilson was a proponent not only of the League of Nations but also of a less punitive peace with Germany.

64. Hendrik A. Lorentz, letter to Einstein, July 27, 1919, AEA [16 471]; also in *CPAE* 9: 76. On the "Manifesto of the 93," or manifesto "To the Civilized World" ("An die Kulturwelt!"), see chapter 1. In the early months of the War, Max Planck had interceded with German authorities in Belgium on behalf of Ernest Solvay. Max Planck, letter to Hendrik A. Lorentz, Nov. 28, 1914, Archief H. A. Lorentz, Rijksarchief Noord-Holland, Haarlem, Netherlands; Heilbron, 79. Marcel Brillouin was a mathematical physicist at the Collège de France. Paul Heger was director of the Brussels Institute of Teaching and Research in the Medical Sciences. Adriaan K. M. Noyons was a physiology professor at the University of Louvain. Ernest Solvay was a Belgian chemical industrialist, philanthropist, and founder in 1911 of the Institut international de physique (IIP), a triannual series of small conferences for top physicists and chemists, which Einstein also attended. Paul Heger and Emile Tassel were members of the IIP's administrative commission. See Hendrik A. Lorentz, "Ernest Solvay," *Naturwissenschaften* 2 (1914): 997–99.

65. Einstein, letter to Auguste Hochberger, July 30, [1919], AEA [43 915]; also in *CPAE* 9: 79.

66. Einstein, postcard to Elsa Einstein, [July 31, 1919], AEA [143 087]; also in *CPAE* 10: 9/79a. Rotten was a signatory to *Lille* in both its original (1919) and revised (1920) versions. Arco et al.; Geiger et al.

67. Einstein, letter to Hendrik A. Lorentz, Aug. 1, 1919, AEA [16 473]; also in *CPAE* 9: 80; and in Nathan and Norden, 33–34. On the creation of the manifesto "To the Civilized World" ("An die Kulturwelt!") and the contributors involved, including Hermann Sudermann, see Jürgen von Ungern-Sternberg and Wolfgang von Ungern-Sternberg, *Der Aufruf "An die Kulturwelt!": Das Manifest der 93 und die Anfänge der Kriegspropaganda im Ersten Weltkrieg* (Stuttgart: Steiner, 1996), 17–49.

68. "Kohlen für Sammelheizung erst am 1. Sept.," *Vossische Zeitung*, July 19, 1919, evening edition; "Die Entente regelt die Kohlenerzeugung. Eine Warning an Deutschland," *Vossische Zeitung*, Aug. 7, 1919, evening edition; and Marc Trachtenberg, *Reparation in World Politics: France and European Economic Diplomacy, 1916–1923* (New York: Columbia Univ. Press, 1980), 144–52.

69. Einstein, postcard to Pauline Einstein, [Aug. 9, 1919], AEA [29 361]; also in *CPAE* 9: 87).

70. On July 31, 1919, the Weimar constitution was adopted. On August 11, it was passed into law. German president Friedrich Ebert accepted the National Assembly's

constitution, which was adopted by a vote of 262 to 75. It established Germany as a parliamentary republic with very strong executive powers in the president to counter the legislative powers of the Reichstag. Kolb, 18–20; Peukert, 36–42; Eyck, 129; MacMillan, 462. For the Weimar Republic's constitution, see *Die Verfassung des Deutschen Reiches vom 11. August 1919*, 7th ed. (Leipzig, 1930); for an English translation of excerpts from it, see "The Constitution of the German Republic," in Kaes et al., 46–51.

On August 19, the German National Assembly established an investigations committee (*Untersuchungsausschuss*) to look into whether the World War could have been avoided, who was responsible for which actions, and related issues. The secretary of the subcommittee to investigate peace overtures during the War, the SPD jurist Hugo Sinzheimer, was attacked in the rightist press for being a Jew; the rightist press also attacked USPD committee member Oscar Cohn for being a Jew and for having received funds from Adolph Joffe, former Russian ambassador to Germany, to organize the November revolution in Berlin. Eyck, 134–39.

71. Between August 16 and 26, a revolt by Poles against Germans in Upper Silesia was suppressed by German troops. On August 24, German army commanders in the Baltics met in Mitau (Jelgava) in Latvia and refused to obey an Allied order to withdraw German Freikorps units from the Baltics; those units had arrived in early 1919 to help the Baltic people drive out the Bolsheviks, then stayed on to the end of 1919 for adventure and in hopes of acquiring land. Peukert, 66; Eyck, 139–42; MacMillan, 222–23.

72. Einstein, letter to Robert Holtzmann, Aug. 17, 1919, AEA [43 946]; also in *CPAE* 9: 91; and in Nathan and Norden, 32.

73. Einstein, letter to Adolf Schmidt, Aug. 17, 1919, AEA [44 960]; also in *CPAE* 9: 92; and in Nathan and Norden, 31–32.

74. Einstein, letter to Auguste Hochberger, Aug. 20, [1919], AEA [43 919]; also in *CPAE* 9: 94.

75. Einstein, letter to Hedwig Born, [Aug. 31, 1919], AEA [8 254]; also in *CPAE* 9: 97. Einstein was referring to the German constitution's prohibition against adding territories, including that of Austria, except through the self-determination of the populations of those territories. He was also referring to a recent amendment to the Treaty of Versailles stipulating that a plebiscite be held in Upper Silesia within eighteen months of Germany's ratification of the treaty to determine whether it would be part of Germany or Poland. Although US president Woodrow Wilson was a great proponent of the League of Nations, he faced isolationist opposition in the US Congress to ratification of the Treaty of Versailles. That ratification was necessary in order for the United States to join the League of Nations.

76. Wilhelm Foerster, letter to Einstein, Sept. 1919, AEA [82 574].

77. Einstein, letter to Pauline Einstein, [Sept. 5, 1919], AEA [29 363]; also in *CPAE* 9: 99.

78. On September 10, 1919, the Treaty of Saint-Germain was signed, ending the War between the Allies and Austria. MacMillan, 256; Kolb, 32; Gilbert, 520.

79. In early October 1919, German troops attacked Riga, Latvia. The Allies responded by prohibiting German shipping in the Baltic. They also threatened to cut off shipments of food and supplies to Germany and sent a commission to investigate. German marine traffic was eventually reopened after German troops had withdrawn from the Baltic countries. Eyck, 141–42; MacMillan, 223; Peukert, 66.

80. On October 2, President Woodrow Wilson, advocate of the League of Nations and various international principles such as open diplomacy, equality of trade, and the sovereignty of populations, who had originally given Germans hope for lenient peace terms, suffered a stroke that left him partly paralyzed and largely incapacitated. He had already encountered difficulties in convincing Americans to support the League of Nations; his direct influence on postwar developments now became virtually nonexistent, notwithstanding his long-term historic contributions to international ideals and aspirations. MacMillan, 441, 491. Many of Wilson's principles pertaining to international relations and his vision for a postwar world were included in his "Fourteen Points" speech to the US Congress on January 8, 1918 (discussed in chapter 3 of this book). For the text of the "Fourteen Points" speech, see appendix C; also in MacMillan, 495–96.

81. Einstein, letter to Paul Ehrenfest, Sept. 12, 1919, AEA [9 433]; also in *CPAE* 9: 103.

82. Einstein, letter to Hendrik A. Lorentz, Sept. 21, 1919, AEA [16 475]; also in *CPAE* 9: 108; and in Nathan and Norden, 34–35. In that final comment, Einstein was referring to the second version of *Lille*.

83. Hendrik A. Lorentz, telegram to Einstein, [Sept. 22, 1919], AEA [16 478]; also in *CPAE* 9: 110.

84. Kartell der deutschen Akademien, attendance list and minutes, 1. Sitzung der Tagung des Kartells der deutschen Akademien, Oct. 3, 1919, II–XII, Bd. 4, B. 169, 170–170d, Archiv der Berlin-Brandenburgischen Akademie der Wissenschaften, Berlin, Germany; also AEA [83 930] and AEA [83 931].

85. Einstein, postcard to Elisabeth Rotten, Oct. 7, 1919, AEA [44 818]; also in *CPAE* 9: 126. Dahomey (present-day Benin) was a French prisoner-of-war camp in which Germans were interned.

86. Wilhelm Schweydar, letter to Einstein, Oct. 10, 1919, AEA [21 567]; also in *CPAE* 9: 130.

87. Einstein, letter to Carl H. Becker, Oct. 15, 1919, AEA [43 222]; also in *CPAE* 9: 133.

88. Einstein, letter to Mileva Einstein-Marić, Oct. 15, 1919, AEA [75 725]; also in *CPAE* 9: 135.

89. Einstein, postcard to Pauline Einstein, Oct. 17, 1919, AEA [29 366]; also in *CPAE* 9: 140. For contemporary newspaper reports on inflation around that time, see "Das internationale Valutaproblem," *Vossische Zeitung*, Oct. 13, 1919, morning edition; and "Geldentwertung und Valuta," *Vossische Zeitung*, Oct. 16, 1919, evening edition.

90. Einstein, letter to Max Born, Oct. 16, 1919, AEA [8 141]; also in *CPAE* 9: 138.

91. Einstein, letter to Otto Lehmann-Rußbüldt, Oct. 17, 1919, AEA [39 724]; also in *CPAE* 9: 141.

92. Heinrich Zangger, letter to Einstein, [after Oct. 17, 1919], AEA [39 722]; also in *CPAE* 9: 143. Bernard Bouvier, a German language and literature professor, and William Rappard, an international law and economic history professor, both at the University of Geneva, were members of the International Committee of the Red Cross (ICRC).

93. Stanley, 57–89. On Einstein's first knowledge of the expedition results in June 1919, see Einstein, postcard to Pauline Einstein, [June 16, 1919], AEA [29 357]; also in *CPAE* 9: 61.

94. Einstein, letter to Herr Dr. Wilouner, Nov. 6, 1919, AEA [44 997]; also in *CPAE* 9: 159. Wilouner was head of the city housing office (Mietseinigungsamt der Stadt Schöneberg).

95. Einstein, letter to Mileva Einstein-Marić, Nov. 16, 1919, AEA [75 728]; also in *CPAE* 9: 166.

96. On November 15, 1919, the former minister Karl Helfferich, a prior liberal turned German Nationalist, testified before the National Assembly's investigations committee regarding the German military's January 1917 decision to conduct unrestricted submarine warfare against the wishes of Chancellor Theobold von Bethmann-Hollweg. Helfferich had originally opposed but later defended that decision. On November 18, General Erich Ludendorff agreed to testify only in General Paul von Hindenburg's presence, then refused to answer questions, instead making personal attacks on committee members. Hindenburg read a prepared statement denying Germany's guilt for the War, in which he also launched the "stab-in-the-back legend." Kolb, 35; Peukert, 68; Eyck, 135–39. For an English translation of Hindenburg's testimony, see Paul von Hindenburg, "The Stab in the Back," in *The Weimar Republic Sourcebook*, ed. Anton Kaes et al. (Berkeley: Univ. of California Press, 1994), 15–16.

97. On November 19, after extensive debate and uncompromising advocacy by President Woodrow Wilson, the US Senate voted against signing the Treaty of Versailles; this was confirmed in a second vote on March 19, 1920. The United States eventually made a separate peace treaty with Germany, which was signed on August 25, 1921. MacMillan, 80, 441, 488–92; Eyck, 126–27; Peukert, 57; Kolb, 57.

98. Einstein, letter to Benjamin de Jong van Beek en Donk, Nov. 9, [1919], AEA [44 032]; also in *CPAE* 9: 163.

99. Arco et al., 3–5, 8–58.

100. Geiger et al., 3–4, 50–61.

101. Einstein, "Leo Arons as Physicist," *Sozialistische Monatshefte* 53 (Nov. 17, 1919): 1055–56; also in *CPAE* 7: 24. A year earlier (Nov. 12, 1918), Einstein had written a letter to Arons complimenting him directly for his advocacy of free speech. Leo Arons,

Universitäten heraus! (Berlin: Verlag der Sozialistischen Monatshefte, 1918), 3; also in *CPAE* 8: 653.

102. Adriaan D. Fokker, letter to Einstein, Nov. 18, 1919, AEA [11 036]; also in *CPAE* 9: 168. After the War, Japan hired hundreds of German naval officers and engineers to teach them how to build submarines. Brayton Harris, *The Navy Times Book of Submarines: A Political, Social, and Military History* (New York: Berkley Books, 1997), 269. Fokker was referring to Rabindranath Tagore, *Nationalismus* (Leipzig: Wolff, [1918]).

103. Adolf Friedrich Lindemann, letter to Einstein, Nov. 23, 1919, AEA [16 407]; also in *CPAE* 9: 174. For instance, Einstein was described as a "Swiss Jew" in "Dr. Albert Einstein," *Times* (London, Nov. 8, 1919).

104. Robert W. Lawson, letter to Einstein, Nov. 26, 1919, AEA [44 253]; also in *CPAE* 9: 177. Felix Ehrenhaft was a physicist at the University of Vienna. Lawson was referring to Arnold Berliner, editor of *Die Naturwissenschaften*; Friedrich A. Paneth, inorganic chemist at the University of Hamburg; Stefan Meyer, Hans Thirring, and Ludwig Flamm, physicists at the University of Vienna; and Heinrich Mache, physicist at the Technical University of Vienna.

105. Robert W. Lawson, letter to Einstein, Nov. 28, 1919, AEA [44 254]; also in *CPAE* 9: 180.

106. Einstein, "Einstein on His Theory. Time, Space, and Gravitation. The Newtonian System," *Times* (London), Nov. 28, 1919, 13–14; also in *CPAE* 7: 26; and in Rowe and Schulmann, 104–5.

107. Einstein, letter to Paul Ehrenfest, Dec. 4, 1919, AEA [9 451]; also in *CPAE* 9: 189.

108. Einstein, letter to Adriaan D. Fokker, [after Dec. 1, 1919], AEA [9 260]; also in *CPAE* 9: 187. Arthur S. Eddington, an astrophysicist with the Cambridge Observatory in England and a Quaker pacifist who had claimed conscientious objector status during the War, had been one of the leaders of the British eclipse expedition that verified Einstein's theory of relativity.

109. Einstein, letter to *Neue Freie Presse*, Dec. 6, 1919, AEA [44 536]; also in *CPAE* 9: 193.

110. At the end of November, health officials declared the influenza pandemic over, after it had caused an estimated 50 million deaths worldwide. The huge toll can be partly attributed to weakened health due to poor nutrition and other War-induced stresses in Germany and elsewhere. While the influenza had caused the highest German civilian mortality rates of the entire War during October and November 1918, by late 1919 these rates had returned to pre-War levels. John M. Barry, *The Great Influenza: The Story of the Deadliest Pandemic in History* (New York and London: Penguin Books, 2005); Richard Bessell, *Germany*, 224.

On November 27, 1919, the Allies signed the Treaty of Neuilly with Bulgaria. MacMillan, 136–42, 364–65; Kolb, 32. By December, the Germans had completed the withdrawal of their troops from the Baltics, although under protest. Eyck, 141–42; MacMillan, 223; Peukert, 66.

111. In December, Social Democrats, Democrats, and Center Party members of the Prussian legislature, over protests from rightist parties, passed by a vote of 210 to 32 a resolution aimed at developing a more centralized German republic through initiatives encouraging discussions between the federal and state governments. That month, the German National Assembly passed a law establishing procedures for prosecuting domestically any Germans whom the Allies wished to indict for violation of the laws of civilized warfare, instead of extraditing them as required in the Treat of Versailles. Eyck, 129–30 and 148.

112. Einstein, letter to Moritz Schlick, Dec. 1, 1919, AEA [21 626]; also in *CPAE* 9: 184.

113. Einstein, letter to Konrad Haenisch, Dec. 6, 1919, AEA [83 282]; also in *CPAE* 9: 194.

114. Einstein, letter to Michele Besso, Dec. 12, 1919, AEA [82 705]; also in *CPAE* 9: 207.

115. Einstein, letter to Heinrich Zangger, Dec. 15 or 22, 1919, AEA [72 152]; also in *CPAE* 9: 217.

116. Einstein, letter to Max Born, [Dec. 8, 1919], AEA [8 143]; also in *CPAE* 9: 198; and in Nathan and Norden, 37.

117. Einstein, letter to Hermann Schüller, Dec. 14, 1919, AEA [45 019]; also in *CPAE* 9: 212. Einstein was probably alluding to a lecture series proposed in the article "Hochschule für Proletarier," *Räte-Zeitung* 1, no. 53 ([Nov.] 1919), 2.

118. Einstein, "Welcoming Address to Paul Colin" ("Begrüßung an Herrn Colin . . .") [Dec. 16, 1919], AEA [28 005]; also in *CPAE* 7: 27; in Nathan and Norden, 38; and in Rowe and Schulmann, 86. The Dutch colleague to whom Einstein was referring was Hendrik A. Lorentz. The Clarté movement, which briefly united communists such as Henri Barbusse and liberal pacifists such as Romain Rolland, shortly afterwards gravitated more definitely into the communist fold. This led to the withdrawal of Rolland from Clarté in 1919. Later, Einstein also distanced himself from the movement, and Colin withdrew from it in 1922. Einstein, letter to Romain Rolland, Apr. 19, 1922, AEA [84 102]; Paul Colin, letter to Einstein, Apr. 22, 1922, AEA [43 467]; Fisher, 88; Jean Albertini, ed., *Romain Rolland. Textes: Politiques, sociaux et philosophiques choisis* (Paris: Editions Sociales, 1970), 182–233.

119. Moritz Schlick, letter to Einstein, Dec. 19, 1919, AEA [21 570]; also in *CPAE* 9: 222.

120. Carl Seelig, letter to Einstein, Dec. 21, 1919, AEA [39 002]; also in *CPAE* 9: 230. Carl Seelig, *Die Zwölf Bücher* (Leipzig, Vienna, and Zürich: E. P. Tal & Co. Verlag, 1919–22). The twelve works ultimately published in Carl Seelig's *Die Zwölf Bücher* were: Romain Rolland, *Die Zeit wird kommen. Drama in drei Akten*; Carl Hauptmann,

Der abtrünnige Zar. Eine Legende in sechs Vorgängen; Hermann Hesse, *Kleiner Garten. Erlebnisse und Dichtungen*; Wilhelm Schmidtbonn, *Die Flucht zu den Hilflosen. Die Geschichte dreier Hunde*; Stefan Zweig, *Fahrten. Landschften und Städte*; Henri Barbusse, *Erste Novellen*; André Suarès, *Cressida*; Otto Zoff, *Gedichte*; Georges Duhamel, *Das Licht. Drama in vier Akten*; Maurice Maeterlinck, *Der Bürgermeister von Stilmonde. Drama in drei Akten*; Wilhelm Schäfer, *Frühzeit*; Ernst Toller, *Die Maschinenstürmer. Ein Drama aus der Zeit der Ludditenbewegung in England in fünf Akten und einem Vorspiel*. For Seelig's editorial processes in putting together this collection, see Ulrich Weinzierl, *Carl Seelig, Schriftsteller* (Vienna: Löcker, 1982), 35–52, 192–93.

121. Einstein, letter to Carl Seelig, Dec. 29, 1919, AEA [39 003]; also in *CPAE* 9: 237.

122. Einstein, letter to Robert W. Lawson, Dec. 26, 1919, AEA [44 257]; also in *CPAE* 9: 234.

123. Einstein, letter to Heinrich Zangger, [Dec. 24, 1919], AEA [86 162]; also in *CPAE* 9: 233.

124. Einstein, letter to Edgar Meyer, Dec. 28, 1919, AEA [75 605]; also in *CPAE* 9: 235.

5. Economic Crisis, Reactionary Politics, and Ongoing Engagement: 1920–1921

1. Peukert, 52–53; Bessell, *Germany*, 254, 283; Eyck, 180. Also see *The Treaty of Peace Between the Allied and Associated Powers and Germany* (London: His Majesty's Stationery Office, 1919); also "The Treaty of Versailles: The Reparations Clauses," in Kaes et al., 8–9.

2. Eyck, 180. On later developments between Germans and Poles regarding Upper Silesia, see MacMillan, 220–21; Kolb, 42, 56; Peukert, 202–3; Eyck, 107, 109.

3. Einstein, letter to Hellmut von Gerlach, Jan. 6, 1920, AEA [43 745]; also in *CPAE* 9: 246. In August 1919, the organ of the New Fatherland League had reported on its advocacy for political prisoners, including writer and philosopher Gustav Landauer and writer Ernst Toller, and protested the execution of Eugen Leviné. *Mitteilungen des Bundes Neues Vaterland*, Neue Folge, Nr. 4 (Aug. 1919), 22–24. Also, in May 1919, Einstein, Gerlach, and fifteen others signed a telegram protesting that execution. Karl A. Aner et al., telegram to Munich Military Tribunal, [May 19, 1919], AEA [86 060]; also in *CPAE* 9: 44).

4. Heinrich von Gleichen Rußworm, for the League of German Scholars and Artists (Bund Deutscher Gelehrter und Künstler), letter to Einstein, Jan. 10, 1920, AEA [43 411]; also in *CPAE* 9: 252.

5. Einstein, letter to League of German Scholars and Artists (Bund Deutscher Gelehrter und Künstler), Jan. 13, 1920, AEA [43 413]; also in *CPAE* 9: 258.

6. Georg Count von Arco letter to Einstein, Jan. 9, 1920, AEA [43 076]; also in *CPAE* 9: 250. Monism is a philosophy that denies the duality of soul and matter, of God and the universe. The German Monist League (Deutscher Monistenbund) propagated

logical extensions of this philosophy in society, such as separation of church and state, a world view based on science, and opposition to religious dogma. See Ernst Haeckel, *Die Welträtsel: Gemeinverständliche Studien über Monistische Philosophie* (Leipzig: Kröner, 1908), 13–14; Paul Ziche, ed., *Monismus um 1900: Wissenschaftskultur und Weltanschauung* (Berlin: Verlag für Wissenschaft und Bildung, 2000); and Roger Chickering, *Imperial Germany and a World Without War*, 89–121. University of Jena zoologist Ernst Haeckel and 1909 Nobel laureate chemist Wilhelm Ostwald were among the signatories of the 1914 manifesto "To the Civilized World," which denied that the German army had committed atrocities in Belgium and France and tended to support German nationalism. See Grete Ostwald, *Wilhelm Ostwald mein Vater* (Stuttgart: Berliner Union, 1953), 137; and Johannes Hemleben, *Ernst Haeckel in Selbstzeugnissen und Bilddokumenten* (Reinbek: Rowohlt, 1964), 139.

7. Einstein, letter to Georg Count von Arco, Jan. 14, 1920, AEA [43 078]; also in *CPAE* 9: 260. Einstein was referring to Ernst Haeckel, who besides being a University of Jena zoology professor was founder of the German Monist League (Deutscher Monistenbund). For more on German Monism, see Haeckel; Ziche.

8. Einstein, letter to Hendrik A. Lorentz, Jan. 12, 1920, AEA [16 498]; also in *CPAE* 9: 256; and in Nathan and Norden, 35. On the development of the two versions of *Lille*, see events of April 1919 in chapter 4.

9. Robert W. Lawson, letter to Einstein, Jan. 8, 1920, AEA [44 259]; also in *CPAE* 9: 249.

10. Einstein, letter to Robert W. Lawson, Jan. 22, 1920, AEA [44 261]; also in *CPAE* 9: 275.

11. On January 16, 1920, the Allies demanded that the Netherlands extradite Kaiser Wilhelm. On January 23, the Netherlands refused. Also on January 16, the US Senate voted against joining the League of Nations. MacMillan, 488–93; Eyck, 126–27; Peukert, 57; Kolb, 25–29.

On January 18, the German National Assembly voted 213 to 56 to pass the Factory Council Law (*Betriebrätegesetz*). This law, which came into effect on February 4, gave workers some rights to negotiate but also required them to act in the interests of the company. Leftists such as the USPD criticized the law as turning workers' councils into company unions. Peukert, 110; Eyck, 133–34.

On January 26, 1920, officer cadet Ottwig von Hirschfeld attempted to murder German finance minister and Center Party member Matthias Erzberger, who had supported a peace resolution in 1917 and had signed the armistice agreement in November 1918 as part of the German delegation. Hirschfeld was sentenced to only eighteen months in prison for this crime. It was the first of three attempts on Erzberger's life, culminating in his murder on August 26, 1921, by Heinrich Tillessen and Heinrich Schulz, former officers of the Ehrhardt Brigade (which implemented the Kapp Putsch in March 1920). Kolb, 44–45; Eyck, 145, 188–89; Bessell, *Germany*, 261; Peukert, 73.

From January 19 through March 12, 1920, the trial of a libel suit brought by Matthias Erzberger against Karl Helfferich for personal attacks in the rightist newspaper *Kreuzzeitung* was conducted. This was a propaganda boon for rightists, even though Helfferich technically lost the case. Kolb, 36; Eyck, 143–44. On newspaper attacks by former German imperial finance minister Karl Helfferich on the new republic's vice-chancellor and finance minister Matthias Erzberger, formerly the German negotiations commissioner for the armistice, see C. Paul Vincent, *A Historical Dictionary of Germany's Weimar Republic, 1918–1933* (Westport, CT: Greenwood, 1997), 113; William S. Halperin, *Germany Tried Democracy* (New York: Norton, 1946), 170; and Siegfried Löwenstein, *Der Prozess Erzberger-Helfferich: Ein Rechtsgutachten* (Ulm: Süddeutsche Verlagsanstalt, 1921).

12. On the disruptions of Nicolai's lectures, see *Deutsche Tageszeitung*, Jan. 13, 1920, [3], and Jan. 15, 1920, [3]; and Zuelzer, 267–83. For his major wartime pacifist published work, see Nicolai, *Biologie des Krieges*. The "Appeal to the Europeans," cosigned by Einstein, is discussed in chapter 1.

13. Einstein, letter to Hans Delbrück, Jan. 26, 1920, AEA [78 123]; also in *CPAE* 9: 282.

14. Einstein, statement in support of Georg Friedrich Nicolai, [Jan. 26, 1920], AEA [78 124]; also in *CPAE* 7: 32; and in Rowe and Schulmann, 84–85. The statement was typed on the reverse side of Einstein's letter to Hans Delbrück, Jan. 26, 1920, AEA [78 123]; also in *CPAE* 9: 282.

15. Einstein, letter to Hedwig and Max Born, Jan. 27, 1920, AEA [8 144]; also in *CPAE* 9: 284. This was a period of military successes by the Russian Red Army. Georges Clemenceau, French prime minister from 1906 to 1909 and 1917 to 1920, the principal advocate of the harsh terms of the Versailles treaty (although accused in France of having been insufficiently tough on Germany), had just been defeated in the January elections. Karl Radek (Sobelsohn), a journalist and member of the Central Committee of the Soviet Union, spent time organizing in Germany (including almost a year in prison) between December 1918 and January 1920, when he returned to Russia. Warren Lerner, *Karl Radek: The Last Internationalist* (Stanford, CA: Stanford Univ. Press, 1970), 76–90.

16. On February 2, consistent with decisions made in Paris, French troops occupied the Memel district, previously a German territory and eventually part of Lithuania. On February 10, 1920, after Polish lieutenant general Jozef Haller de Hallenburg captured Pomerania and Gdansk (Danzig), a German naval yard industrial center during the War, Poles celebrated their country's recovery of access to the sea. On the decision at Paris and other context regarding Gdansk, see Kolb, 27–28; Bessel, *Germany*, 146, note 85; Eyck, 106–8; Peukert, 44. On February 17, the Allies agreed to allow Germany to prosecute Germans domestically for crimes against the laws of civilized warfare instead of extraditing them. MacMillan, 165; Eyck, 148, 175, 187–88.

17. Hendrik A. Lorentz, letter to Einstein, February 11, 1920, AEA [16 502]; also in *CPAE* 9: 308. On the Lille commission and the two versions of its published booklet, see discussion of the events of April 1919 in chapter 4.

18. Death Certificate, Pauline Einstein, February 20, 1920, AEA [29 164]; Burial of Pauline Einstein in Maxstrasse cemetery, Berlin Schöneberg, Feb. 23, 1920, AEA [29 164.1].

19. Einstein, letter to Marcel Grossmann, Feb. 27, 1920, AEA [11 457]; also in *CPAE* 9: 330. On Moritz Schlick, see Einstein, letter to Moritz Schlick, Dec. 1, 1919, AEA [21 626]; also in *CPAE* 9: 184; and Moritz Schlick, letter to Einstein, Dec. 19, 1919, AEA [21 570]; also in *CPAE* 9: 222. Paul Kammerer was an adjunct (*Adjunkt*) in zoology at the Research Institute for Biology (Biologische Versuchsanstalt) in Vienna.

20. Einstein, letter to Heinrich Zangger, Feb. 27, 1920, AEA [86 168]; also in *CPAE* 9: 332.

21. Einstein, letter to Moritz Schlick, Feb. 27, 1920, AEA [21 630]; also in *CPAE* 9: 331.

22. On February 29, 1920, Walther von Lüttwitz, a general in charge of two military brigades in central and eastern Germany, protested to German minister of defense Gustav Noske his order that such brigades be disbanded. On March 9, the German National Assembly considered a petition from the German National People's Party (DNVP) to dissolve on May 1 and hold elections for the Reichstag and president; it rejected the petition by a vote of 176 to 60. Also on March 9, Lüttwitz presented various plans for violence to Oskar Hergt, head of the DNVP. On March 10, Lüttwitz met with German president Friedrich Ebert and Noske to make various political and military demands, including an immediate Reichstag election, to which Ebert and Noske made counterarguments. On March 11, Noske removed Lüttwitz from his position and ordered the arrests of Wolfgang Kapp and Waldemar Pabst, but the police delayed and the two found safe hiding. On March 12, the day before the putsch, Noske described the situation to the German cabinet, but the generals present, especially General Hans von Seeckt, refused to promise to defend the government; thus, the army failed as a shield of the constitution and the republic. Eyck, 149–50; Kolb, 37; Peukert, 68–69.

23. "Haenisch zum Fall Nicolai. Aufhebung der Senatsentscheidung," *Vossische Zeitung*, Mar. 10, 1920, evening edition.

24. Einstein, letter to Konrad Haenisch, [Mar. 10–12, 1920], AEA [83 264]; also in *CPAE* 9: 349. On Leo Arons, see the discussion in chapter 4.

25. Georg Count von Arco, letter to Einstein, Mar. 12, 1920, AEA [43 080]; also in *CPAE* 9: 350.

26. Konrad Haenisch, letter to Einstein, Mar. 12, 1920, AEA [43 824]; also in *CPAE* 9: 351.

27. Moritz Schlick, letter to Einstein, Mar. 13, 1920, AEA [21 572]; also in *CPAE* 9: 352.

28. In the Kapp Putsch, troops of the Ehrhardt naval brigade, which opposed the reduction of the military required by the Versailles treaty, marched through the Brandenburg Gate in Berlin, and the national government of Friedrich Ebert and Gustav Bauer left Berlin for Dresden and later Stuttgart. Wolfgang Kapp, provincial director-general of East Prussia, seized authority and the title of chancellor. Kapp, son of Friedrich Kapp, the 1848 revolutionary and liberal member of the Reichstag, had become a conservative agrarian, nationalist, and militarist, and was a founding member of the German Fatherland Party (a short-lived rightist party active during late World War I). That day, the SPD called for a general strike; the following day, March 14, in the largest recorded general strike in Germany, all public services ceased throughout Berlin. "Aufruf der sozialdemokratischen Mitglieder der Reichsregierung und des Parteivorstandes der SPD zum Generalstreik," in Karl Brammer, ed., *Fünf Tage Militärdiktatur* (Berlin: Verlag für Politik und Wissenschaft, 1920), 65; for an English translation, see Social Democratic Party (SPD), "Appeal of the Social Democratic Party for a General Strike," in *The Weimar Republic Sourcebook*, ed. Anton Kaes et al. (Berkeley: Univ. of California Press, 1994), 16. The Kapp regime was disorganized, did not have popular support among Berliners, and was defeated by a general strike of workers and the refusal of civil servants to support it (e.g., to authorize banks to provide funding). On March 17, the putsch collapsed after four days. Kapp declared his resignation as chancellor and fled to Sweden. When crowds at the Brandenburg Gate cheered as the Ehrhardt brigade marched away, officers ordered machine guns fired on the people. In the aftermath of the putsch, Gustav Bauer resigned as chancellor, former foreign minister Hermann Müller became the new chancellor, and the government for the time being remained a coalition of the SPD, Center Party, and DDP. On March 18, in a National Assembly session, Philipp Scheidemann pushed through the resignation of Gustav Noske from the SPD, despite Ebert's defense of him, due to Noske's more aggressive actions against the striking communists (KPD) than against the Kapp generals. Eyck, 150–52, 156–57, 159; Kolb, 36–37; Peukert, 68–69.

29. Midway through the four days of the Kapp Putsch, on March 15, in the industrial region of the Ruhr, workers' organizations formed a "Red Army" of 60,000 workers. Over the following days, after various Red Army units had occupied most of the region, the government used both army and Freikorps militias to suppress these units violently, which alienated a large proportion of workers from the Weimar Republic and the SPD. The Bielefeld Agreement, concluded on March 24, disbanded the Red Army units and met union demands but also granted immunity to troops that had broken laws in violently suppressing the rebellion. In response to this use of German troops, the French, under socialist Prime Minister Alexandre Millerand, occupied Frankfurt, Darmstadt, Homburg, and Hanau. The British government dissociated itself from this French action. Kolb, 38; Peukert, 69–70; Eyck, 155–56, 159–60; Bessel, *Germany*, 261; George Eliasberg, *Der Ruhrkrieg von 1920* (Bonn-Bad Godesberg: Verlag Neue Gesellschaft, 1974).

30. MacMillan, 492; Eyck, 127; Peukert, 57; Kolb, 25.

31. Einstein, letter to Hendrik A. Lorentz, Mar. 18, 1920, AEA [16 506]; also in *CPAE* 9: 356; and in Nathan and Norden, 36–37, 39.

32. Einstein, letter to Hans Albert and Eduard Einstein, Mar. 26, [1920], AEA [75 730]; also in *CPAE* 9: 360. For the exchange rate of 13 marks to 1 Swiss franc on that date, see *Berliner Tageblatt*, Mar. 26, 1920, evening edition. On the limitation of rail traffic, see *Vossische Zeitung*, Mar. 24, 1920, morning edition, 1st supplement. On the closing and reopening of the University of Berlin, see *Vossische Zeitung*, Mar. 25, 1920, evening edition.

33. Einstein, letter to Heinrich Zangger, Mar. 26, 1920, AEA [40 001]; also in *CPAE* 9: 361. For a contemporaneous detailed report and analysis of the rampant political murders being committed mostly by rightist militias known as the Freikorps, largely composed of demobilized soldiers retaining their weapons from the War, see Emil J. Gumbel, *Zwei Jahre Mord* (Berlin: Verlag Neues Vaterland, 1921); and Emil J. Gumbel, *Vier Jahre politischer Mord* (Berlin-Fichtenau: Verlag der Neuen Gesellschaft, 1922). For more on the Freikorps, see Nigel H. Jones, *Hitler's Heralds: The Story of the Freikorps, 1918–1923* (New York: Dorset Press, 1992); and Hagen Schulze, *Freikorps und Republik, 1918–1920* (Boppard am Rhein: Boldt, 1969). For a newspaper article two days earlier alleging that government officials were transferring money out of the country, see "Geldvershiebung nach dem Ausland?" *Vossische Zeitung*, Feb. 24, 1920, evening edition.

34. On March 29, 1920, at a meeting of the German National Assembly, the new chancellor, Hermann Müller, called for speedy elections to the Reichstag, moving them to early June. Eyck, 161; Kolb, 38; Peukert, 70.

By April, 15,000 officers had been dismissed from the German army, with a reduction of another 5,000 required by September under the Versailles treaty. In the change of government, DDP member of the Reichstag Otto Gessler took over from Gustav Noske as minister of defense and Hans von Seekt became head of the Army Command (*Heeresleitung*). Seeckt resisted compliance with the Versailles treaty, authorizing the army to collaborate with the Freikorps in suppressing workers' uprisings in Saxony, Thuringia, and the Ruhr. On March 26, the German government asked France for permission to use German troops against not-yet-disbanded Red Army units in the French-occupied area of the Ruhr, and on April 2 the German army entered the Ruhr for this purpose. During April, communists led by Max Hölz took control of the Saxon Vogtland for several weeks. On April 8, German minister of interior Erich Koch, as directed by the Allies, ordered German provincial governments to dissolve their Home Guards (*Einwohnerwehren*). When Prussia complied, it was attacked by rightists in the Prussian legislature (*Landtag*) and the press; when Bavaria refused, it was praised by rightists. Kolb, 30, 36–38; Peukert, 46, 68–70; Eyck, 155–60, 164–66; and Eliasberg. On the size of the German army and limitation of it, see Bessell, *Germany*, 5, 127, 161, 180. On the German army during the early years of the Weimar Republic, see Vincent, *Historical Dictionary*, 390–91. On communist activity in Saxony three years later in 1923, and the contrasting treatment of leftist and rightist insurgencies by the federal government at that time, see Kolb, 48; Peukert, 72, 75. On the growth of trade union membership in Germany after the War, see Bessell, *Germany*, 135–38.

35. On April 19, after prisoner-of-war exchanges with England, the United States, and France had taken place in late 1919 and early 1920, Germany and the Soviet Union reached a prisoner-of-war exchange agreement. On German use of foreign prisoners of war for labor during the War, the loss of this at the end of the War, and the return of Germans who had been prisoners of war, see Bessell, *Germany*, 22–23, 59–61, 72–73, 114, 126–27, 144–45, 161–62, 180, 197–203.

36. Einstein wrote the appeal on the request of the *Neue Zürcher Zeitung*'s editor. Eduard Korrodi, letter to Einstein, Mar. 23, 1920, AEA [44 546]; also in *CPAE* 9: 359.

37. Einstein, untitled article advocating an organized international exchange of scientific literature, published in *Neue Zürcher Zeitung*, Apr. 4, 1920, 1st Sunday edition, section 2: [1]; also in *CPAE* 7: 36.

38. Einstein, letter to Hans Albert Einstein, Apr. 5, 1920, AEA [75 731]; also in *CPAE* 9: 369.

39. Einstein, letter to Carl H. Unthan, Apr. 5, 1920, AEA [45 162]; also in *CPAE* 9: 370.

40. Einstein, letter to Paul Ehrenfest, Apr. 7, 1920, AEA [9 488]; also in *CPAE* 9: 371; and in Nathan and Norden, 39. On the Allies' blockade of Germany and its extension past the end of the War, see Roerkohl; Vincent, *Politics of Hunger*; and Bane and Lutz.

41. Einstein, letter to Ulrich von Wilamowitz-Moellendorff, Apr. 19, 1920, AEA [81 833]; also in *CPAE* 9: 379; and in Nathan and Norden, 39.

42. Einstein, letter to Heinrich Zangger, Apr. 19, 1920, AEA [86 096]; also in *CPAE* 9: 380.

43. On April 27, 1920, the new German Franchise Law established a proportional ballot, in which voters elected parties rather than persons into power; the aim of this was to attain greater equity in the weight of votes. On May 4, consistent with a longer-range provision of the new German constitution, a Temporary National Economic Council was formed, intended both to review and to initiate economic and social legislation. Eyck, 69–70; 76–77; Peukert, 38–39. For broader context on German economic and industrial relations policy in the early Weimar Republic, see Peukert, 108–12. On the Demobilization Ministry that existed between November 1918 and April 1919, and continued in regional and local branches until 1921 and 1924, see Bessell, *Germany*, 91–124.

44. Eliasberg; Eyck, 156; Kolb, 38; Peukert, 69.

45. Gottlieb Haberlandt, letter to Einstein, May 1, 1920, AEA [43 819]; also in *CPAE* 10: 3.

46. Moritz Schlick, letter to Einstein, May 10, 1920, AEA [21 574]; also in *CPAE* 10: 12. University of Rostock professors who supported the Kapp Putsch included Rudolf Helm (classics), Gerhard Hilbert (theology), and Ernst Schwalbe (pathology), as well as an instructor, Friedrich Dettweiler (animal husbandry). Among other nationalist or

reactionary Rostock professors were university rector Gustav Herbig (comparative literature), Johannes Reinmöller (dentistry), Otto Staude (mathematics), and Wilhelm Walther (church history). Gerhard Heitz et al., *Geschichte der Universität Rostock 1419–1969. Festschrift zur Fünfhundertfüngzig-Jahr-Feier der Universität. Band I. Die Universität von 1419–1945* (Berlin: Deutscher Verlag der Wissenschaft, 1969), 166–74.

47. Carl H. Unthan, letter to Einstein, May 23, 1920, AEA [45 163]; also in *CPAE* 10: 28. Unthan mischaracterized Fried, who held the former government of Germany responsible for the War but explicitly declared the current people of Germany not guilty of this. Alfred Hermann Fried, *Auf hartem Grund. Offene Antwort auf den an mich gerichteten Brief von Dr. Jur. Hermann M. Popert im "Vortrupp" vom 1. März 1919* (Hamburg: Pfadweiser, 1919). Fried also publicly argued against the Versailles treaty's unfair treatment of Germany. Fried, *Mein Kampf gegen Versailles und St. Germain vom Nov. 1918 bis Juni 1919* (Leipzig: Die Neue Geist, 1920).

48. "Gegen die 'unfruchtbare Ablehnung des neuen politischen Zustandes,'" *Berliner Tageblatt*, May 30, 1920, morning edition; "Für die demokratische Verfassung: Ein Aufruf der Hochschullehrer," *Vossische Zeitung*, May 30, 1920; Christian Nottmeier, *Adolf von Harnack und die deutsche Politik 1890–1930. Ein biographische Studie zum Verhältnis von Protetstantismus, Wissenschaft und Politik* (Tübingen: Siebbeck, 2004), 476.

49. Paul Epstein, letter to Einstein, May 30, 1920, AEA [10 558]; also in *CPAE* 10: 38.

50. On May 21, 1920, Hans Paasche, a former German naval officer turned pacifist, was assassinated. During the summer of 1920, many political murders were committed by radical rightists, who during this period were shifting from a prior focus exclusively on revolutionary leftists to target pacifists and others portrayed as "traitors." Hellmut von Gerlach, Emil J. Gumbel, and Georg Friedrich Nicolai all received death threats, and Gumbel was physically attacked and beaten up. Zuelzer, 263, 265, 293.

51. On June 6, in Reichstag elections, Germans showed their frustration with the ineffectiveness of the moderate Weimar coalition of SPD and DDP by voting it out. Those two parties' total share of the vote dropped from 19 million to 11 million, while votes for more extreme rightist and leftist parties increased, on the right from 5.6 to 9.1 million, on the left from 2.1 to 5.3 million. The SPD, the DDP, and the Center Party, which together had held 78 percent of seats (331 out of 423), now held only 44.6 percent of seats (205 out of 459). The SPD was still the strongest party in the Reichstag with 113 seats, but this was more than 50 seats less than before; DDP seats in the Reichstag dropped from 75 to 45. Seats of the more leftist USPD rose from 22 to 81, and the KPD won 2 seats for first time ever. The government of Hermann Müller gave way to a coalition of the DVP, the Center Party, and the DDP, led by Konstantin Fehrenbach of the Center Party. In the Reichstag, the new coalition controlled only 176 of 466 seats and could function only with the SPD's cooperation. This meant that the original Weimar coalition had lost its window of opportunity to establish meaningful democratic reform at the outset of the Weimar Republic. Henceforth it would be far more difficult for the

Weimar government to take decisive action. Kolb, 38–39, 194–95; Peukert, 70–71; Eyck, 161–63. After a temporary retirement during the Helfferich case, Matthias Erzberger was re-elected to the Reichstag. Eyck, 188; Kolb, 36.

52. The Allies and Hungary signed the Treaty of Trianon on June 4, 1920. It ceded two-thirds of Hungary's territory to Rumania, Czechoslovakia, and Yugoslavia, evoking great bitterness among Hungarians. MacMillan, 268–70; Kolb, 32; Peukert, 45. On June 15, a new border treaty between Germany and Denmark, based on a plebiscite held in February, gave northern Schleswig to Denmark. MacMillan, 169–70; Kolb, 28; Eyck, 87, 99, 108. On July 20, Germany declared neutrality in the war between Poland and Soviet Union, which meant that Allied aid to Poland could no longer be transported through Germany. Meanwhile, the Russian Red Army continued an offensive into Poland in response to a Polish expansion of its boundaries into Russia that started on April 25. Eyck, 169–70. For some context to the conflict, and for the German perspective on Poland, see Kolb, 55–56; Peukert, 56, 202–3.

53. Einstein, letter to Frank D. Fackenthal, Columbia University, June 6, 1920, AEA [30 130].

54. Einstein, letter to Arthur S. Eddington, June 11, 1920, AEA [73 839]; also in *CPAE* 10: 52.

55. Einstein, letter to Hendrik A. Lorentz, June 15, 1920, AEA [16 516]; also in *CPAE* 10: 56.

56. Eduard Sthamer, letter to Konrad Burdach and Einstein, June 23, 1920 AEA [79 295]; and Einstein, "Vereidigungsnachweis," July 1, 1920, AEA [79 297]; also in Kirsten and Treder, docs. 20 and 21, 112.

57. Gaston Moch, letter to Einstein, July 3, 1920, AEA [44 468]; also in *CPAE* 10: 69.

58. Einstein, letter to Gaston Moch, July 19, 1920, AEA [44 469]; also in *CPAE* 10: 78.

59. On July 5 to 16, 1920, a conference between the Allies and the Germans, represented by Otto Gessler and General Hans von Seeckt, took place at Spa, Belgium. The main topic of discussion was the Versailles treaty's provisions for German disarmament. The Allies refused Germany's request for a larger army (of 200,000 rather than 100,000). They also refused Germany's request for more time to comply; the Germans would have to do so by January 1, 1921. The Allies threatened to occupy new parts of Germany, possibly the industrial Ruhr region, if the new deadline was not met and coal deliveries were not made. The outcome of the conference was received with anger and indignation in Germany, reinforcing the sentiments that were driving this summer of violence and threats of violence. Eyck, 165–66. On related conferences focused on reparations, see Kolb, 41. On some of the broader implications for Weimar culture of the limitation of the size of the German military, see Bessell, *Germany*, 272–73.

60. German League for the League of Nations (Deutsche Liga für Völkerbund), letter to Einstein, July 8, 1920, AEA [44 335]; also in *CPAE* 10: 73.

61. Einstein, letter to German League for the League of Nations, July 23, 1920, AEA [44 336]; also in *CPAE* 10: 82.

62. Elsa Hermann, German Central Committee for Foreign Relief (Deutscher Zentralausschuss für die Auslands Hilfe), letter to Einstein, July 9, 1920, AEA [44 694]; also in *CPAE* 10: 74.

63. Einstein, handwritten draft of statement praising American and English Quaker relief work in Germany (in Ilse Einstein's handwriting), [July 11, 1920], AEA [44 695]; also in *CPAE* 7: 40; and in Nathan and Norden, 39–40. To add a personal anecdote, when I was engaged in research in Germany during 1997–98, an elderly lifelong member of the Berlin Quaker Meeting told me that the Anglo-American Quaker hunger relief effort in Germany after World War I made such a profound impression on Germans that, throughout the period of the Nazi regime, despite Nazis' professed hostility toward pacifism, they allowed the Berlin Quaker Meeting to continue its Meeting for Worship (although always with an SS officer present).

64. Einstein, typed statement on the Quaker relief effort in Central Europe, [after July 11, 1920], AEA [40 006]; also in *CPAE* 7: 41.

65. Einstein, letter to Michele Besso, July 26, 1920, AEA [70 366]; also in *CPAE* 10: 85.

66. On August 4, 1920, a law was passed in Germany that granted amnesty to participants in the Kapp Putsch who were not instigators or leaders. (Most of the leaders had disappeared; only Traugott von Jagow was convicted.) Eyck, 159–60; Kolb, 38. On August 11, when the disarmament law under the Versailles treaty came into force, the Bavarian government refused to disarm the Bavarian home guards. Eyck, 165, 175–76; Kolb, 38.

67. Between August 13 and 25, the Poles defeated the Russian Red Army in the battle of Warsaw. Meanwhile, from August 19 to 25, Poles in Upper Silesia repeated their uprising of a year earlier against the Germans. This conflict would continue into 1921, requiring a new German-Polish agreement, which was reached in May 1922. Kolb, 56; Eyck, 169, 180–81, 191–93; Bessell, *Germany*, 96, 209; Peukert, 56, 66, 202–3.

68. On August 10, the Ottoman sultan's representatives signed the Treaty of Sèvres with the Allies. This treaty, which included protections for the Armenian and Kurdish minorities, quickly proved ineffective. In 1920, and continuing into 1923, Kemal Ataturk's Turkish nationalist forces were driving out Greek forces and taking military control over most of Asia Minor. This led to an armistice in October 1922 and the Treaty of Lausanne in July 1923 (of which Turkey dictated most of the terms), which did not retain protections for minorities. MacMillan, 447–55; Kolb, 32.

69. Einstein, letter to Norwegian Students' Association (Norwegische Studentenverein), Sept. 9, 1920, AEA [44 597]; also in *CPAE* 10: 141; and in Nathan and Norden, 40. Einstein had done a brief lecture tour in Kristiania (Oslo), Norway, for a few days in mid-June 1920.

70. Einstein, letter to Hendrik A. Lorentz, [after Sept. 25, 1920], AEA [16 531]; also in *CPAE* 10: 155. On the ideal of an international federation of nations, which had been part of the public discourse among pacifists since before the War, see Alfred Hermann Fried, *Kurzgefasste Darstellung der Pan-Amerikanischen Bewegung* (Berlin: Verlag der "Friedens-Warte," 1912). On the harsh terms of the Versailles treaty as exemplifying a failure of leadership on the part of France and England, a view shared by both German nationalists and prominent German pacifists, see Fried, *Mein Kampf gegen Versailles*.

71. Otto Glogau et al., letter to Einstein, Sept. 29, 1920, AEA [43 536].

72. Einstein, "On the Contribution of Intellectuals to International Reconciliation," [after Sept. 29, 1920], in *Thoughts on Reconciliation* (New York: Deutscher Gesellig-Wissenschaftlicher Verein von New York, 1920), 10–11; AEA [43 537]; also in *CPAE* 7: 47; and in Rowe and Schulmann, 87–88. Einstein's concern about reconciling international divisions in the scientific community arose in the context of an Allied boycott, via the International Research Council (founded in July 1919), which was attempting to control international scientific exchanges and to limit German and Austrian participation in international scientific activity. For more on the International Research Council, see, for example, Ulrich Marsch, *Notgemeinschaft der Deutschen Wissenschaft. Gründung und frühe Geschichte 1920–1925* (Frankfurt am Main: Lang, 1994), 38; Brigitte Schröder-Gudehus, "Deutsche Wissenschaft und Internationale Zusammenarbeit 1914–1928. Ein Beitrag zum Studium kultureller Beziehungen in politischen Krisenzeiten" (PhD diss., University of Geneva, 1966), 96–120; and "Die wissenschaftliche Blockade Deutschlands," *Tägliche Rundschau*, Feb. 22, 1921, morning edition, [7]. On the boycott's economic implications, see Friedrich Schmidt-Ott and Eduard Wildhagen, "Die Not der deutschen Wissenschaft," in *Internationale Monatsschrift für Wissenschaft, Kunst und Technik* 15, no. 1 (Oct. 1920): 1–36; and in *Forschungsförderung in drei Epochen: Deutschen Forschungsgemeinschaft—Geschichte, Arbeitsweise, Kommentar*, ed. Kurt Zierold (Wiesbaden: Steiner, 1968), 564–65.

73. During October 1920, Russian communists in Germany, including Gregory Zinoviev, convinced a number of USPD members to join the KPD, which increased that party's influence. Eyck, 179; Peukert, 71–72; Kolb, 35, 39, 44. In November, Prussia finished drafting its state constitution, and the Prussian constitutional convention voted to facilitate a unified federal state of Germany. Prussia accounted for three-fifths of Germany's territory, including Berlin, and was home to three-fifths of its population. Throughout the Weimar years, Prussia's state government would be more stable and effective (under an SPD-centered coalition) than the national one. Kolb, 68–69; Eyck, 129–30; Peukert, 213, 225.

74. On October 9, Polish troops occupied Vilna, Lithuania. On October 12, after the Polish army captured Tarnopol, Dubno, Minsk, and Dryssa, a ceasefire was called and took hold. For context, see Kolb, 56; Eyck, 108.

75. Paul Hertz, letter to Einstein, Oct. 28, 1920, AEA [12 210]; also in *CPAE* 10: 186. For more on the charges on which Kelen and nine other defendants were convicted on Dec. 25, 1920, and sentenced to life imprisonment, see *Arbeiter-Zeitung* (Vienna), Dec. 29, 1920; and Erika Rév, *A népbiztosok pere* (Budapest: Kossuth, 1969).

76. Einstein, letter to Jolán Kelen-Fried, Nov. 8, 1920, AEA [86 857]; published in *Neue Freie Presse* (Vienna), Nov. 19, 1920, morning edition, 7; also in *CPAE* 10: 194. Einstein's earlier appeal on behalf of Friedrich Adler is discussed in chapter 2 of this book.

77. Friedrich Adler, letter to Einstein, Nov. 9, 1920, AEA [6 015]; also in *CPAE* 10: 196. Otto Bauer and Sigmund Kunfy (editors of the *Arbeiter-Zeitung*), circular letter sent to Einstein, Nov. 9, 1920, AEA [34 407]. The petition to pardon ten officials of the Hungarian Soviet Republic (with Einstein among signatories) was published in *Arbeiter-Zeitung* (Vienna), Dec. 15, 1920.

78. Paul Hertz, letter to Einstein, Nov. 11, 1920, AEA [12 211]; also in *CPAE* 10: 20). For Kármán's article, see *Frankfurter Zeitung*, Nov. 4, 1920, 1st morning edition.

79. Jolán Kelen-Fried, letter to Einstein, Nov. 12, 1920, AEA [12 215]; also in *CPAE* 10: 202.

80. The League of Nations moved its headquarters to Geneva, Switzerland, on October 27 and held its first League of Nations assembly there on November 15. On the League of Nations, see Anique H. M. van Ginneken, *Historical Dictionary of the League of Nations* (Lanham, MD: Scarecrow Press, 2006); F. S. Northedge, *The League of Nations: Its Life and Times, 1920–1946* (New York: Holmes and Meier, 1986); and James Avery Joyce, *Broken Star: The Story of the League of Nations* (Swansea: C. Davies, 1978). For a summary of the vision of the League of Nations, see Brower, 143–44. On November 17, the Council of the League of Nations accepted the constitution for the "international" Free City of Danzig (Gdansk), which had been established under the Treaty of Versailles in conjunction with Poland's corridor to the Baltic Sea. Kolb, 27; Peukert, 44; Eyck, 86, 106–7; Brower, 52; Duiker, 135. On December 23, Britain and France ratified the border between French-administered Syria and British-administered Palestine, part of the mandate system under the auspices of the League of Nations. For broader context, see Brower, 59; and Duiker, 84, 107.

81. Paul Mühsam, letter to Einstein, Dec. 7, 1920, AEA [44 502]; also in *CPAE* 10: 221. Paul Mühsam, *Aus dem Schicksalsbuch der Menschheit* (Dresden: Rödel, 1919). Einstein retained a copy of the book, with the author's autograph dated December 7, 1920, in his personal library for the rest of his life.

82. Einstein, letter to Ernest Pickworth Farrow, Dec. 28, 1920, AEA [43 662]; also in *CPAE* 10: 245. This was in reply to Ernest Pickworth Farrow, letter to Einstein, Dec. 17, 1920, AEA [43 661].

83. In January 1921 in Paris, the Allies established a reparations figure of 269 billion marks, to be paid in gradually increasing amounts across forty-two years in tandem

with the delivery of 12 percent of German exports to the Allied nations. The Germans resented having been forced to sign the treaty before the amount of reparations had been determined, and at a Reichstag session on February 2, indignant protests were expressed against the Allies' January decisions in Paris. At the London conference on April 27 (discussed below), the Allied Reparations Commission reduced the reparations figure to 132 billion gold marks (6.6 billion or $33 billion pounds) and established a system of payments to extend for thirty years. MacMillan, 479–80; Kolb, 41; Peukert, 52–54, 58; Eyck, 173–80.

84. On January 19, there was a debate in the Reichstag between Bavarian and federal officials over the former's refusal to obey the law requiring it to disarm its Bavarian Home Guards. In March 1921, Bavarian officials indicated that their Home Guards included 320,000 men and held 240,000 rifles, 2,780 machine guns, 44 light cannons, and other weapons. Eyck, 175–79; Hans Mommsen, *The Rise and Fall of Weimar Democracy* (Chapel Hill: Univ. of North Carolina Press, 1996), 112; Gerald D. Feldman, *The Great Disorder: Politics, Economics, and Society in the German Inflation, 1914–1924* (New York: Oxford Univ. Press, 1997), 328–29. For context, see Kolb, 38; Peukert, 73.

85. Paul Zacharias, letter to Einstein, Jan. 6, 1921, AEA [45 293]; also in *CPAE* 12: 10.

86. Otto Lehmann-Rußbüldt's article, published in the Swiss newspaper *Neue Schweizer Zeitung* and the French paper *Le Matin* in early January 1921, was later reprinted in Otto Lehmannn-Rußbüldt, *"Die Brücke über den Abgrund": Für die Verständigung zwischen Deutschland und Frankreich* (Berlin: Verlag Neues Vaterland, E. Berger u. Co., [1922]).

87. Rudolf Lebius, "Ungesühnter Landesverrat," *Deutsche Staatsbürger-Zeitung*, Jan. 9, 1921, 1.

88. Otto Lehmann-Rußbüldt, letter to Einstein, Jan. 18, 1921, AEA [44 306]. The other fifty-two recipients were: Georg Count von Arco; Ernst L. von Aster, philosopher and historian of philosophy; Martin H. Baege, co-founder of the journal *Neuen Erziehung. Zeitschrift für entschiedene Schulreform und freiheitliche Schulpolitik*; Carl Ballad, economist; Hans Georg von Beerfelde, journalist, writer, and former military officer turned pacifist; Eduard Bernstein, a German SPD leader; Alfred Blaschko, dermatologist and cofounder the German Society for the Fight Against Sexually Transmitted Diseases (Liga zur Bekämpfung der Geschlechtskrankheiten); Ludwig Otto Blumenthal, mathematician; Wilhelm Bölcke, World War I pilot; Walther Borgius, anarchist writer on economics and international trade; Wilhelm Börner, Ethical Culture Society leader in Washington, DC, and in Vienna; Rudolf Breitscheid, publisher, politician, and an SPD leader; Hermann Eckardtstein, diplomat; Viktor Engelhardt, historian; Hermann Fernau, journalist and writer; Friedrich Wilhelm Foerster, historian; Alfred Hermann Fried; Hellmuth von Gerlach; Kurt Albert Gerlach, sociologist and economist; Julius Gerson, one-time associate of Karl Kautsky; Martin Gerson, one-time associate of Leo Baeck; Rudolf Breitscheid, Austrian sociologist; Kurt Goldstein, physician, psychiatrist, and

director of the Lazarett Hospital in Frankfurt; Richard Grelling, writer; Emil J. Gumbel, mathematical statistician; Alfons Herzberger; Magnus Hirschfeld, physician, sexologist, and homosexual rights advocate; Alfons Horten, business manager and writer; Morikatsu Inagaki, journalist associated with the League of Nations Association of Japan and proponent of world federation; Lilli Jannasch, head of the publisher Verlag Neues Vaterland; Harry Count Kessler, diplomat, patron of the arts, writer, publisher, and activist; Auguste Kirchhof, jurist; Johann Georg Koenigsberger, mineralogist; Felix Linke, astronomer, SPD member, and prominent figure in the International Hotel Owners Association (Internationalen Hotelbesitzer-Vereins); Heinrich Mann, writer; Professor Mohner; Willy Meyer, board member of the German Professional Pilots Association (Berufsvereinigung Deutscher Flugzeugfuhrer); Ilse Müller-Oestrich; Paul Oestrich, writer and educational reformer; Otto Opet, writer and jurist; Kurt von Reibnitz, an SPD leader in Mecklenburg-Strelitz; Elisabeth Rotten; Rene Schickele, writer and French-German translator; Hugo Simon, writer, banker, and former Prussian finance minister; Georg Steindorff, Egyptologist; Helene Stöcker; Heinrich Ströbel, journalist and SPD politician; Kurt von Tepper-Laski, head of the New Fatherland League from 1914 to 1918; Kuno Tiemann, writer on German foreign policy; Professor Timm; Kurt Tucholsky, poet, writer, satirist, and editor; and Hans Wehberg, international lawyer and pacifist writer and editor.

89. George B. Jeffery, letter to Einstein, Jan. 23, 1921, AEA [13 424]; also in *CPAE* 12: 31. For Einstein's letter accompanying the book that he sent a month earlier, see Einstein, letter to George B. Jeffery, Dec. 14, 1920, AEA [13 436]; also in *CPAE* 10: 230.

90. On Emil J. Gumbel in this period, see Arthur D. Brenner, *Emil J. Gumbel: Weimar German Pacifist and Professor* (Boston and Leiden: Brill, 2001), 4–61. During 1921, Gumbel published the first of two works on political murders in Germany during the early Weimar Republic: *Zwei Jahre Mord* and *Vier Jahre politischer Mord*.

91. Einstein, letter to George B. Jeffery, Mar. 18, 1921, AEA [13 426]; also in *CPAE* 12: 104.

92. Einstein, letter to Max Born, Jan. 31, 1921, AEA [8 158]; also in *CPAE* 12: 37. The punitive terms of the Versailles treaty had been advocated more intensely by the French than the British due to the more extreme suffering of the French in the War. For the previous day's newspaper account in Berlin, see "Das Entschädigungs-Abkommen der Alliierten," *Berliner Tageblatt*, Jan. 30, 1921, morning edition.

93. Otto Neurath, letter to Einstein, Jan. 12, 1921, AEA [18 396]; also in *CPAE* 12: 14.

94. Einstein, draft of letter to Otto Neurath, Mar. 3, 1921, AEA [18 399]; also in *CPAE* 12: 76; and Einstein, letter to Otto Neurath, Mar. 3, 1921, AEA [18 401]; also in *CPAE* 12: 77.

95. Einstein, letter to Romain Rolland, Mar. 1, 1921, AEA [84 171]; also in *CPAE* 12: 72. Romain Rolland, *Clerambault: Histoire d'une conscience libre pendant la guerre* (Paris: P. Ollendorff, 1920); for the passage in which Rolland praised Einstein, see page

336. Einstein retained his copy of this book, with autograph by Rolland dated February 1921, in his personal library through the end of his life.

96. On March 1, 1921, a German delegation arrived on invitation to confer with the Allies in London. On March 3, British prime minister Lloyd George told German foreign minister Walter Simons that various sanctions, including the occupation of three Rhine port cities in the Ruhr, would go into effect on March 7, unless the Germans accepted the Paris proposals of January or made appropriate counterproposals. On March 8, the Allies occupied three Rhine ports, Düsseldorf, Duisburg, and Ruhrort; in response, the German delegation protested and left the London conference. In Germany, Walter Simons and German army chief Hans von Seeckt were greeted by cheering crowds. On March 12, the Reichstag voted approval of Simons's and Seeckt's rejection of the Allies' ultimatum in London, but only after lengthy debate expressing a wide variety of perspectives. Kolb, 41; Eyck, 177–78; MacMillan, 479–80.

According to the Versailles treaty, Germany's first reparation payment of 20 billion marks was due on May 1, 1921. On May 5, Lloyd George handed a new ultimatum to the German ambassador in London, Friedrich Sthamer, which consisted of the following: disarmament; 12 billion marks still due (of the 20 billion due by May 1); war crimes trials; a schedule of payment of the reduced figure of 132 billion marks; and 2 billion marks with a statement of acceptance by May 12. If these demands were not met, Allied troops would occupy the Ruhr industrial region. On May 9, the British government received questions from Gustav Stresemann regarding concessions if Germany accepted the Allies' ultimatum. On May 11, the cabinet of Chancellor Joseph Wirth and the Reichstag officially accepted the reparations plan in the London ultimatum. On May 12, the British government responded favorably to almost all of Stresemann's requests. Kolb, 41–42; Peukert, 52–54, 58; Eyck, 179–80.

97. On March 21, 1921, in the plebiscite in Upper Silesia promised by the Versailles treaty, 479,000 voted to join Poland, while 707,000 voted to remain with Germany, a result partly due to patriotic Upper Silesians (*heimattreuen Oberschlesier*), including many Jews who traveled from other parts of Germany back to Silesia to vote. In response, Poles in Upper Silesia started an uprising in early May with help from French occupying troops. On May 13, speaking in the House of Commons, Lloyd George declared it unfair not to allow Germany to use its army to restore order in Upper Silesia, but France disagreed. Between May 21 and June 6, German volunteer militias successfully battled against Poles around Annaberg, although the German government, pressured by the Allies, tried to restrain them. At the end of June, an Allied commission negotiated a compromise settlement in Upper Silesia, and on October 20 the Allies, in consultation with the League of Nations, decided to partition the plebiscite area, giving the more industrial eastern portion to Poland and the larger, less industrial western area to Germany. This loss of the industrial region and mostly German cities of Kattowicz and Königshütte provoked a governmental crisis, in which Joseph Wirth and his cabinet resigned, although

they reformed a new government and cabinet four days later. Kolb, 42; Peukert, 202–3; MacMillan, 220–21; Eyck, 180–81.

98. On May 23, 1921, the first war crimes trials before German Supreme Court (Reichsgericht) began. In a case of mistreatment of British prisoners of war, three of four defendants were found guilty, but British observers found their sentences too light. For actions against France, most defendants were set free. Two submarine officers who sank a British hospital ship and then its lifeboats were sentenced to four years but escaped a few weeks later. Eyck, 187–88.

99. Rudolph Goldscheid, letter to Einstein, May 10, 1921, AEA [43 768]).

100. Hellmut von Gerlach, Alfons Horten, and Heinrich Ströbel, letter to members of the New Fatherland League (Bund Neues Vaterland), [after June 17, 1921], AEA [44 310]; also in *CPAE* 12: 151. Horten, a former member of the Thyssenwerke steel mills, advocated the socialization of the coal and steel industries. See Alfons Horten, *Sozialisierung und Wiederaufbau; praktische Vorschläge zur Sozialisierung und zur Wiederaufrichtung unseres Wirtschaftslebens* (Berlin: Verlag Neues Vaterland, [1920]). Ströbel was a politician, publicist, former SPD member, and former editor of its organ, *Vorwärts*. At this time, he was a member of the USPD.

101. On July 2, 1921, the state of war between Germany and the United States (which had not ratified the Versailles treaty) was formally ended by a resolution of the US Congress, signed by President Warren G. Harding. On August 25, the United States and Germany concluded a separate peace treaty; in it, the United States claimed all rights and privileges in the Versailles treaty but indicated that it was not part of the League of Nations, nor was it a participant in the Reparations Commission (although, as the chief creditor of the European nations, it ultimately had great influence over reparations payments). Kolb, 26, 29, 57; Eyck, 126–27; MacMillan, 492; Peukert, 57.

102. On August 26, German Undersecretary of State Matthias Erzberger, who had signed the Armistice agreement, was assassinated by Heinrich Schultz and Heinrich Tillessen, former officers in the Ehrhardt Brigade (which had led the Kapp Putsch in March 1920) and members of the secret rightist Organisation Consul. They evaded justice by escaping to Hungary. On August 29, in response to the assassination of Erzberger, Chancellor Joseph Wirth made a proclamation condemning the republic's enemies, and he used the Weimar constitution's Article 48 to empower authorities against those who encouraged sedition. This state of emergency lasted almost four months, until December 16, although the Bavarian government refused to execute the decree. Kolb, 44–45; Eyck, 188–89; Peukert, 73. For the commentary of a political contemporary on the Erzberger murder in the context of other political murders in that era, see Gumbel, *Vier Jahre politischer Mord*; for an English translation of part of the book, see Emil J. Gumbel, "Four Years of Political Murder," in Kaes et al., 100–104.

103. Käthe Kollwitz, letter to Einstein, Sept. 11, 1921, AEA [34 653]; also in *CPAE* 12: 234.

104. Arthur Holitscher, letter to Einstein, August 16, 1921, AEA [34 647]; also in *CPAE* 12: 209. Willi Münzenberg, letter to Einstein, Sept. 1, 1921, AEA [44 841]; also in *CPAE* 12: 222. Otto Gäbel, letter to Einstein, Sept. 3, 1921, AEA [44 794]; also in *CPAE* 12: 227. Max Barthel, letter to Einstein, Sept. 6, 1921, AEA [44 845]; also in *CPAE* 12: 231. On Willi Münzenberg and his role in this campaign, see Babette Gross, *Willi Münzenberg: Eine politische Biographie* (Stuttgart: Deutsche Verlags-Anstalt, 1967); Stephen Koch, *Double Lives: Stalin, Willi Münzenberg, and the Seduction of the Intellectuals*, rev. ed. (New York: Enigma Books, 2004); and Sean McMeekin, *The Red Millionaire: A Political Biography of Willi Münzenberg, Moscow's Secret Propaganda Tsar in the West* (New Haven: Yale Univ. Press, 2003).

105. Einstein, draft of letter to Käthe Kollwitz, Sept. 19, 1921, AEA [34 654]; also in *CPAE* 12: 240.

106. Central Committee, Incorporated for the Relief of Distress in Germany and Austria, letter to Einstein, Sept. 9, 1921, AEA [43 438]; also in *CPAE* 12: 232.

107. Einstein, statement on the suffering of children in Germany due to the unstable economy, Oct. 7, 1921, AEA [43 440]; also in *CPAE* 7: 65.

108. Einstein, draft of letter to French League for Human Rights, [between Sept. 1921 and Jan. 2, 1922], AEA [28 012]; also in *CPAE* 12: 220.

109. Lehmann-Rußbüldt, "*Die Brücke über den Abgrund*," 4–28; Lehmann-Rußbüldt, *Der Kampf der Deutschen Liga*, 101; Brenner, 48–49; Norman Ingram, *The Politics of Dissent: Pacifism in France, 1919–1939* (Oxford: Oxford Univ. Press, 1991), 40–46; Henri Sée, *Histoire de la Ligue des Droits de l'Homme (1898–1926)* (Paris: Ligue des Droits de l'Homme, 1927), 179.

6. Social Justice Advocacy for Jews, and Confident Jewish Identity: 1914–1921

1. For a study focused primarily on Einstein's Jewish identity, see Ze'ev Rosenkranz, *Einstein Before Israel: Zionist Icon or Iconoclast?* (Princeton: Princeton Univ. Press, 2011).

2. In October 1916, in response to antisemitic agitation blaming Jews for the German army's failure to win a quick victory in the war, the Prussian War Ministry ordered a census of frontline soldiers by religion. Although its results were never published, this census provoked a storm of protest from the German Jewish community. They responded by conducting a separate census (which they did publish) and by founding the German League of Jewish Frontline Soldiers (Reichsbund jüdischer Frontsoldaten), an organization of Jewish veterans of combat in World War I and the second-largest Jewish organization in Weimar Germany. On the Reichsbund jüdischer Frontsoldaten and similar Jewish organizations, see Carl Jeffrey Rheins, "German Jewish Patriotism 1918–1935: A Study of the Attitudes and Actions of the 'Reichsbund Jüdischer Frontsoldaten,' the 'Verband Nationaldeutscher Juden,' the 'Schwarzes Fahnlein, Jungenschaft,' and the 'Deutscher Vortrupp, Gefolgschaft Deutscher Juden'" (PhD diss., State University of New York at Stony

Brook, 1978); Rheins, "The Verband nationaldeutscher Juden 1921–1933," *Leo Baeck Institute Yearbook* 25 (1980); and Ulrich Dunker, *Der Reichsbund jüdischer Frontsoldaten 1919–1938: Geschichte eines jüdischen Abwehrvereins* (Düsseldorf: Droste Verlag, 1977).

3. Einstein, letter to Petr Petrovich Lazarev, May 16, 1914, AEA [84 163]; also in *CPAE* 8: 7; and in Kirsten and Treder, 354. On the trial of Mendel Beilis, a Jew from Kiev accused of ritual murder of a Gentile in 1911, then acquitted in late 1913, see *Neue Zürcher Zeitung* 134, no. 1581 (Nov. 11, 1913), 1st evening supplement, [1]; and Paul Mendes-Flohr and Jehuda Reinharz, *The Jew in the Modern World: A Documentary History*, 2nd ed. (Oxford: Oxford Univ. Press, 1995), 412–13. On the outbreak of pogroms in the Russian Empire from 1903 to 1906, see John D. Klier and Schlomo Lambrosa, *Pogroms: Anti-Jewish Violence in Modern Russian History* (Cambridge: Cambridge Univ. Press, 1992), chapter 8.

4. Zionist Association of Germany (Zionistische Vereinigung für Deutschland), letter to Einstein, May 23, 1918, AEA [45 330]; also in *CPAE* 8: 547. They were referring to Rabbi Markus Braude. On Jewish secondary-level teaching in Poland, see Arjeh Tartakower, "Das jüdische Mittelschulwesen in Polen," *Der Jude* 9, separatum no. 2: "Erziehung" (1926): 79–89.

5. Zionist Association of Germany, letter to Einstein, Dec. 9, 1918, AEA [45 331]; also in *CPAE* 8: 666.

6. Zionist Association of Germany, letter to Einstein, Dec. 12, 1918, AEA [45 334]; also in *CPAE* 8: 671.

7. Einstein, letter to Paul Ehrenfest, Mar. 22, 1919, AEA [9 427]; also in *CPAE* 9: 10; and in Nathan and Norden, 29. The University of Berlin was closed from early January to March 17, 1919; see *Vossische Zeitung*, Mar. 15, 1919, evening edition; and Mar. 17, 1919, evening edition.

8. Jonathan Wright, *Gustav Stresemann: Weimar's Greatest Statesman* (Oxford: Oxford Univ. Press, 2002). On Stresemann's horror at the murder of Walther Rathenau, the Weimar Republic's foreign minister, who was of Jewish background, on June 24, 1922, see Eyck, 218. On the quality of Stresemann's leadership more broadly, see Kolb, 57–58; and Thomas Mann, "Appell an die Vernunft," *Berliner Tageblatt* (Oct. 18, 1930); for an English translation, see Thomas Mann, "An Appeal to Reason," in *The Weimar Republic Sourcebook*, ed. Anton Kaes et al. (Berkeley: Univ. of California Press, 1994), 150–59.

9. Paul Epstein, letter to Einstein, Sept. 11, 1919, AEA [10 554]; also in *CPAE* 9: 102.

10. Einstein, letter to Paul Epstein, Oct. 5, 1919, AEA [75 315]; also in *CPAE* 9: 122.

11. Relief and Works Agency for Palestine (Hilfswerk für Palästina), letter to Einstein, [Oct. 13, 1919], AEA [86 057]; also in *CPAE* 9: 132. The letter was signed with the initial "B," probably by the Zionist Organization's Berlin secretary Julius Berger. The Relief and Works Agency for Palestine contacted Einstein on the recommendation of graphic artist Hermann Struck. Struck, among the founders of the Orthodox-Zionist Misrachi Party, was one of the earliest people who introduced Einstein to Zionism, along

with Kurt Blumenfeld, with whom Einstein had been in contact since he attended meetings of the Zionist Association of Germany in December 1918.

12. Wertheimer.

13. Einstein, "Die Zuwanderung aus dem Osten" ("Immigration from the East"), *Berliner Tageblatt*, Dec. 30, 1919, morning edition, [2]; also in *CPAE* 7: 29; and in Rowe and Schulmann, 139–41.

14. *Deutsche Zeitung*, Feb. 17, 1920, [5]; and "Antisemitischer Studentenpöbel gegen Einstein," *Vorwärts*, Feb. 19, 1920, evening edition, [2].

15. Fölsing, 519.

16. *Berliner Tageblatt*, Feb. 14, 1920, morning edition, [2].

17. *Tägliche Rundschau*, Feb. 18, 1920, morning edition, [6].

18. Albert Einstein, "Tumultszenen bei einer Einstein-Vorlesung. Eine Erklärung Professor Einsteins," *8-Uhr Abendblatt* (Berlin) 73, no. 38, Feb. 13, 1920, 2–3; also in *CPAE* 7: 33. On February 17, Einstein publicly stated his conditions for people to participate further in his classes, guaranteeing seats for registered students, but allowing nonmatriculated students to occupy any empty and available seats. *Berliner Tageblatt*, Feb. 17, 1920, morning edition.

19. University of Berlin students (291 signatures), declaration to Einstein, Feb. 19, 1920, AEA [45 156]; also in *CPAE* 9: 320. For contemporaneous newspaper accounts of this event, see *Berliner Tageblatt*, Feb. 20, 1920, morning edition; and *Vossische Zeitung*, Feb. 20, 1920, evening edition, supplement.

20. *Vossische Zeitung*, Feb. 20, 1920, evening edition, supplement.

21. Einstein and Leopold Landau, letter to Konrad Haenisch, Feb. 19, 1920, AEA [70 952]; also in *CPAE* 9: 317. A list of eighteen names of professors offering to participate and their courses were enclosed with the letter.

22. Einstein, typed essay entitled "Assimilation und Antisemitismus" ("Assimilation and Antisemitism"), [Apr. 3, 1920], AEA [36 625]; also in *CPAE* 7: 34; and in Rowe and Schulmann, 142–45.

23. Einstein, typed essay entitled "Antisemitismus. Abwehr durch Erkenntnis" ("Anti-Semitism. Defense through Knowledge"), [after Apr. 3, 1920], AEA [36 626]; also in *CPAE* 7: 35.

24. German News Agency for Foreign University and Student Affairs (Deutsches Korrespondenzbüro für ausländisches Universitäts- und Studentenangelegenheiten), letter to Einstein, [before July 27, 1920], AEA [43 541].

25. Einstein, letter to German News Agency for Foreign University and Student Affairs (Deutsches Korrespondenzbüro für ausländisches Universitäts- und Studentenangelegenheiten), July 27, 1920, AEA [43 543]; also in *CPAE* 10: 86.

26. Heinrich Stern, letter to Einstein, Aug. 3, 1920, AEA [2 083] and AEA [2 082.1].

27. On the events leading to Nicolai's emigration, see Zuelzer, chapters 15–20.

28. *8-Uhr-Abendblatt*, Aug. 25, 1920; and *Vorwärts*, Aug. 25, 1920, evening edition, 2.

29. On these events, see “Einstein’s Encounter with German Anti-Relativists” in *CPAE* 7, 101–13; “Introduction,” part 2, *CPAE* 10, xxix–lii; and Jeroen van Dongen, “On Einstein’s Opponents, and Other Crackpots,” *Studies in History and Philosophy of Modern Physics* 41, no. 1 (Jan. 2010): 78.

30. Einstein, “Meine Antwort. Ueber die anti-relativitätstheoretische G.m.b.H.,” *Berliner Tageblatt*, Aug. 27, 1920, morning edition, [1–2]; also in *CPAE* 7: 45; and in Rowe and Schulmann, 105–9. Paul Weyland was a rightist journalist and publicist. On the context of this article, see Hubert Goenner, “The Reaction to Relativity Theory I: The Anti-Einstein Campaign in Germany in 1920,” *Science in Context* 6 (1993): 107–33; and Fölsing, 520–29.

The 86th meeting of the Society of German Natural Scientists and Doctors (Gesellschaft Deutscher Naturforscher und Ärzte, or GDNA), which Einstein mentioned, took place in Bad Nauheim about a month later on September 23–24, 1920. For a printed text of discussions following six lectures on the theory of relativity at that meeting, see *Physikalische Zeitschrift* 21 (Nov. 1 and 15, 1920): 650–51, 662, 666–68; and *CPAE* 7: 46. For contemporary newspaper accounts of the debate between Philipp Lenard and Einstein at that meeting, see *Vossische Zeitung*, Sept. 24, 1920, evening edition, 1; and *Berliner Tageblatt*, Sept. 24, 1920, evening edition, [3]. See also Jeroen van Dongen, “Reactionaries and Einstein’s Fame: ‘German Scientists for the Preservation of Pure Science,’ Relativity, and the Bad Nauheim Meeting,” *Physics in Perspective* 9 (2007): 212–30.

31. *Berliner Tageblatt*, Aug. 28, 1920, morning edition.

32. Hellmut von Gerlach, “Hakenkreuz, hurra hoch!” *Die Welt am Montag*, Aug. 30, 1920.

33. Oscar Bie et al., telegram to Einstein, Aug. 30, 1920, AEA [36 049]; published as, “Um Einstein,” *Berliner Tageblatt*, Aug. 31, 1920, evening edition; “Der Kampf um Einstein,” *Deutsche Allgemeine Zeitung*, Sept. 1, 1920; “Eine Sympathiekundgebung für Professor Einstein,” *Neue Freie Presse*, Vienna, Sept. 1, 1920; also in *CPAE* 10: 117. The telegram was signed by Oscar Bie, publicist Joseph Chapiro, actor Werner Krauss, writer Andreas Latzko, actor Alexander Mossji, actress Johanna Terwin, actress Helene Thimig, theater director Max Reinhart, and writer and playwright Stefan Zweig. On the Festspiele, see Joseph Chapiro and Andreas Latzko, “Die Salzburger Festspiele,” *Freiheit*, Sept. 6, 1920, evening edition.

34. Helmut Bloch, letter to Einstein, Aug. 30, 1920, AEA [36 009]; also in *CPAE* 10: 118.

35. Arnold Sommerfeld, letter to Einstein, Sept. 3, 1920, AEA [21 339]; also in *CPAE* 10: 131. Friedrich von Müller was a professor of internal medicine at the University of Munich and the presiding officer of the Society of German Natural Scientists and Doctors (Gesellschaft Deutscher Naturforscher und Ärzte) at its meeting in Bad Nauheim. Jean Jaurès, a socialist member of the French Chamber of Deputies, was killed due to his pacifism. Bertrand Russell, an English mathematician and philosopher, was relieved

of his lectureship at Trinity College, Cambridge, in 1916 for his advocacy for conscientious objectors, and sentenced to six months in prison in 1918 for writing a pacifist article. Joseph Caillaux, a French statesman, was imprisoned in 1920 for his opposition to the War. However, options for dissent during the War were probably even fewer in Germany; moreover, given Einstein's renown as a scientist, his not being of draft age, and his moderate, conciliatory approach to activism, he probably would have been treated somewhat similarly in any country.

36. Konrad Haenisch, letter to Einstein, Sept. 6, 1920, AEA [43 826]; also in *CPAE* 10: 135.

37. Einstein, letter to Konrad Haenisch, Sept. 8, 1920, AEA [83 330]; also in *CPAE* 10: 137; and in Kirsten and Treder, 204.

38. Minna Cauer, letter to Einstein, Sept. 19, 1920, AEA [43 436]; also in *CPAE* 10: 151. Cauer's absence was probably due to her work on an electoral campaign in Upper Silesia. For more on Cauer, see Gerlinde Naumann, *Minna Cauer: Eine Kämpferin für Frieden, Demokratie und Empanzipation* (Berlin: Sekretariat des Zentral Vorstandes der Liberal-Demokratischen Partei Deutschlands im Buchverlag Der Morgen, 1988); and Else Lüders, ed., *Minna Cauer, Leben und Werk, dargestellt an Hand ihrer Tagebücher und nachgelassenen Schriften* (Gotha: Klotz, 1925). The April 1919 appeal to the German government is discussed in chapter 4.

39. Stefan Zweig, letter to Einstein, Sept. 22, 1920, AEA [34 305]; also in *CPAE* 10: 152.

40. Einstein, letter to Stefan Zweig, Nov. 10, 1920, AEA [72 150]; also in *CPAE* 10: 198.

41. Einstein, letter to [Minna Cauer], Nov. 19, 1920, AEA [71 453]; also in *CPAE* 10: 205.

42. Rudolf Goldscheid, letter to Einstein, Dec. 13, 1920, AEA [43 766]; also in *CPAE* 10: 228.

43. Ilse Einstein, letter to the Association for Combating Antisemitism (Verein zur Abwehr des Antisemitismus), Sept. 14, 1920, AEA [43 063]; also in *CPAE* 10: 150.

44. Albert Einstein, "Ein Erkenntnisbrief Einsteins," *Israelitisches Wochenblatt für die Schweiz*, no. 9 (Sept. 24, 1920), 10; typed drafts dated Apr. 5, 1920, AEA [43 443] and AEA [35 060]; also in *CPAE* 7: 37; and in Rowe and Schulmann, 146–47. The Central Association of German Citizens of the Jewish Faith (Centralverein deutscher Staatsbürger jüdischen Glaubens) was the largest, most prominent Jewish liberal (as they termed themselves; later called "assimilationist") association in Weimar Germany.

45. Albert Einstein, "Wie ich Zionist wurde," *Jüdische Rundschau* 26, no. 49 (July 21, 1921), 351–52; also in *CPAE* 7: 57; and in Rowe and Schulmann, 149–53.

46. Einstein, handwritten draft of a speech delivered to a Zionist meeting in Berlin, June 27, 1921, AEA [28 008]; also in *CPAE* 7: 59.

47. Einstein, speech delivered on June 27, 1921 to a Zionist meeting in Berlin, *Jüdische Rundschau*, July 1, 1921, 37; also in *CPAE* 7: 60; and in Rowe and Schulmann, 153–55.

48. Einstein, "Zur Errichtung der hebräischen Universität in Jerusalem," *Jüdische Pressezentrale Zürich*, Aug. 26, 1921, [1]; also in *CPAE* 7: 62; and in Rowe and Schulmann, 156–58.

Epilogue

1. For a more comprehensive study of Einstein's lifelong commitment to pacifism, on which this epilogue is based, see Holmes, "Was Einstein Really a Pacifist?"

2. The evidence of Einstein's strong Jewish identity is extensive, including his firm commitment to the founding of the Hebrew University in Jerusalem, the fact that his first trip to the United States in 1921 was a fundraising campaign alongside the Zionist leader Chaim Weizmann, and that he bequeathed all of his papers, upon his death, to the Hebrew University.

3. Einstein, letter to Franklin D. Roosevelt, Aug. 2, 1939 [38 088]; also in Rowe and Schulmann, 359–61; and in Nathan and Norden, 294–96. For the start of the public perception of Einstein as "grandfather of the bomb," see *Time*, July 1, 1946; and "Einstein, the Man Who Started It All," *Newsweek*, Mar. 10, 1947.

4. See, for example, Rowe and Schulmann; and Walter Isaacson, *Einstein: His Life and Universe* (New York: Simon & Schuster, 2007). For arguments against Einstein's identity as a pacifist, see Hubert Goenner and Giuseppe Castagnetti, "Albert Einstein as Pacifist and Democrat during World War I," *Science in Context* 9, no. 4 (winter 1996): 325–86; Evelyn Wilcock, *Pacifism and the Jews* (Lansdown: Hawthorn Press, 1994), 121; and Fölsing, 399–402, 446–49. These sources are also discussed in Holmes, "Was Einstein Really a Pacifist?"

5. See, for example, Nathan and Norden, v; Ronald W. Clark, *Einstein: The Life and Times* (New York: World Publishing, 1971), 620; Jamie Sayen, *Einstein in America: The Scientist's Conscience in the Age of Hitler and Hiroshima* (New York: Crown Publishers, 1985), 268; and Reiner Braun and David Krieger, eds., *Einstein—Peace Now! Visions and Ideas* (Weinheim: WILEY-VCH Verlag, 2005), 1. These sources are also discussed in Holmes, "Was Einstein Really a Pacifist?"

6. Einstein, "Impact of Science on the Development of Pacifism," [before Dec. 9, 1921], in *CPAE* 7: 69; also in *Die Friedensbewegung. Ein Handbuch der Weltfriedensströmungen der Gegenwart*, ed. Kurt Lenz and Walter Fabian, 78–79 (Berlin: Schwetschke, 1922).

7. In the spring of 1923, after France had refused to submit the war reparations issue to arbitration and the French military had occupied the German industrial Ruhr region, Einstein resigned, stating that the League of Nations had neither the strength or good

will to implement its great task, and, as a serious-minded pacifist, he did not consider it right to be in connected with it. *Vossische Zeitung*, Mar. 22, 1923; and Einstein, letter to Maurice Solovine, Oct. 30, 1924, AEA [80 847]. He later rejoined the Committee on Intellectual Cooperation and attended its meetings from 1924 to 1927. On his fluctuating involvement, see Nathan and Norden, 58–89.

8. Einstein, statement to Women's International League for Peace and Freedom, Jan. 4, 1928, AEA [48 818]; also in Nathan and Norden, 90.

9. Einstein, statement for the World Peace League, 1928, "Golden Book of Peace" (Geneva: World Peace League, 1928), Albert Einstein Collection, AR 4858 (I: 24), Leo Baeck Institute, New York.

10. *Die Menschenrechte* 3, no. 1 (Jan. 31, 1928).

11. Nathan and Norden, 91. Einstein signed this along with almost seventy signatories from fifteen countries.

12. Einstein, letter to No More War Movement (British section of War Resisters International, London), Nov. 25, 1928, in Nathan and Norden, 91.

13. Einstein, letter to the League for the Organization of Progress, Yellow Springs, Ohio, Dec. 26, 1928, AEA [47 363].

14. Einstein, letter to No More War Movement, Apr. 1929, AEA [28 076].

15. Einstein, speech to New History Society at New York Ritz-Carlton Hotel, Dec. 14, 1930. The original of the speech is not extant, but published versions appeared in the *New York Times*, Dec. 21, 1930; and *World Tomorrow*, Jan. 1931.

16. "Report of the German Consulate-General to Chicago" (presented in Berlin, March 6, 1931), 236; *New York World-Telegram* (Mar. 4, 1931); also in *New York Times* (Mar. 5, 1931).

17. Einstein, statement to Belgians gathering at Dixmude, Belgium, on August 23, 1931 (sent to Jan de Bondt on August 8, 1931), AEA [45 637]; also in Nathan and Norden, 144–45.

18. Einstein, letter to *Nieuw Europa*, Sept. 17, 1931, AEA [47 747]; also in Nathan and Norden, 145.

19. Einstein, "My Opinion on the War," [Oct. 23–Nov. 11, 1915], AEA [70 457]; also in *CPAE* 6: 20; and Einstein, "Meine Meinung über den Krieg," in *Das Land Goethes 1914–1916. Ein vaterländisches Gedenkbuch. Herausgegeben vom Berliner Goethebund* (Stuttgart and Berlin: Deutsche Verlags-Anstalt, 1916), 30.

20. Einstein and Sigmund Freud, *Why War? A Correspondence Between Albert Einstein and Sigmund Freud*, trans. by Stuart Gilbert (London: Peace Pledge Union, 1939; c1933).

21. Einstein, "Wie ich Zionist wurde," *Jüdische Rundschau* 26, no. 49 (June 21, 1921): 351–52; also in *CPAE* 7: 57.

22. "Einstein über die Friedensmission des Zionismus," *Vossische Zeitung*, July 5, 1921, Morning Edition, 4.

23. Einstein, "Botschaft," *Jüdische Rundschau* 30, no. 14 (Feb. 17, 1925), 129. Einstein wrote, "*Daher glaube ich behaupten zu können, daß der Zionismus, eine dem Anschein nach nationalistische Bewegung, letzten Endes ein bedeutendes Verdienst um die Menschheit hat.*"

24. On the Jüdischer Friedensbund, see Virginia Iris Holmes, "Integrating Diversity, Reconciling Contradiction: The Jüdischer Friedensbund in Late Weimar Germany," *Leo Baeck Institute Year Book* 47 (2002): 175–94.

25. Einstein, letter to Leo Baeck, July 1931, Albert Einstein Collection, AR 4858 (I: II: b: 1: 1), Leo Baeck Institute, New York. Einstein wrote, "*Ich brauche Ihnen wohl nicht zu sagen, dass mir der Versuch, die Geistlichkeit fuer das pazifistische Ziel zu mobilisieren, ausserordentlich am Herzen liegt.*"

26. Stephen M. Poppel, *Zionism in Germany, 1897–1933: The Shaping of a Jewish Identity* (Philadelphia: Jewish Publication Society of America, 1976), 149–50. In this violence of August 1929, 133 Jews were killed and 339 wounded, and 116 Arabs were killed and 232 wounded, although the Arab casualties were inflicted by British Mandate forces. For more on the events of August 1929 in Palestine, see *Report of the Commission on the Palestine Disturbances of August, 1929* (London: His Majesty's Stationery Office, 1930); and Maurice Samuel, *What Happened in Palestine: The Events of August, 1929, Their Background and Their Significance* (Boston: Stratford, 1929). On Einstein's and other Jewish pacifists' responses to the issue of Jews and Arabs in Palestine, see Holmes, "Integrating Diversity." On other responses to the violence of August 1929 in Palestine, see Horace B. Samuel, *Beneath the Whitewash: A Critical Analysis of the Report of the Commission on the Palestine Disturbances of August, 1929* (London: Hogarth Press, 1930); and Naomi W. Cohen, *The Year after the Riots: American Responses to the Palestine Crisis of 1929–30* (Detroit: Wayne State Univ. Press, 1988).

27. Albert Einstein, "Einsteins Brief," *Jüdische Rundschau* 34, no. 69 (Sept. 3, 1929), 447. A couple of months later, Einstein stated that, just as Jews wished to be understood by the Arabs, so they must try to understand the Arabs, and responsibility to seek peace and to work together with the Arab people rested "on us and on us alone." "Einstein über jüdisch-arabische Verständigung," *Jüdische Rundschau* 34, no. 91 (Nov. 19, 1929), 610.

28. "Einsteins Brief," *Jüdische Rundschau* 35, no. 15 (Feb. 21, 1930), 98.

29. Nathan and Norden, 127; Albert Einstein, *The Fight Against War*, ed. Alfred Lief (New York: John Day, 1933), 26; Sayen, 105; and *New York Times*, June 1, 1930.

30. Albert Einstein, "Einstein an die arabische Welt," *Jüdische Rundschau* 35, no. 9 (Jan. 31, 1930), 57; for the same letter, see Einstein, "Letter to the Editor of *Falastin*, Jan. 28, 1930," *Falastin* (Feb. 1, 1930); also in Rowe and Schulmann, 181–82.

31. Albert Einstein, "Einstein und die Araberfrage. Ein zweiter Brief Einsteins an den Redakteur des 'Falastin,'" *Jüdische Rundschau* 35, no. 23 (Mar. 21, 1930), 155. See also Einstein, "Letter to the Editor of *Falastin*," *Falastin*, Mar. 15, 1930, AEA [46 154]; also in Rowe and Schulmann, 183–84. Two Jewish newspapers in Palestine found Einstein's

proposal well-meaning but overly simplistic, naive, unrealistic, and unworkable. The Jewish peace group in Palestine, Brith Schalom, by contrast, greeted Einstein's proposal with enthusiasm. "Zum Brief von Professor Einstein," *Jüdische Rundschau* 35, no. 27 (Apr. 4, 1930), 187. On Einstein's power-sharing structure, as modeled after the Swiss constitution, see *New York Times*, Mar. 5, 1931.

32. Einstein, "Einstein über die jüdisch-arabische Verständigung," *Jüdische Rundschau* 35, no. 81/82 (Oct. 17, 1930), 528. Einstein expressed similar sentiments in New York a couple of months later, and in Los Angeles over a year after that, in March 1932, while fundraising for Jewish settlements in Palestine. "Einstein in New York," *Jüdische Rundschau* 35, no. 100 (Dec. 19, 1930), 673; and "Einstein-Bankett für den Keren Kajemeth," *Jüdische Rundschau* 37, no. 22 (Mar. 18, 1932), 106.

33. Einstein, *Fight Against War.*

34. Einstein, letter to Joseph Merory, May 1, 1933, AEA [51 148].

35. Einstein, letter to Rev. J. B. Th. Hugenholtz, July 1, 1933, AEA [50 320]. Einstein had been collaborating with Hugenholtz for over a year to create an international pacifist organization.

36. Einstein, letter to King Albert of Belgium, July 14, 1933, AEA [32 361].

37. Einstein, letter to Alfred Nahon, July 20, 1933, AEA [51 231]; also in Rowe and Schulmann, 283; and in Nathan and Norden, 229; and Einstein, letter to David Davies, of the New Commonwealth Society, July 20, 1933, AEA [49 450].

38. Clark, 491.

39. International Antimilitaristic Commission, "Einstein an Apostate," *Press Service* (Haarlem, Netherlands), Sept. 7, 1933; also in Nathan and Norden, 232.

40. Romain Rolland, diary entry of September 1933, *Journal des années de guerre* (Paris: Oeuvres libres, 1950); also in Nathan and Norden, 232–33.

41. Einstein, letter to G. C. Heringa, Sept. 11, 1933, AEA [50 199].

42. Rabbi Philip S. Bernstein, letter to Einstein, Mar. 22, 1934, AEA [49 275].

43. Einstein, letter to Rabbi Philip S. Bernstein, Apr. 5, 1934, AEA [49 276].

44. Einstein, "Antwort auf einen Artikel von B. D. Allinson," handwritten draft, [after Nov. 8, 1934], AEA [28 296]; Einstein, "A Re-examination of Pacifism," *Polity* (Jan. 1935); and Brent Dow Allinson, "Speak, Einstein, for the Peace of Europe," *Polity* (Nov. 1934). Einstein was not alone among pacifists in addressing this quandary. During pacifism's zenith in the decade and a half after the First World War, a slowly growing acceptance of the option of conscientious objection had spread from England to initially less receptive countries such as France. However, the rise of the Nazi regime, Italy's invasion of Ethiopia, and the Spanish Civil War deprived pacifism of the support of internationalists, who relied on international law and disarmament, and leftists and trade unionists, who conflated war with unjust wars of imperial expansion, and also created divisions inside of groups of primarily pacifist identity. Norman Ingram, *The Politics of Dissent: Pacifism in France 1919–1939* (Oxford: Clarendon Press, 1991), 56–69, 93–100; Brock and

Young, 117–26; Brock, 126–33; and Catherine Foster, *Women for All Seasons: The Story of the Women's International League for Peace and Freedom* (Athens: Univ. of Georgia Press, 1989), 21–22.

45. Einstein, letter to Antiwar Committee at New York University, Mar. 22, 1934, AEA [51 268]; also in Nathan and Norden, 248–49. Einstein, address to New Jersey State Conference on "The Cause and Cure of War," Princeton, New Jersey, Apr. 1934, AEA [72 635]; also in *New York Times* (Oct. 23, 1935); and in Nathan and Norden, 250–51.

46. Einstein and Alfred E. Smith, radio address, Columbia Broadcasting System (CBS), Oct. 22, 1935; also in *New York Times*, Oct. 23, 1935; and in Nathan and Norden, 262–63. Boris Schwarz, Contribution to Jerusalem Symposium in *Albert Einstein, Historical and Cultural Perspectives: Centennial Symposium in Jerusalem*, ed. Gerald Holton and Yehuda Elkana (Princeton: Princeton Univ. Press, 1982), 409; and Einstein, letter to Charles C. Burlingham, Apr. 9, 1938, AEA [28 419] and AEA [28 423]; also in Nathan and Norden, 278–79.

47. Einstein, letter to Franklin Delano Roosevelt, Aug. 2, 1939, AEA [38 088]; also in Rowe and Schulmann, 359–61; and in Nathan and Norden, 294–96. (On October 30, 1940, Einstein became a US citizen.)

48. Vannevar Bush, letter to Frank Aydelotte, Dec. 30, 1941; Clark, 684; and Richard A. Schwartz, "Einstein and the War Department," *Isis* 80 (1989): 281–84. See also Fred Jerome, *The Einstein File: J. Edgar Hoover's Secret War Against the World's Most Famous Scientist* (New York: St. Martin's Press, 2002).

49. Einstein, letter to Roy Kepler, Aug. 8, 1948, AEA [58 969]; for the letter to which he was responding, see Roy Kepler, letter to Einstein, July 28, 1948, AEA [58 967].

50. Einstein, letter to Ann Morrisett, Mar. 21, 1952, AEA [60 595].

51. Einstein, Interview at Saranac Lake, September 14, 1945, *New York Times*, Sept. 15, 1945.

52. Einstein, "The War Is Won, but the Peace Is Not," address to Nobel Anniversary Dinner, New York, Dec. 10, 1945; also in *New York Times*, Dec. 11, 1945; in Einstein, *Ideas and Opinions* (New York: Bonanza Books, 1954), 115–17; and in Rowe and Schulmann, 381–82.

53. Leo Szilard was the chief organizer of the Emergency Committee of Atomic Scientists, Einstein became its chairperson, and other members included Harold C. Urey, Hans A. Bethe, Torfin R. Hogness, Philip M. Morse, Linus Pauling, and Victor Weisskopf. See Emergency Committee of Atomic Scientists, "Certification of Incorporation of the Emergency Committee of Atomic Scientists, Incorporated," Aug. 2, 1946, AEA [40 462]; and "The Emergency Committee of Atomic Scientists, Incorporated: Its Purposes, Its Program, Its Requirements," [after Aug. 2, 1946], AEA [40 479].

54. "Einstein, the Man Who Started It All," *Newsweek*, Mar. 10, 1947. Most historians of the Manhattan project would agree with Einstein's assessment, according to Fölsing, 725.

55. *Kaizo* (Tokyo), Sept. 20, 1952; Einstein, letter to Seiei Shinohara, June 23, 1953, AEA [61 298]; also in Rowe and Schulmann, 491–92; and in Nathan and Norden, 589.

56. Einstein, letter to Henry Usborne, Jan. 9, 1948, AEA [58 922]; also in Rowe and Schulmann, 398–99; and in Nathan and Norden, 463–64.

57. Einstein, letter to John Dudzik, Mar. 8, 1948, AEA [58 108]; also in Rowe and Schulmann, 454–55; and in Nathan and Norden, 468–69.

58. Einstein, statement for the Senate Committee on Armed Services, Mar. 18, 1948, read into the Congressional Record on Mar. 24, 1948, in *U.S. Congress, Senate, Committee on Armed Services, Hearing on Universal Military Training,* 267 (Washington, DC, Government Printing Office, 1948).

59. Einstein, television statement to Eleanor Roosevelt, Feb. 12, 1950, in "Einstein Warns World: Outlaw H-Bomb or Perish!" *New York Post,* Feb. 13, 1952.

60. Albert Einstein, letter to Max von Laue, Mar. 19, 1955, AEA [16 205]; also in Nathan and Norden, 621. Einstein wrote, ". . . *ich . . . unterschrieb angesichts der Gefahr, dass Hitler zuerst in den Besitz der Bombe kommen könnte. Wenn ich gewusst hätte, dass diese Befurchtung grundlos war, hätte ich mich an der Öffnung der Pandora-Büchse nicht beteiligt* . . ."

61. "Declaration," *New York Times,* July 10, 1955; also in Nathan and Norden, 632–35. It was signed by eight scientists, in addition to Einstein and Russell: Percy W. Bridgman and Hermann J. Muller of the United States, Cecil F. Powell and Joseph Rotblat of England, Frédéric Joliot-Curie of France, Leopold Infeld of Poland, Hideki Yukawa of Japan, and Max Born of Germany.

62. Einstein, letter to Seiei Shinohara, June 23, 1953, AEA [61 297]; also in Rowe and Schulmann, 491–92; and in Nathan and Norden, 589.

Bibliography

All Einstein documents (including correspondence listed in endnotes) are located at the Albert Einstein Archives at the Hebrew University in Jerusalem, unless otherwise indicated. Archive numbers of documents in the Albert Einstein Archives (AEA) are indicated in brackets, e.g., AEA [38 088]. Brackets around dates indicate educated guesses of dates that are uncertain.

Archives and Collections

Albert Einstein Archives (AEA), Hebrew University, Jerusalem.
Albert Einstein Collection, Leo Baeck Institute, New York.
Archief H. A. Lorentz, Rijksarchief Noord-Holland, Haarlem, Netherlands.
Berlin-Brandenburgischen Akademie der Wissenschaften, Archiv, Berlin.
Bundesarchiv, Koblenz, Germany.
Eidgenössische Technische Hochschule (ETH) Bibliothek, Zürich.
Institut für Zeitgeschichte, Archiv, Munich.
Library of Congress, Prints & Photographs Division, Washington, DC.
Mary Evans Picture Library, London.
Nachlaß Hans Delbrück, Handschriftenabteilung, Deutsche Staatsbibliothek, Berlin.
Nachlaß Hans Wehberg, Bundesarchiv, Koblenz, Germany.
Nachlaß Nicolai, Archiv, Institut für Zeitgeschichte, Munich, Germany.
Nachlaß Paul Natorp, Universitätsbibliothek Marburg, Germany.
Nachlaß Wilhelm Wien, Archiv, Deutschen Museum, Munich, Germany.
Schweizerische Nationalbibliothek, Prints and Drawings Department, Bern, Switzerland.
Schweizerisches Bundesarchiv, Bern, Switzerland.
Stadtarchiv, Zürich.

Newspapers Cited

8-Uhr Abendblatt (Berlin)
Arbeiter-Zeitung (Vienna)
Berliner Tageblatt
Deutsche Allgemeine Zeitung
Deutsche Staatsbürger-Zeitung
Deutsche Tageszeitung
Deutsche Zeitung
Falastin
Frankfurter Zeitung
Freiheit
Die Friedens-Warte. Blätter für zwischenstaatliche Organisation
Jüdische Pressezentrale Zürich
Jüdische Rundschau
Die Menschenrechte
Mitteilungen des Bundes Neues Vaterland
Neue Freie Presse
Neue Zürcher Zeitung
Newsweek
New York Post
New York Times
New York World-Telegram
Press Service
Räte-Zeitung: Organ der Hand- und Kopfarbeiter-Räte Deutschlands
Republik
Die Rote Fahne
Sozialistische Monatshefte
Tägliche Rundschau
Time
Times (London)
Volksrecht
Vorwärts
Vossische Zeitung
Die Welt am Montag

Books, Articles, and Public Documents

Adler, Friedrich, and Johann Wolfgang Brügel, eds. *Friedrich Adler vor dem Ausnahmegericht. 18. u. 19. Mai 1917.* Vienna: Europa Verlag, 1967.

Albertini, Jean, ed. *Romain Rolland. Textes: Politiques, sociaux et philosophiques choisis.* Paris: Editions Sociales, 1970.

Allinson, Brent Dow. "Speak, Einstein for the Peace of Europe." *Polity,* Nov. 1934.

"An die Kulturwelt!" ("To the Civilized World"). In Nicolai, *Die Biologie des Krieges,* 7–10; Nicolai, *Biology of War,* xi–xiv; Ungern-Sternberg, *Der Aufruf "An die Kulturwelt!"* 144–47.

"Antisemitischer Studentenpöbel gegen Einstein." *Vorwärts,* Feb. 19, 1920, evening edition, [2].

"Appeal of the German Intellect for Socialism" ("*Aufruf des deutschen Geistes zum Sozialismus*"). Attached to a circular signed by Paul Natorp, Wilhelm Schäfer, Hermann Herrigel, Alfons Paquet, and Leopold Ziegler, dated Apr. 1919. Nachlaß Paul Natorp, Ms. 831. Universitätsbibliothek Marburg, Germany.

"Appeal to the Europeans." *See* "Aufruf an die Europäer."

"Un Appel: Fière Déclaration d'Intellectuels." *L'Humanité* (Paris), June 26, 1919.

Arco, Georg Count von, Albert Einstein, Walburga Geiger, Hellmut von Gerlach, Maximilian Harden, Max Hodann, Luise Kautsky, Elisabeth Rotten, Erich Schlesinger, and Helene Stöcker. *Lille: Beiträge zur Naturgeschichte des Krieges.* Berlin: Engelmann, 1919. [First/original version.]

Ardelt, Rudolf G. *Friedrich Adler. Probleme einer Persönlicheitsentwicklung um die Jahrhundertwende.* Vienna: Österreichischer Bundesverlag, 1984.

Arons, Leo. *Universitäten heraus!* Berlin: Verlag der Sozialistische Monatshefte, 1918.

"Aufruf an die Europäer" ("Appeal to the Europeans"). Mid-Oct. 1914. In Nicolai, *Die Biologie des Krieges,* 12–14; Nicolai, *The Biology of War,* xvii–xix; Nicolai, *Aufruf an die Europäer,* 21–25.

"Aufruf der sozialdemokratischen Mitglieder der Reichsregierung und des Parteivorstandes der SPD zum Generalstreik." *Fünf Tage Militärdiktatur.* Edited by by Karl Brammer. Berlin: Verlag für Politik und Wissenschaft, 1920.

Bane, Suda Lorena, and Ralph Haswell Lutz, eds. *The Blockade of Germany after the Armistice 1918–1919: Selected Documents of the Supreme Economic*

Council, Superior Blockade Council, American Relief Administration, and Other Wartime Organizations. Stanford: Stanford Univ. Press, 1942.

Barry, John M. *The Great Influenza: The Story of the Deadliest Pandemic in History*. New York and London: Penguin Books, 2005.

Belke, Ingrid. *Die sozialreformischen Ideen von Josef Popper-Lynkeus (1838–1921) im Zusammenhang mit allgemeinen Reformbestrebungen des Wiener Bürgertums um die Jahrhundertwende*. Tübingen: Mohr, 1978.

Bernhard, Georg. "Berlin Proteststreik." *Vossische Zeitung*, June 7, 1919, evening edition.

Bessell, Richard. *Germany after the First World War*. Oxford: Oxford Univ. Press, 1993.

——. "Violence as Propaganda: The Role of the Storm Troopers in the Rise of National Socialism." In *The Formation of the Nazi Constituency, 1919–1933*, edited by Thomas Childers, 131–46. London: Croom Helm, 1986.

Born, Max. *My Life: Recollections of a Nobel Laureate*. New York: Scribner's Sons, 1978.

Brammer, Karl, ed. *Fünf Tage Militärdiktatur*. Berlin: Verlag für Politik und Wissenschaft, 1920.

Braun, Reiner, and David Krieger, eds. *Einstein—Peace Now! Visions and Ideas*. Weinheim: WILEY-VCH Verlag, 2005.

Brenner, Arthur D. *Emil J. Gumbel: Weimar German Pacifist and Professor*. Boston and Leiden: Brill, 2001.

Brock, Peter. *Twentieth-Century Pacifism*. New York: Van Nostrand Reinhold, 1970.

Brock, Peter, and Nigel Young. *Pacifism in the Twentieth Century*. Syracuse, New York: Syracuse Univ. Press, 1999.

Brockdorff-Rantzau, Ulrich von. "Ansprache des Reichsaußenministers Grafen Brockdorff-Rantzau bei Überreichung des Friedensvertrags-Entwurfs durch die Alliierten und Assoziierten Mächte." *Dokumente*. Charlottenburg: Deutsche Verlasgesellschaft für Politik und Geschichte, 1920. 113 ff.

———. "Speech of the German Delegation, Versailles, May 7, 1919." In *The Weimar Republic Sourcebook*, edited by Anton Kaes et al., 9–12. Berkeley: Univ. of California Press, 1994.

Brocke, Bernhard vom, and Hubert Laitko, eds. *Die Kaiser-Wilhelm-/Max-Planck-Gesellschaft und Ihre Institute. Das Harnack-Prinzip*. Berlin: Walter de Gruyter, 1996.

Brower, Daniel R. *The World in the Twentieth Century: From Empires to Nations*, 4th ed. Upper Saddle River, NJ: Prentice Hall, 1999.

Chapiro, Joseph, and Andreas Latzko. "Die Salzburger Festspiele." *Freiheit*, Sept. 6, 1920, evening edition.

Chickering, Roger. *Imperial Germany and a World Without War: The Peace Movement and German Society, 1892–1914*. Princeton: Princeton Univ. Press, 1975.

———. *Imperial Germany and the Great War*. Cambridge: Cambridge Univ. Press, 1998.

Chickering, Roger, and Stig Förster, eds. *Great War, Total War: Combat and Mobilization on the Western Front, 1914–1918*. Cambridge: Cambridge Univ. Press, 2000.

Clark, Ronald W. *Einstein: The Life and Times*. New York: World Publishing, 1971.

Cohen, Naomi W. *The Year after the Riots: American Responses to the Palestine Crisis of 1929–30*. Detroit: Wayne State Univ. Press, 1988.

Cohen, Richard A. "The German League for Human Rights in the Weimar Republic." PhD diss., State University of New York at Buffalo, 1989.

The Collected Papers of Albert Einstein. Vol. 1, *The Early Years, 1879–1902*. Edited by John Stachel et al. Princeton: Princeton Univ. Press, 1987.

———. Vol. 5, *The Swiss Years, 1902–1914*. Edited by Martin J. Klein et al. Princeton: Princeton Univ. Press, 1993.

———. Vol. 6, *The Berlin Years: Writings, 1914–1917*. Edited by A. J. Kox et al. Princeton: Princeton Univ. Press, 1996.

———. Vol. 7, *The Berlin Years: Writings, 1918–1921*. Edited by Michel Jannsen et al. Princeton: Princeton Univ. Press, 2002.

———. Vol. 8, *The Berlin Years: Correspondence, 1914–1918*. Edited by Robert Schulmann et al. Princeton: Princeton Univ. Press, 1998.

———. Vol. 9, *The Berlin Years: Correspondence, January 1919–April 1920*. Edited by Diana Kormos Buchwald et al. Princeton: Princeton Univ. Press, 2004.

———. Vol. 10, *The Berlin Years: Correspondence, May–Dec. 1920*. Edited by Diana Kormos Buchwald et al. Princeton: Princeton Univ. Press, 2006.

———. Vol. 12. *The Berlin Years: Correspondence, 1921*. Edited by Diana Kormos Buchwald et al. Princeton: Princeton Univ. Press, 2009.

Craig, Gordon A. *Germany, 1866–1945*. New York: Oxford Univ. Press, 1978.

Dalton, Dennis. *Mahatma Gandhi: Nonviolent Power in Action*. New York: Columbia Univ. Press, 1993.

Delbrück, Hans. "Die Differenzen über die Kriegsziele hüben und drüben." *Preußische Jahrbücher* 162 (1915): 167–72.

Delbrück, Hans, and Bernhard Dernburg (with 140 signatories). Petition to German Chancellor Theobold von Bethmann Hollweg. July 27, 1915.

"Die wissenschaftliche Blockade Deutschlands." *Tägliche Rundschau*, Feb. 22, 1921, morning edition, [7]

Dongen, Jeroen van. "On Einstein's Opponents, and Other Crackpots." *Studies in History and Philosophy of Modern Physics* 41, no. 1 (Jan. 2010): 78.

———. "Reactionaries and Einstein's Fame: 'German Scientists for the Preservation of Pure Science,' Relativity, and the Bad Nauheim Meeting." *Physics in Perspective* 9 (2007): 212–30.

Dorst, Tankred. *Die Münchner Räterepublik: Zeugnisse und Kommentar.* Frankfurt am Main: Suhrkamp, 1967.

"Dr. Albert Einstein." *Times* (London), Nov. 8, 1919.

Duiker, William J. *Twentieth-Century World History.* Belmont, CA: Wadsworth Publishing, 1999.

Dunker, Ulrich. *Der Reichsbund jüdischer Frontsoldaten 1919–1938: Geschichte eines jüdischen Abwehrvereins.* Düsseldorf: Droste Verlag, 1977.

Einstein, Albert. Address to New Jersey State Conference on "The Cause and Cure of War." Princeton, New Jersey. Apr. 1934. *New York Times*, Oct. 23, 1935.

———. "Antisemitismus. Abwehr durch Erkenntnis." Typed essay. [After Apr. 3, 1920]. AEA [36 626].

———. "Antwort auf einen Artikel von B. D. Allinson." Handwritten draft. [After Nov. 8, 1934]. AEA [28 296].

———. "Assimilation und Antisemitismus." Typed essay. [Apr. 3, 1920]. AEA [36 625].

———. "Botschaft." *Jüdische Rundschau* 30, no. 14 (Feb. 17, 1925): 129.

———. "Ein Erkenntnisbrief Einsteins." *Israelitisches Wochenblatt für die Schweiz*, no. 9 (Sept. 24, 1920): 10.

———. "Einstein an die arabische Welt." *Jüdische Rundschau* 35, no. 9 (Jan. 31, 1930): 57.

———. "Einstein on His Theory. Time, Space, and Gravitation. The Newtonian System." *Times* (London), Nov. 28, 1919.

———. "Einsteins Brief." *Jüdische Rundschau* 34, no. 69 (Sept. 3, 1929): 447.

———. "Einsteins Brief." *Jüdische Rundschau* 35, no. 15 (Feb. 21, 1930): 98.

———. "Einstein über die Friedensmission des Zionismus." *Vossische Zeitung*, July 5, 1921, morning edition, 4.

———. "Einstein über die jüdisch-arabische Verständigung." *Jüdische Rundschau* 35, no. 81/82 (Oct. 17, 1930): 528.

———. "Einstein über jüdisch-arabische Verständigung." *Jüdische Rundschau* 34, no. 91 (Nov. 19, 1929): 610.

———. "Einstein und die Araberfrage. Ein zweiter Brief Einsteins an den Redakteur des 'Falastin.'" *Jüdische Rundschau* 35, no. 23 (Mar. 21, 1930): 155.

———. *The Fight Against War.* Edited by Alfred Lief. New York: John Day, 1933.

———. "Friedrich Adler als Physiker." (Interview of Einstein.) *Vossische Zeitung,* no. 259 (May 23, 1917): 2–3.

———. Handwritten draft of statement praising American and English Quaker relief work in Germany (in Ilse Einstein's handwriting). [July 11, 1920]. AEA [44 695].

———. Handwritten text of a speech supporting the German Revolution delivered by Einstein at a meeting the New Fatherland League on November 13, 1918. [Nov. 13, 1918]. AEA [28 001].

———. *Ideas and Opinions.* New York: Bonanza Books, 1954.

———. "Impact of Science on the Development of Pacifism." In *Die Friedensbewegung. Ein Handbuch der Weltfriedensströmungen der Gegenwart,* edited by Kurt Lenz and Walter Fabian, 78–79. Berlin: Schwetschke, 1922.

———. Interview at Saranac Lake. Sept. 14, 1945. *New York Times,* Sept. 15, 1945.

———. "Leo Arons as Physicist." *Sozialistische Monatshefte* 53 (Nov. 17, 1919): 1055–56.

———. "Letter to the Editor of *Falastin,* Jan. 28, 1930." *Falastin,* Feb. 1, 1930.

———. "Letter to the Editor of *Falastin.*" *Falastin,* Mar. 15, 1930.

———. "Meine Antwort. Ueber die anti-relativitätstheoretische G.m.b.H." *Berliner Tageblatt,* Aug. 27, 1920, morning edition, [1–2].

———. "Meine Meinung über den Krieg." In *Das Land Goethes 1914–1916. Ein vaterländisches Gedenkbuch. Herausgegeben vom Berliner Goethebund,* 30. Stuttgart and Berlin: Deutsche Verlags-Anstalt, 1916.

———. "My Opinion on the War" ("*Meine Meinung über den Krieg*"). Oct. 23–Nov. 11, 1915. AEA [70 457].

———. "On the Contribution of Intellectuals to International Reconciliation." [After Sept. 29, 1920]. In *Thoughts on Reconciliation,* 10–11. New York: Deutscher Gesellig-Wissenschaftlicher Verein von New York, 1920.

———. "On the Questionnaire Concerning the Right of National Self-Determination" ("*Zum Fragebogen über das Selbstbestimmungsrecht der Völker*"). [July 1917–before Mar. 10, 1918]. AEA [28 095].

———. "A Re-examination of Pacifism." *Polity,* Jan. 1935.

———. Speech delivered on June 27, 1921 to a Zionist meeting in Berlin. *Jüdische Rundschau*, July 1, 1921, 37.

———. Speech to New History Society at New York Ritz-Carlton Hotel. Dec. 14, 1930. *New York Times*, Dec. 21, 1930; also in *World Tomorrow*, Jan. 1931.

———. Statement for the Senate Committee on Armed Services. Mar. 18, 1948. Read into the Congressional Record on Mar. 24, 1948. In *U.S. Congress, Senate, Committee on Armed Services, Hearing on Universal Military Training*, 267. Washington, DC: Government Printing Office, 1948.

———. Statement for the World Peace League. "Golden Book of Peace." Geneva: World Peace League, 1928. Albert Einstein Collection. AR 4858 (I: 24). Leo Baeck Institute, New York.

———. Statement in support of Georg Friedrich Nicolai. [Jan. 26, 1920]. AEA [78 124].

———. Statement on the suffering of children in Germany due to the unstable economy. Oct. 7, 1921. AEA [43 440].

———. Statement to Belgians gathering at Dixmude, Belgium. Aug. 23, 1931. (Sent to Jan de Bondt on Aug. 8, 1931.) AEA [45 637].

———. Statement to Women's International League for Peace and Freedom. Jan. 4, 1928. AEA [48 818].

———. Television statement to Eleanor Roosevelt, Feb. 12, 1950. In "Einstein Warns World: Outlaw H-Bomb or Perish!" *New York Post*, Feb. 13, 1952.

———. "To the Society 'A Guaranteed Subsistence for All'" ("*An den Verein 'Allgemeine Nährpflicht'*"). Dec. 12, 1918. AEA [32 754].

———. "Tumultszenen bei einer Einstein-Vorlesung. Eine Erklärung Professor Einsteins." *8-Uhr Abendblatt* (Berlin) 73, no. 38 (Feb. 13, 1920): 2–3.

———. Typed statement on the Quaker relief effort in Central Europe. [After July 11, 1920]. AEA [40 006].

———. Untitled article. *Neue Zürcher Zeitung*, Apr. 4, 1920, 1st Sunday edition, section 2: [1].

———. "The War Is Won, but the Peace Is Not." Address to Nobel Anniversary Dinner, New York. Dec. 10, 1945. *New York Times*, Dec. 11, 1945.

———. "Welcoming Address to Paul Colin" ("*Begrüßung an Herrn Colin . . .*"). [Dec. 16, 1919]. AEA [28 005].

———. "Wie ich Zionist wurde." *Jüdische Rundschau* 26, no. 49 (June 21, 1921): 351–52.

———. "Zur Errichtung der hebräischen Universität in Jerusalem." *Jüdische Pressezentrale Zürich*, Aug. 26, 1921, [1].

———. "Die Zuwanderung aus dem Osten" ("Immigration from the East"). *Berliner Tageblatt*, Dec. 30, 1919, morning edition, [2].

Einstein, Albert, Bertrand Russell, Percy W. Bridgman, Hermann J. Muller, Cecil F. Powell, Joseph Rotblat, Frédéric Joliot-Curie, Leopold Infeld, Hideki Yukawa, and Max Born. "Declaration." *New York Times*, July 10, 1955.

Einstein, Albert, and Alfred E. Smith. Radio address. Columbia Broadcasting System (CBS). Oct. 22, 1935. Also in *New York Times*, Oct. 23, 1935.

Einstein, Albert, and Sigmund Freud. *Why War? A Correspondence between Albert Einstein and Sigmund Freud*. Translated by Stuart Gilbert. London: Peace Pledge Union, 1939, c1933.

"Einstein-Bankett für den Keren Kajemeth." *Jüdische Rundschau* 37, no. 22 (Mar. 18, 1932): 106.

"Einstein, the Man Who Started It All." *Newsweek*, Mar. 10, 1947.

"Einstein in New York." *Jüdische Rundschau* 35, no. 100 (Dec. 19, 1930): 673.

Eliasberg, George. *Der Ruhrkrieg von 1920*. Bonn-Bad Godesberg: Verlag Neue Gesellschaft, 1974.

Emergency Committee of Atomic Scientists. "Certification of Incorporation of the Emergency Committee of Atomic Scientists, Incorporated." Aug. 2, 1946. AEA [40 462].

———. "The Emergency Committee of Atomic Scientists, Incorporated: Its Purposes, Its Program, Its Requirements." [After Aug. 2, 1946.] AEA [40 479].

Eyck, Erich. *A History of the Weimar Republic*. Translated by Harlan P. Hanson and Robert G. L. Waite. Cambridge, MA: Harvard Univ. Press, 1967.

Feldman, Gerald D. *The Great Disorder: Politics, Economics, and Society in the German Inflation, 1914–1924*. New York: Oxford Univ. Press, 1997.

Ferro, Marc. *The Great War 1914–1918*. London: Routledge, 1973; orig. Paris: Éditions Gallimard, 1969.

Fisher, David J. *Romain Rolland and the Politics of Intellectual Engagement*. Berkeley: Univ. of California Press, 1988.

Florence, Ronald. *Fritz: The Story of a Political Assassin*. New York: Dial Press, 1971.

Fölsing, Albrecht. *Albert Einstein: A Biography*. Translated by Ewald Osers. New York: Viking Penguin, 1997.

Foster, Catherine. *Women for All Seasons: The Story of the Women's International League for Peace and Freedom*. Athens, GA: Univ. of Georgia Press, 1989.

Frank, Philipp. *Einstein: His Life and Times*. New York: A. A. Knopf, 1947.

Freundes-Rat des Internationalen Jugend-Bundes et al. "Aufruf an die freie Jugend aller Stände und Völker." *Republik* (Feb. 8, 1919). Reprinted in *Die Friedens-Warte. Blätter für zwischenstaatliche Organisation* 21, no. 3 (Apr. 1919): 74–75.

Fried, Alfred Hermann. *Auf hartem Grund. Offene Antwort auf den an mich gerichteten Brief von Dr. Jur. Hermann M. Popert im "Vortrupp" vom 1. März 1919.* Hamburg: Pfadweiser, 1919.

———. *Kurzgefasste Darstellung der Pan-Amerikanischen Bewegung.* Berlin: Verlag der "Friedens-Warte," 1912.

———. *Mein Kampf gegen Versailles und St. Germain vom Nov. 1918 bis Juni 1919.* Leipzig: Die Neue Geist, 1920.

Fussell, Paul. *The Great War and Modern Memory.* Oxford and New York: Oxford Univ. Press, 2000, c1975.

Galison, Peter L., Gerald Holton, and Silvan S. Schweber, eds. *The Assassin of Relativity.* DVD video. University of California, Berkeley, Department of Physics, 2006.

———. *Einstein for the 21st Century: His Legacy in Science, Art, and Modern Culture.* Princeton: Princeton Univ. Press, 2008.

Gautschi, Willi, ed. *Dokumente zum Landesstreik 1918.* Zurich: Benziger, 1971.

"Gegen den Länderraub." *Volksrecht* 18, no. 195 (Aug. 23, 1915): [1–2]

Geiger, Walburga, Max Hodann, Elisabeth Rotten, and Erich Schlesinger. *Lille: Beiträge zur Naturgeschichte des Krieges.* 2nd ed. Berlin: Engelmann, 1920. [Second/revised version.]

Gerlach, Hellmut von. "Hakenkreuz, hurra hoch!" *Die Welt am Montag,* Aug. 30, 1920.

Gesellschaft Deutscher Naturforscher und Ärzte (GDNA), 86th meeting, Bad Nauheim, Sept. 23–24, 1920. *Physikalische Zeitschrift* 21 (Nov. 1 and 15, 1920): 650–51, 662, 666–68.

Gilbert, Martin. *The First World War: A Complete History.* New York: Henry Holt, 1994.

Ginneken, Anique H. M. van. *Historical Dictionary of the League of Nations.* Lanham, MD: Scarecrow Press, 2006.

Goenner, Hubert. *Einstein in Berlin.* Munich: Verlag C. H. Beck, 2005.

———. "The Reaction to Relativity Theory I: The Anti-Einstein Campaign in Germany in 1920." *Science in Context* 6 (1993): 107–33.

Goenner, Hubert, and Giuseppe Castagnetti. "Albert Einstein as Pacifist and Democrat during World War I." *Science in Context* 9, no. 4 (winter 1996): 325–86.

Goldscheid, Rudolf. *Reine Vernunft und Staatsvernunft*. Leipzig: Suschitzky, 1918.

Graham, Frank D. *Exchange, Prices, and Production in Hyperinflation: Germany, 1920–1923*. New York: Russell and Russell, 1967.

Gross, Babette. *Willi Münzenberg: Eine politische Biographie*. Stuttgart: Deutsche Verlags-Anstalt, 1967.

Gumbel, Emil J. "Four Years of Political Murder." In *The Weimar Republic Sourcebook*, edited by Anton Kaes et al., 100–104. Berkeley: Univ. of California Press, 1994.

———. *Vier Jahre politischer Mord*. Berlin-Fichtenau: Verlag der Neuen Gesellschaft, 1922.

———. *Zwei Jahre Mord*. Berlin: Verlag Neues Vaterland, 1921.

Haeckel, Ernst. *Die Welträtsel: Gemeinverständliche Studien über Monistische Philosophie*. Leipzig: Kröner, 1908.

Halperin, William S. *Germany Tried Democracy*. New York: Norton, 1946.

Harris, Brayton. *The Navy Times Book of Submarines: A Political, Social, and Military History*. New York: Berkley Books, 1997.

Hausenstein, Wilhelm. "Erinnerungen an Eisner." *Der neue Merkur* 3 (1919–20): 56–57.

———. "Remembering Eisner." In *The Weimar Republic Sourcebook*, edited by Anton Kaes et al., 52–53. Berkeley: Univ. of California Press, 1994.

Heilbron, John L. *The Dilemmas of an Upright Man: Max Planck as Spokesman for German Science*. Berkeley: Univ. of California Press, 1986.

Heitz, Gerhard, et al. *Geschichte der Universität Rostock 1419–1969. Festschrift zur Fünfhundertfüngzig-Jahr-Feier der Universität. Band I. Die Universität von 1419–1945*. Berlin: Deutscher Verlag der Wissenschaft, 1969.

Hemleben, Johannes. *Ernst Haeckel in Selbstzeugnissen und Bilddokumenten*. Reinbek: Rowohlt, 1964.

Hiller, Kurt. *Geist werde Herr. Kundgebungen eines Aktivisten vor, in und nach dem Kriege*, 3rd ed. Berlin: Reiß, 1920.

Hiller, Kurt, ed. "Anhang: Dokumente." In *Das Ziel: Jahrbücher für geistige Politik*. Vol. 3, part 1. Leipzig: Wolff, 1919.

Hindenburg, Paul von. "The Stab in the Back." In *The Weimar Republic Sourcebook*, edited by Anton Kaes et al., 15–16. Berkeley: Univ. of California Press, 1994.

"Hochschule für Proletarier." *Räte-Zeitung* 1, no. 53 ([Nov.] 1919): 2.

Holl, Karl. *Pazifismus in Deutschland*. Frankfurt am Main: Suhrkamp, 1988.

Holmes, Virginia Iris. "Integrating Diversity, Reconciling Contradiction: The Jüdischer Friedensbund in Late Weimar Germany." *Leo Baeck Institute Year Book* 47 (2002): 175–94.

———. "'The Inviolability of Human Life': Pacifism and the Jews in Weimar Germany." PhD diss., Binghamton University, 2001.

———. "Was Einstein Really a Pacifist? Einstein's Forward-Thinking, Pragmatic, Persistent Pacifism." *Peace and Change* 33, no. 2 (Apr. 2008): 274–307.

Horne, John N., and Alan Kramer. *German Atrocities, 1914: A History of Denial.* New Haven: Yale Univ. Press, 2001.

———. "War between Soldiers and Civilians, 1914–1915." *Great War, Total War: Combat and Mobilization on the Western Front, 1914–1918.* Edited by Roger Chickering and Stig Förster, 153–68. Cambridge: Cambridge Univ. Press, 2000.

Horten, Alfons. *Sozialisierung und Wiederaufbau; praktische Vorschläge zur Sozialisierung und zur Wiederaufrichtung unseres Wirtschaftslebens.* Berlin: Verlag Neues Vaterland, [1920].

Ingram, Norman. *The Politics of Dissent: Pacifism in France, 1919–1939.* Oxford: Oxford Univ. Press, 1991.

International Antimilitaristic Commission. "Einstein an Apostate." *Press Service* (Haarlem, Netherlands), Sept. 7, 1933.

Isaacson, Walter. *Einstein: His Life and Universe.* New York: Simon & Schuster, 2007.

Jegelka, Norbert. *Paul Natorp: Philosophie, Pädagogik, Politik.* Würzburg: Könighausen und Neumann, 1992.

Jerome, Fred. *The Einstein File: J. Edgar Hoover's Secret War against the World's Most Famous Scientist.* New York: St. Martin's Press, 2002.

Jones, Nigel H. *Hitler's Heralds: The Story of the Freikorps, 1918–1923.* New York: Dorset Press, 1992.

Joyce, James Avery. *Broken Star: The Story of the League of Nations.* Swansea: C. Davies, 1978.

Kaes, Anton, et al., eds. *The Weimar Republic Sourcebook.* Berkeley: Univ. of California Press, 1994.

"Der Kampf um Einstein." *Deutsche Allgemeine Zeitung,* Sept. 1, 1920.

Kármán, Theodor von. Article supporting József Kelen. *Frankfurter Zeitung,* Nov. 4, 1920, 1st morning edition.

Keegan, John. *The First World War.* New York: Vintage Books, Random House, 1998.

Kern, Robert William. *Friedrich Adler and the Austrian Social Democratic Party.* AM thesis, University of Chicago, 1961.

Kessler, Harry Graf. *Tagebücher.* Frankfurt am Main: Insel-Verlag, 1961.

Keynes, John M. *The Economic Consequences of the Peace.* New York: Harcourt, 1920.

Kirsten, Christa, and Hans-Jürgen Treder, eds. *Albert Einstein in Berlin 1913–1933. Part 1, Darstellung und Dokumente.* Berlin: Akademie-Verlag, 1979.

Klier, John D., and Schlomo Lambrosa. *Pogroms: Anti-Jewish Violence in Modern Russian History.* Cambridge: Cambridge Univ. Press, 1992.

Koch, Stephen. *Double Lives: Stalin, Willi Münzenberg, and the Seduction of the Intellectuals.* Rev. ed. New York: Enigma Books, 2004.

Kolb, Eberhard. *The Weimar Republic.* Translated by P. S. Falla. London and New York: Routledge, 1988, c1984.

Kramer, Alan. *Dynamic of Destruction: Culture and Mass Killing in the First World War.* Oxford: Oxford Univ. Press, 2007.

Lebius, Rudolf. "Ungesühnter Landesverrat." *Deutsche Staatsbürger-Zeitung,* Jan. 9, 1921, 1.

Lehmannn-Rußbüldt, Otto. *"Die Brücke über den Abgrund": Für die Verständigung zwischen Deutschland und Frankreich.* Berlin: Verlag Neues Vaterland, E. Berger u. Co., [1922].

———. *Der Kampf der Deutschen Liga für Menschenrechte, vormals Bund Neues Vaterland, für den Weltfrieden, 1914–1927.* Berlin: Hensel & Co. Verlag, 1927.

Lenz, Kurt and Walter Fabian, eds. *Die Friedensbewegung. Ein Handbuch der Weltfriedensströmungen der Gegenwart.* Berlin: Schwetschke, 1922.

Lerner, Warren. *Karl Radek: The Last Internationalist.* Stanford, CA: Stanford Univ. Press, 1970.

Levenson, Thomas. *Einstein in Berlin.* New York: Bantam Dell, 2003.

Levin, N. Gordon, Jr. *Woodrow Wilson and World Politics: America's Response to War and Revolution.* Oxford: Oxford Univ. Press, 1968.

Lichem, Heinz von. *Gebirgskrieg, 1915–1918.* Bozen, Germany: Athesia, 1980–82.

Liddle, Peter H. *The 1916 Battle of the Somme, A Reappraisal.* London: Leo Cooper, 1992.

Liga zur Beförderung der Humanität (Menschheitsbund). "Erklärung in Sachen Liebknecht-Luxemburg." [Jan. 23, 1919]. Nachlaß Nicolai, ED 184/15. Institut für Zeitgeschichte, Munich, Germany. (Dated by letter and draft from Georg Friedrich Nicolai to Hans Delbrück, Jan. 23, 1919. Nachlaß Hans

Delbrück, Mappe G. Nicolai, Bl. 1. Handschriftenabteilung, Deutsche Staatsbibliothek, Berlin).

Lipkes, J. *Rehearsals: The German Army in Belgium, August 1914*. Leuven, Belgium: Leuven Univ. Press, 2007.

Lorentz, Hendrik A. "Ernest Solvay." *Naturwissenschaften* 2 (1914): 997–99.

Löwenstein, Siegfried. *Der Prozess Erzberger-Helfferich: Ein Rechtsgutachten*. Ulm: Süddeutsche Verlagsanstalt, 1921.

Lüders, Else, ed. *Minna Cauer, Leben und Werk, dargestellt an Hand ihrer Tagebücher und nachgelassenen Schriften*. Gotha: Klotz, 1925.

Lüttgemeier-Davin, Reinhold. *Pazifismus zwishcen Kooperation und Konfrontation: Das Deutche Friedenskartell in der Weimarer Republik*. Köln: Pahl-Rugenstein Verlag, 1982.

Luxemburg, Rosa. "Founding Manifesto of the Communist Party of Germany (KPD)." In *The Weimar Republic Sourcebook*, edited by Anton Kaes et al., 40–46. Berkeley: Univ. of California Press, 1994.

———. "Konstituierung der Kommunistische Partei." *Die Rote Fahne*, no. 45 (Dec. 31, 1918).

MacMillan, Margaret. *Paris 1919: Six Months that Changed the World*. New York: Random House, 2001.

Mann, Thomas. "An Appeal to Reason." In *The Weimar Republic Sourcebook*, edited by Anton Kaes et al., 150–59. Berkeley: Univ. of California Press, 1994.

———. "Appell an die Vernunft." *Berliner Tageblatt*, Oct. 18, 1930.

March, Harold. *Romain Rolland*. New York: Twayne, 1971.

Marsch, Ulrich. *Notgemeinschaft der Deutschen Wissenschaft. Gründung und frühe Geschichte 1920–1925*. Frankfurt am Main: Lang, 1994.

Massart, Jean. "Les intellectuels allemands et la recherche de la verité." *Revue de Paris* 25 (1918): 643–72.

McMeekin, Sean. *The Red Millionaire: A Political Biography of Willi Münzenberg, Moscow's Secret Propaganda Tsar in the West*. New Haven: Yale Univ. Press, 2003.

Mendes-Flohr, Paul, and Jehuda Reinharz. *The Jew in the Modern World: A Documentary History*. 2nd ed. Oxford: Oxford Univ. Press, 1995.

Mommsen, Hans. *The Rise and Fall of Weimar Democracy*. Chapel Hill: Univ. of North Carolina Press, 1996.

Mommsen, Wolfgang J. "Die deutsche öffentliche Meinung und der Zusammenbruch des Regierungssystems Bethmann Hollweg im Juli 1917." *Geschichte in Wissenschaft und Unterricht* 19 (1968): 656–71.

Mühsam, Paul. *Aus dem Schicksalsbuch der Menschheit*. Dresden: Rödel, 1919.

Nathan, Otto, and Heinz Norden. *Einstein on Peace*. New York: Schocken Books, 1968, c1960.

Natorp, Paul. "Ein Weg der Rettung." *Kunstwart und Kulturwart. Halbmonatschau für Ausdruckskultur auf allen Lebensgebieten* 32, no. 3 (Apr.–June 1919): 100–107.

———. "Ein Weg der Rettung." *Frankfurter Zeitung*, Mar. 30, 1919, morning edition.

Naumann, Gerlinde. *Minna Cauer: Eine Kämpferin für Frieden, Demokratie und Empanzipation*. Berlin: Sekretariat des Zentral Vorstandes der Liberal-Demokratischen Partei Deutschlands im Buchverlag Der Morgen, 1988.

New Fatherland League (Bund Neues Vaterland). "Appeal Signed by Cultural Associations of All of Germany and by Individuals" ("*Aufruf der von Kulturvereinen ganz Deutschlands und von Einzelpersonen unterzeichnet werden soll*"). Postdated May 7, 1919. ZSg. 1–13/1 (8). Bundesarchiv, Koblenz, Germany. Also AEA [82 572].

Nicolai, Georg Friedrich. *Aufruf an die Europäer: gesammelte Aufsätze zum Wiederaufbau Europas*. Edited by Hans Wehberg. Leipzig, Vienna, and Zurich: Verlag der Wiener Graphischen Werkstätte, 1921.

———. *Die Biologie des Krieges: Betrachtungen eines deutschen Naturforschers*. Zurich: Orell Füssli, 1917.

———. *The Biology of War*. Translated by Constance A. Grande and Julian Grande. New York: Century, 1919.

Nicolai, Georg Friedrich, ed. *Romain Rollands Manifest und die deutschen Antworten*. Charlottenburg: Mundus, [1920].

Northedge, F. S. *The League of Nations: Its Life and Times, 1920–1946*. New York: Holmes and Meier, 1986.

Nottmeier, Christian. *Adolf von Harnack und die deutsche Politik 1890–1930. Ein biographische Studie zum Verhältnis von Protetstantismus, Wissenschaft und Politik*. Tübingen: Siebbeck, 2004.

Ostwald, Grete. *Wilhelm Ostwald mein Vater*. Stuttgart: Berliner Union, 1953.

Petition to pardon ten officials of the Hungarian Soviet Republic (Einstein among signatories). *Arbeiter-Zeitung* (Vienna), Dec. 15, 1920.

Peukert, Detlev J. K. *The Weimar Republic: The Crisis of Classical Modernity*. Translated by Richard Deveson. New York: Hill and Wang, 1993, c1987.

Poppel, Stephen M. *Zionism in Germany, 1897–1933: The Shaping of a Jewish Identity*. Philadelphia: Jewish Publication Society of America, 1976.

Popper-Lynkeus, Josef. *Nach dem Kriege. Ein Auszug aus dem Werke Die allgemeine Nährpflicht als Lösung der sozialen Frage*, 2nd ed. Dresden: Reißner, [1915].

Quidde, Ludwig. *Der deutsche Pazifismus während des Weltkrieges 1914–1918*. Aus dem Nachlaß Ludwig Quiddes. Edited by Karl Holl and Helmut Donat. Boppard am Rhein: Harald Boldt Verlag, 1979.

Rathenau, Walther. *Von kommenden Dingen*. Berlin: Fischer Verlag, 1917.

Reiser, Anton (pseudonym of Rudolf Kayser). *Albert Einstein: A Biographical Portrait*. New York: A. & C. Boni, 1930.

Report of the Commission on the Palestine Disturbances of August, 1929. London: His Majesty's Stationery Office, 1930.

Rév, Erika. *A népbiztosok pere*. Budapest: Kossuth, 1969.

Rheins, Carl Jeffrey. "German Jewish Patriotism 1918–1935: A Study of the Attitudes and Actions of the 'Reichsbund Jüdischer Frontsoldaten,' the 'Verband Nationaldeutscher Juden,' the 'Schwarzes Fahnlein, Jungenschaft,' and the 'Deutscher Vortrupp, Gefolgschaft Deutscher Juden.'" PhD diss., State University of New York at Stony Brook, 1978.

———. "The Verband nationaldeutscher Juden 1921–1933." *Leo Baeck Institute Yearbook* 25 (1980).

Riesenberger, Dieter. *Geschichte der Friedensbewegung in Deutschland: Von den Anfängen bis 1933*. Göttingen: Vandenhoeck & Ruprecht, 1985.

Roerkohl, Anne. *Hungerblockade und Heimatfront: Die kommunale Lebensmittelversorgung in Westfalen während des ersten Weltkrieges*. Stuttgart: Steiner, 1991.

Rolland, Romain. *Clerambault: Histoire d'une conscience libre pendant la guerre*. Paris: P. Ollendorff, 1920.

———. Diary entry of September 1933. *Journal des années de guerre*. Paris: Oeuvres libres, 1950.

———. *Liluli*. Paris: Ollendorff, 1919.

———. *Pierre et Luce*. Geneva: Sablier, 1920.

Rosenkranz, Ze'ev. *Einstein before Israel: Zionist Icon or Iconoclast?* Princeton: Princeton Univ. Press, 2011.

Rowe, David E., and Robert Schulmann, eds. *Einstein on Politics: His Private Thoughts and Public Stands on Nationalism, Zionism, War, Peace, and The Bomb*. Princeton, NJ: Princeton Univ. Press, 2007.

Samuel, Horace B. *Beneath the Whitewash: A Critical Analysis of the Report of the Commission on the Palestine Disturbances of August, 1929*. London: Hogarth Press, 1930.

Samuel, Maurice. *What Happened in Palestine: The Events of August, 1929, Their Background and Their Significance.* Boston: Stratford, 1929.

Sayen, Jamie. *Einstein in America: The Scientist's Conscience in the Age of Hitler and Hiroshima.* New York: Crown Publishers, 1985.

Schmidt-Ott, Friedrich, and Eduard Wildhagen. "Die Not der deutschen Wissenschaft." In *Internationale Monatsschrift für Wissenschaft, Kunst und Technik* 15, no. 1 (Oct. 1920): 1–36. Also in *Forschungsförderung in drei Epochen: Deutschen Forschungsgemeinschaft—Geschichte, Arbeitsweise, Kommentar,* edited by Kurt Zierold Wiesbaden: Steiner, 1968.

Schröder-Gudehus, Brigitte. "Deutsche Wissenschaft und Internationale Zusammenarbeit 1914–1928. Ein Beitrag zum Studium kultureller Beziehungen in politischen Krisenzeiten." PhD diss., University of Geneva, 1966.

Schulze, Hagen. *Freikorps und Republik, 1918–1920.* Boppard am Rhein: Boldt, 1969.

Schwabe, Klaus. "Deutsche Professoren und die politischen Grundfragen des Ersten Weltkrieges." PhD diss., Philosophische Fakultät, Albert Ludwigs Universität, Freiburg im Breisgau, 1958.

———. "Zur politischen Haltung der deutschen Professoren im Ersten Weltkrieg." *Historische Zeitschrift* 193 (1961).

Schwartz, Richard A. "Einstein and the War Department." *Isis* 80 (1989): 281–84.

Schwarz, Boris. Contribution to Jerusalem Symposium. *Albert Einstein Historical and Cultural Perspectives: Centennial Symposium in Jerusalem.* Edited by Gerald Holton and Yehuda Elkana. Princeton: Princeton Univ. Press, 1982.

Sea Sorrow: The Story of the Iolaire Disaster. Stornoway: Stornoway Gazette, 1972.

Sée, Henri. *Histoire de la Ligue des Droits de l'Homme (1898–1926).* Paris: Ligue des Droits de l'Homme, 1927.

Seeberg, Reinhold (with 1,347 signatories). "Petition of the Intellectuals" ("*Intellektuelleneingabe*"). July 8, 1915.

Seelig, Carl. *Die Zwölf Bücher.* Leipzig, Vienna, and Zürich: E.P. Tal & Co. Verlag, 1919–22.

Sitter, Willem de. "On Einstein's Theory of Gravitation, and Its Astronomical Consequences. First Paper." *Royal Astronomical Society. Monthly Notices* 76 (1915–16): 699–728.

———. "On Einstein's Theory of Gravitation, and Its Astronomical Consequences. Second Paper." *Royal Astronomical Society. Monthly Notices* 77 (1916–1917): 155–84.

Skalweit, August. *Die deutsche Kriegsnährungswirtschaft.* Stuttgart: Deutsche Verlags-Anstalt, 1927.

Social Democratic Party (SPD). "Appeal of the Social Democratic Party for a General Strike." In *The Weimar Republic Sourcebook,* edited by Anton Kaes et al., 16. Berkeley: Univ. of California Press, 1994.

Sorel, Georges. *Reflections on Violence.* 3rd ed. New York: Huebsch, 1912.

———. *Réflexions sur la Violence.* Paris: Librairie de "Pages libres," 1908.

"Spartacus Manifesto." In "Richtlinien für die Arbeiter– und Soldatenräte Deutschlands," *Die Rote Fahne* (Nov. 26, 1918); *The Weimar Republic Sourcebook,* edited by Anton Kaes et al. (Berkeley: Univ. of California Press, 1994), 37–38.

Stanley, Matthew. "'An Expedition to Heal the Wounds of War': The 1919 Eclipse and Eddington as Quaker Adventurer." *Isis* 94 (2003): 57–89.

Starr, William T. *Romain Rolland, One against All: A Biography.* The Hague: Mouton, 1971.

Steglich, Wolfgang. *Die Friedenspolitik der Mittelmächte 1917/18.* Vol 1. Wiesbaden: Steiner, 1964.

Stokesbury, James L. *A Short History of World War I.* New York: Harper Collins, 1981).

"Eine Sympathiekundgebung für Professor Einstein." *Neue Freie Presse* (Vienna), Sept. 1, 1920.

Tagore, Rabindranath. *Nationalismus.* Leipzig: Wolff, [1918].

Tartakower, Arjeh. "Das jüdische Mittelschulwesen in Polen." *Der Jude* 9, separatum no. 2: "Erziehung" (1926): 79–89.

Thimme, Annelise. *Hans Delbrück als Kritiker der wilhelminischen Epoche.* Düsseldorf: Droste, 1955.

Trachtenberg, Marc. *Reparation in World Politics: France and European Economic Diplomacy, 1916–1923.* New York: Columbia Univ. Press, 1980.

The Treaty of Peace between the Allied and Associated Powers and Germany. London: His Majesty's Stationery Office, 1919. *See* "The Treaty of Versailles: The Reparations Clauses."

"The Treaty of Versailles: The Reparations Clauses." In *The Weimar Republic Sourcebook,* edited by Anton Kaes et al., 8–9. Berkeley: Univ. of California Press, 1994.

"To the Civilized World." *See* "An die Kulturwelt!"

"To the Intellectuals of the Proletariat" ("*An die Geistigen des Proletariats*"). Attached to a circular signed by Paul Natorp, Wilhelm Schäfer, Hermann

Herrigel, Alfons Paquet, and Leopold Ziegler, dated Apr. 1919. Nachlaß Paul Natorp, Ms. 831. Universitätsbibliothek Marburg, Germany. Also in *Räte-Zeitung: Organ der Hand- und Kopfarbeiter-Räte Deutschlands*, June 29, 1919.

Ulrich, Bernd, and Benjamin Ziemann, eds. *Frontalltag im Ersten Weltkrieg: Wahn und Wirklichkeit*. Frankfurt am Main: Fischer Taschenbuch Verlag, 1994.

"Um Einstein." *Berliner Tageblatt*, Aug. 31, 1920, evening edition.

Ungern-Sternberg, Jürgen von, and Wolfgang von Ungern-Sternberg. *Der Aufruf "An die Kulturwelt!" Das Manifest der 93 und die Anfänge der Kriegspropaganda im Ersten Weltkrieg*. Stuttgart: Franz Steiner Verlag, 1996.

University of Berlin students (291 signatures). Declaration to Einstein. Feb. 19, 1920. AEA [45 156].

Vellacott, Jo. *Bertrand Russell and the Pacifists in the First World War*. Bath: Harvester Press, 1980.

Die Verfassung des Deutschen Reiches vom 11. August 1919, 7th ed. Leipzig, 1930. *See* "Constitution of the German Republic."

Vincent, C. Paul. *A Historical Dictionary of Germany's Weimar Republic, 1918–1933*. Westport, CT: Greenwood, 1997.

———. *The Politics of Hunger: The Allied Blockade of Germany, 1915–1919*. Athens: Ohio Univ. Press, 1985.

Wehberg, Hans. *Wider den Aufruf der 93! Das Ergebnis einer Rundfrage an die Intellektuellen über die Kriegsschuld*. Charlottenburg: Deutsche Verlagsgesellschaft für Politik und Geschichte, 1920.

Weinzierl, Ulrich. *Carl Seelig, Schriftsteller*. Vienna: Löcker, 1982.

Wells, H. G. *The War That Will End War*. London: Frank and Cecil Palmer, 1914.

Wertheimer, Jack. *Unwelcome Strangers: East European Jews in Imperial Germany*. Oxford: Oxford Univ. Press, 1987.

Wilcock, Evelyn. *Pacifism and the Jews*. Lansdown: Hawthorn Press, 1994.

Wilson, Woodrow. "Fourteen Points." President Wilson's Message to Congress, Jan. 8, 1918. Records of the United States Senate. RG 46. National Archives.

Winteler-Einstein, Maja. "Albert Einstein—Beitrag für sein Lebensbild" (Excerpt). In *The Collected Papers of Albert Einstein*. Vol. 1, *The Early Years, 1879–1902*, lxiii–lxiv. Edited by John Stachel et al. Princeton: Princeton Univ. Press, 1987.

"Wir unterzeichneten Deutschen . . ." Enclosure in Georg Friedrich Nicolai and Hans Wehberg, letter to Wilhelm Mühlon. June 27, 1919. Nachlaß Wilhelm

Mühlon, ED 142, Band 26, Blatt 4753. Institut für Zeitgeschichte, Munich, Germany.

Wolff, Theodor. "Hunger, Angst und Verrohung." *Berliner Tageblatt*, Mar. 10, 1919.

Wright, Jonathan. *Gustav Stresemann: Weimar's Greatest Statesman*. Oxford: Oxford Univ. Press, 2002.

Wurgaft, Lewis D. "The Activists: Kurt Hiller and the Politics of Action on the German Left 1914–1933." *American Philosophical Society: Transactions* 67 (1977): 51–52.

Ziche, Paul, ed. *Monismus um 1900: Wissenschaftskultur und Weltanschauung*. Berlin: Verlag für Wissenschaft und Bildung, 2000.

Zimmel, Brigitte, and Gabriele Kerber, eds. *Hans Thirring. Ein Leben für Physik und Frieden*. Vienna: Böhlau, 1992.

Zuckerman, Larry. *The Rape of Belgium: The Untold Story of World War I*. New York: New York Univ. Press, 2004.

Zuelzer, Wolf. *The Nicolai Case: A Biography*. Detroit: Wayne State Univ. Press, 1982.

Zu Max Plancks sechzigstem Geburtstag. Ansprachen gehalten am 26. April 1918 in der Deutschen Physikalischen Gesellschaft von E. Warburg, M. v. Laue, A. Sommerfeld und A. Einstein. Karlsruhe: C. F. Müllersche Hofbuchhandlung, 1918.

"Zum Brief von Professor Einstein." *Jüdische Rundschau* 35, no. 27 (Apr. 4, 1930): 187.

Index

Italic page numbers denote illustrations.

Virginia Iris Holmes earned an MA in international relations from the Fletcher School at Tufts University, an MA in history from the University of Michigan, and a PhD in history from Binghamton University. Dr. Holmes was a Senior Research Editor with the Einstein Papers Project from 2002 to 2008. She has been a vegan for twenty years and counting, and she served in the United States Peace Corps in the Republic of Georgia from 2014 to 2016. Current projects include revising her dissertation into a book, and another monograph on Jewish communities in Weimar Germany.